Springer Uncertainty Research

Series editor

Baoding Liu, Beijing, China

Springer Uncertainty Research

Springer Uncertainty Research is a book series that seeks to publish high quality monographs, texts, and edited volumes on a wide range of topics in both fundamental and applied research of uncertainty. New publications are always solicited. This book series provides rapid publication with a world-wide distribution.

More information about this series at http://www.springer.com/series/13425

Yuanguo Zhu

Uncertain Optimal Control

Springer

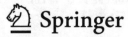

Yuanguo Zhu
Department of Mathematics
Nanjing University of Science
and Technology
Nanjing, China

ISSN 2199-3807 ISSN 2199-3815 (electronic)
Springer Uncertainty Research
ISBN 978-981-13-4737-5 ISBN 978-981-13-2134-4 (eBook)
https://doi.org/10.1007/978-981-13-2134-4

This Springer imprint is published by the registered company Springer Nature Singapore Pte Ltd.
The registered company address is: 152 Beach Road, #21-01/04 Gateway East, Singapore 189721, Singapore

Preface

If a dynamical system is disturbed by uncertain factors, it may be described by an uncertain differential equation. A problem of optimizing an index subject to an uncertain differential equation is called an uncertain optimal control problem. It is a novel topic on optimal control based on the uncertainty theory.

This book is to introduce the theory and applications of uncertain optimal control. Two types of models including expected value uncertain optimal control and optimistic value uncertain optimal control are established. These models which have continuous-time forms and discrete-time forms are dealt with by dynamic programming. The uncertain optimal control theory concerns on establishing models based on expected value and optimistic value criterions, equation of optimality, bang–bang optimal control, optimal control for switched uncertain system, optimal control for uncertain system with time delay, and parametric optimal control. The applications of uncertain optimal control are shown in portfolio selection, engineering, and management.

The book is suitable for researchers, engineers, and students in the field of mathematics, cybernetics, operations research, industrial engineering, artificial intelligence, economics, and management science.

Acknowledgement

This work was partially supported by the National Natural Science Foundation of China (Grant Nos. 61273009, 61673011).

Nanjing, China
June 2018

Yuanguo Zhu

Contents

1 Basics on Uncertainty Theory 1
1.1 Uncertainty Space 1
1.1.1 Uncertain Variable 3
1.1.2 Independence 4
1.2 Expected Value 4
1.2.1 Distribution of Function of Uncertain Variable 10
1.2.2 Expected Value of Function of Uncertain Variable 11
1.3 Optimistic Value and Pessimistic Value 16
1.4 Uncertain Simulation 20
1.5 Uncertain Process 21
1.5.1 Liu Process 22
1.5.2 Liu Integral 23
1.6 Uncertain Differential Equation 23
References 25

2 Uncertain Expected Value Optimal Control 27
2.1 Problem of Uncertain Optimal Control 27
2.2 Principle of Optimality 28
2.3 Equation of Optimality 29
2.4 Equation of Optimality for Multidimension Case 30
2.5 Uncertain Linear Quadratic Model 33
2.6 Optimal Control Problem of the Singular Uncertain System 39
References 46

3 Optimistic Value-Based Uncertain Optimal Control 47
3.1 Optimistic Value Model 47
3.2 Equation of Optimality 49
3.3 Uncertain Optimal Control Model with Hurwicz Criterion 52

3.4 Uncertain Linear Quadratic Model Under Optimistic Value Criterion . . . 57
3.5 Optimistic Value Optimal Control for Singular System . . . 60
3.5.1 Example . . . 65
References . . . 68

4 Optimal Control for Multistage Uncertain Systems . . . 69
4.1 Recurrence Equation . . . 69
4.2 Linear Quadratic Model . . . 71
4.3 General Case . . . 75
4.3.1 Hybrid Intelligent Algorithm . . . 75
4.3.2 Finite Search Method . . . 76
4.3.3 Optimal Controls for Any Initial State . . . 77
4.4 Example . . . 79
4.5 Indefinite LQ Optimal Control with Equality Constraint . . . 83
4.5.1 Problem Setting . . . 83
4.5.2 An Equivalent Deterministic Optimal Control . . . 85
4.5.3 A Necessary Condition for State Feedback Control . . . 88
4.5.4 Well Posedness of the Uncertain LQ Problem . . . 92
4.5.5 Example . . . 95
References . . . 97

5 Bang–Bang Control for Uncertain Systems . . . 99
5.1 Bang–Bang Control for Continuous Uncertain Systems . . . 99
5.1.1 An Uncertain Bang–Bang Model . . . 101
5.1.2 Example . . . 102
5.2 Bang–Bang Control for Multistage Uncertain Systems . . . 105
5.2.1 Example . . . 109
5.3 Equation of Optimality for Saddle Point Problem . . . 110
5.4 Bang–Bang Control for Saddle Point Problem . . . 113
5.4.1 A Special Bang–Bang Control Model . . . 115
5.4.2 Example . . . 116
References . . . 119

6 Optimal Control for Switched Uncertain Systems . . . 121
6.1 Switched Uncertain Model . . . 122
6.2 Expected Value Model . . . 122
6.2.1 Two-Stage Algorithm . . . 123
6.2.2 Stage (a) . . . 123
6.2.3 Stage (b) . . . 127
6.2.4 An Example . . . 128
6.3 LQ Switched Optimal Control Problem . . . 129

6.4 MACO Algorithm for Optimal Switching Instants 133
6.4.1 Example 135
6.5 Optimistic Value Model 137
6.5.1 Two-Stage Approach 138
6.5.2 Stage (a) 138
6.5.3 Stage (b) 142
6.5.4 Example 142
6.6 Discrete-Time Switched Linear Uncertain System 144
6.6.1 Analytical Solution 145
6.6.2 Two-Step Pruning Scheme 149
6.6.3 Local Pruning Scheme 150
6.6.4 Global Pruning Scheme 151
6.6.5 Examples 152
References 154

7 Optimal Control for Time-Delay Uncertain Systems 157
7.1 Optimal Control Model with Time-Delay 158
7.2 Uncertain Linear Quadratic Model with Time-Delay 161
7.2.1 Example 165
7.3 Model with Multiple Time-Delays 168
7.3.1 Example 173
References 175

8 Parametric Optimal Control for Uncertain Systems 177
8.1 Parametric Optimization Based on Expected Value 177
8.1.1 Parametric Optimal Control Model 179
8.1.2 Parametric Approximation Method 180
8.2 Parametric Optimization Based on Optimistic Value 183
8.2.1 Piecewise Optimization Method 184
References 186

9 Applications 187
9.1 Portfolio Selection Models 187
9.1.1 Expected Value Model 187
9.1.2 Optimistic Value Model 189
9.2 Manufacturing Technology Diffusion Problem 190
9.3 Mitigation Policies for Uncertain Carbon Dioxide Emissions 193
9.4 Four-Wheel Steering Vehicle Problem 200
References 205

Index 207

Chapter 1
Basics on Uncertainty Theory

For modeling indeterminacy, there exist many ways. Roughly speaking, there are two representative theories: one is probability theory and the other is uncertainty theory [1]. Probability is interpreted as frequency, while uncertainty is interpreted as personal belief degree. When the sample size is large enough, probability theory is the unique method to deal with the problem on the basis of estimated probability distributions. However, in many cases, no samples are available to estimate a probability distribution. We have to invite some domain experts to evaluate the belief degree that each event will happen. By the Nobelist Kahneman and his partner Tversky [2], human tends to overweight unlikely events, and the belief degree has a much larger range than the true frequency as a result. In this case, probability theory does not work [3], so uncertainty theory is founded to deal with this type of indeterminacy. In order to rationally deal with belief degrees, uncertainty theory was founded in 2007 [1]. Nowadays, uncertainty theory has become a branch of axiomatic mathematics for modeling belief degrees [4]. Theory and practice have shown that uncertainty theory is an efficient tool to deal with some nondeterministic information, such as expert data and subjective estimations, which appears in many practical problems. During the past years, there have been many achievements in uncertainty theory, such as uncertain programming, uncertain statistics, uncertain logic, uncertain inference, and uncertain process.

1.1 Uncertainty Space

To begin with, some basic concepts in the uncertainty theory [1, 4] are listed. Let Γ be a nonempty set, and $\mathcal{L}$ a σ-algebra over Γ. Each element $A \in \mathcal{L}$ is called an event. A set function $\mathcal{M}$ defined on the σ-algebra $\mathcal{L}$ is called an uncertain measure if it satisfies (i) $\mathcal{M}(\Gamma) = 1$; (ii) $\mathcal{M}(A) + \mathcal{M}(A^c) = 1$ for any event A; (iii) $\mathcal{M}(\bigcup_{i=1}^{\infty} A_i) \le \sum_{i=1}^{\infty} \mathcal{M}(A_i)$ for every countable sequence of events A_i.

Y. Zhu, *Uncertain Optimal Control*, Springer Uncertainty Research,
https://doi.org/10.1007/978-981-13-2134-4_1

Definition 1.1 ([1]) Let Γ be a nonempty set, let $\mathcal{L}$ be a σ-algebra over Γ, and let $\mathcal{M}$ be an uncertain measure. Then the triplet $(\Gamma, \mathcal{L}, \mathcal{M})$ is called an uncertainty space.

Product uncertain measure was defined to produce an uncertain measure of compound event by Liu [5] in 2009, thus producing the fourth axiom of uncertainty theory. Let $(\Gamma_k, \mathcal{L}_k, \mathcal{M}_k)$ be uncertainty spaces for $k = 1, 2, \ldots$ Write

$$\Gamma = \Gamma_1 \times \Gamma_2 \times \cdots \tag{1.1}$$

that is the set of all ordered tuples of the form $(\gamma_1, \gamma_2, \ldots)$, where $\gamma_k \in \Gamma_k$ for $k = 1, 2, \ldots$ A measurable rectangle in Γ is a set

$$\Lambda = \Lambda_1 \times \Lambda_2 \times \cdots \tag{1.2}$$

where $\Lambda_k \in \mathcal{L}_k$ for $k = 1, 2, \ldots$ The smallest σ-algebra containing all measurable rectangles of Γ is called the product σ-algebra, denoted by

$$\mathcal{L} = \mathcal{L}_1 \times \mathcal{L}_2 \times \cdots \tag{1.3}$$

Then the product uncertain measure $\mathcal{M}$ on the product σ-algebra $\mathcal{L}$ is defined by the following product axiom [5].

Axiom 4 (*Product Axiom*) Let $(\Gamma_k, \mathcal{L}_k, \mathcal{M}_k)$ be uncertainty spaces for $k = 1, 2, \ldots$ The product uncertain measure $\mathcal{M}$ is an uncertain measure satisfying

$$\mathcal{M}\left\{\prod_{k=1}^{\infty} \Lambda_k\right\} = \bigwedge_{k=1}^{\infty} \mathcal{M}_k\{\Lambda_k\} \tag{1.4}$$

where Λ_k are arbitrarily chosen events from $\mathcal{L}_k$ for $k = 1, 2, \ldots$, respectively.

For each event $\Lambda \in \mathcal{L}$, we have

$$\mathcal{M}\{\Lambda\} = \begin{cases} \sup\limits_{\Lambda_1 \times \Lambda_2 \times \cdots \subset \Lambda} \min\limits_{k \ge 1} \mathcal{M}_k\{\Lambda_k\} \\ \quad \text{if} \sup\limits_{\Lambda_1 \times \Lambda_2 \times \cdots \subset \Lambda} \min\limits_{k \ge 1} \mathcal{M}_k\{\Lambda_k\} > 0.5, \\ 1 - \sup\limits_{\Lambda_1 \times \Lambda_2 \times \cdots \subset \Lambda^c} \min\limits_{k \ge 1} \mathcal{M}_k\{\Lambda_k\} \\ \quad \text{if} \sup\limits_{\Lambda_1 \times \Lambda_2 \times \cdots \subset \Lambda^c} \min\limits_{k \ge 1} \mathcal{M}_k\{\Lambda_k\} > 0.5, \\ 0.5, \text{ otherwise.} \end{cases} \tag{1.5}$$

Definition 1.2 Assume that $(\Gamma_k, \mathcal{L}_k, \mathcal{M}_k)$ are uncertainty spaces for $k = 1, 2, \ldots$ Let $\Gamma = \Gamma_1 \times \Gamma_2 \times \cdots$, $\mathcal{L} = \mathcal{L}_1 \times \mathcal{L}_2 \times \cdots$ and $\mathcal{M} = \mathcal{M}_1 \wedge \mathcal{M}_2 \wedge \cdots$ Then the triplet $(\Gamma, \mathcal{L}, \mathcal{M})$ is called a product uncertainty space.

1.1.1 Uncertain Variable

Definition 1.3 ([1]) An uncertain variable is a function ξ from an uncertainty space $(\Gamma, \mathcal{L}, \mathcal{M})$ to the set of real numbers R such that $\{\xi \in B\}$ is an event for any Borel set B.

Definition 1.4 ([1]) The uncertainty distribution Φ of an uncertain variable ξ is defined by

$$\Phi(x) = \mathcal{M}\{\xi \le x\} \tag{1.6}$$

for any real number x.

Theorem 1.1 ([6]) *A function $\Phi(x) : R \to [0, 1]$ is an uncertainty distribution if and only if it is a monotone increasing function except $\Phi(x) \equiv 0$ and $\Phi(x) \equiv 1$.*

Example 1.1 An uncertain variable ξ is called linear if it has a linear uncertainty distribution

$$\Phi(x) = \begin{cases} 0, & \text{if } x \le a \\ (x-a)/(b-a), & \text{if } a \le x \le b \\ 1, & \text{if } x \ge b \end{cases} \tag{1.7}$$

denoted by $\mathcal{L}(a, b)$ where a and b are real numbers with $a < b$.

Example 1.2 An uncertain variable ξ is called zigzag if it has a zigzag uncertainty distribution

$$\Phi(x) = \begin{cases} 0, & \text{if } x \le a \\ (x-a)/2(b-a), & \text{if } a \le x \le b \\ (x+c-2b)/2(c-b), & \text{if } b \le x \le c \\ 1, & \text{if } x \ge c \end{cases} \tag{1.8}$$

denoted by $\mathcal{Z}(a, b, c)$ where a, b, c are real numbers with $a < b < c$.

Example 1.3 An uncertain variable ξ is called normal if it has a normal uncertainty distribution

$$\Phi(x) = \left(1 + \exp\left(\frac{\pi(e-x)}{\sqrt{3}\sigma}\right)\right)^{-1}, \quad x \in R \tag{1.9}$$

denoted by $\mathcal{N}(e, \sigma)$ where e and σ are real numbers with $\sigma > 0$.

Example 1.4 An uncertain variable ξ is called empirical if it has an empirical uncertainty distribution

$$\Phi(x) = \begin{cases} 0, & \text{if } x < x_1 \\ \alpha_i + \dfrac{(\alpha_{i+1} - \alpha_i)(x - x_i)}{x_{i+1} - x_i}, & \text{if } x_i \le x \le x_{i+1},\ 1 \le i < n \\ 1, & \text{if } x > x_n \end{cases} \tag{1.10}$$

where $x_1 < x_2 < \cdots < x_n$ and $0 \le \alpha_1 \le \alpha_2 \le \cdots \le \alpha_n \le 1$.

1.1.2 Independence

Definition 1.5 ([5]) The uncertain variables $\xi_1, \xi_2, \ldots, \xi_n$ are said to be independent if

$$\mathcal{M}\left\{\bigcap_{i=1}^{n}(\xi_i \in B_i)\right\} = \bigwedge_{i=1}^{n}\mathcal{M}\{\xi_i \in B_i\} \tag{1.11}$$

for any Borel sets $B_1, B_2, \ldots, B_n$.

Theorem 1.2 ([5]) *The uncertain variables $\xi_1, \xi_2, \ldots, \xi_n$ are independent if and only if*

$$\mathcal{M}\left\{\bigcup_{i=1}^{n}(\xi_i \in B_i)\right\} = \bigvee_{i=1}^{n}\mathcal{M}\{\xi_i \in B_i\} \tag{1.12}$$

for any Borel sets $B_1, B_2, \ldots, B_n$.

Theorem 1.3 ([5]) *Let $\xi_1, \xi_2, \ldots, \xi_n$ be independent uncertain variables, and let $f_1, f_2, \ldots, f_n$ be measurable functions. Then $f_1(\xi_1), f_2(\xi_2), \ldots, f_n(\xi_n)$ are independent uncertain variables.*

1.2 Expected Value

Expected value is the average value of uncertain variable in the sense of uncertain measure and represents the size of uncertain variable.

Definition 1.6 ([1]) Let ξ be an uncertain variable. Then the expected value of ξ is defined by

$$E[\xi] = \int_0^{+\infty}\mathcal{M}\{\xi \geq x\}\mathrm{d}x - \int_{-\infty}^{0}\mathcal{M}\{\xi \leq x\}\mathrm{d}x \tag{1.13}$$

provided that at least one of the two integrals is finite.

Theorem 1.4 ([1]) *Let ξ be an uncertain variable with uncertainty distribution Φ. Then*

$$E[\xi] = \int_0^{+\infty}(1 - \Phi(x))\mathrm{d}x - \int_{-\infty}^{0}\Phi(x)\mathrm{d}x. \tag{1.14}$$

Definition 1.7 ([4]) An uncertainty distribution $\Phi(x)$ is said to be regular if it is a continuous and strictly increasing function with respect to x at which $0 < \Phi(x) < 1$, and

$$\lim_{x \to -\infty}\Phi(x) = 0, \quad \lim_{x \to +\infty}\Phi(x) = 1.$$

Theorem 1.5 ([4]) *Let ξ be an uncertain variable with regular uncertainty distribution Φ. Then*

$$E[\xi] = \int_0^1 \Phi^{-1}(\alpha)\mathrm{d}\alpha. \tag{1.15}$$

Theorem 1.6 ([7]) *Assume $\xi_1, \xi_2, \ldots, \xi_n$ are independent uncertain variables with regular uncertainty distributions $\Phi_1, \Phi_2, \ldots, \Phi_n$, respectively. If $f(x_1, x_2, \ldots, x_n)$ is strictly increasing with respect to $x_1, x_2, \ldots, x_m$ and strictly decreasing with respect to $x_{m+1}, x_{m+2}, \ldots, x_n$, then the uncertain variable $\xi = f(\xi_1, \xi_2, \ldots, \xi_n)$ has an expected value*

$$E[\xi] = \int_0^1 f(\Phi_1^{-1}(\alpha), \cdots, \Phi_m^{-1}(\alpha), \Phi_{m+1}^{-1}(1-\alpha), \cdots, \Phi_n^{-1}(1-\alpha))\mathrm{d}\alpha. \tag{1.16}$$

Theorem 1.7 ([4]) *Let ξ and η be independent uncertain variables with finite expected values. Then for any real numbers a and b, we have*

$$E[a\xi + b\eta] = aE[\xi] + bE[\eta]. \tag{1.17}$$

Definition 1.8 ([1]) Let ξ be an uncertain variable with finite expected value e. Then the variance of ξ is

$$V[\xi] = E[(\xi - e)^2]. \tag{1.18}$$

Let ξ be an uncertain variable with expected value e. If we only know its uncertainty distribution Φ, then the variance

$$\begin{aligned}
V[\xi] &= \int_0^{+\infty} \mathcal{M}\{(\xi - e)^2 \geq x\}\mathrm{d}x \\
&= \int_0^{+\infty} \mathcal{M}\{(\xi \geq e + \sqrt{x}) \cup (\xi \leq e - \sqrt{x})\}\mathrm{d}x \\
&\leq \int_0^{+\infty} (\mathcal{M}\{\xi \geq e + \sqrt{x}\} + \mathcal{M}\{\xi \leq e - \sqrt{x}\})\mathrm{d}x \\
&= \int_0^{+\infty} (1 - \Phi(e + \sqrt{x}) + \Phi(e - \sqrt{x}))\mathrm{d}x.
\end{aligned}$$

Thus the following stipulation is introduced.
Stipulation. Let ξ be an uncertain variable with uncertainty distribution Φ and finite expected value e. Then

$$V[\xi] = \int_0^{+\infty} (1 - \Phi(e + \sqrt{x}) + \Phi(e - \sqrt{x}))\mathrm{d}x. \tag{1.19}$$

Now let us give an estimation for the expected value of $a\xi + \xi^2$ if ξ is a normal uncertain variable [8].

Theorem 1.8 ([8]) *Let ξ be a normal uncertain variable with expected value* 0 *and variance σ^2 ($\sigma > 0$), whose uncertainty distribution is*

$$\Phi(x) = \left(1 + \exp\left(\frac{-\pi x}{\sqrt{3}\sigma}\right)\right)^{-1}, \quad x \in R.$$

Then for any real number a,

$$\frac{\sigma^2}{2} \le E[a\xi + \xi^2] \le \sigma^2. \tag{1.20}$$

Proof We only need to verify the conclusion under the case that $a > 0$ because the similar method is suitable to the case that $a \le 0$. Let

$$x_1 = \frac{-a - \sqrt{a^2 + 4r}}{2}, \quad x_2 = \frac{-a + \sqrt{a^2 + 4r}}{2}$$

which is derived from the solutions of the equation $ax + x^2 = r$ for any real number $r \ge -a^2/4$ (Denote $y_0 = -a^2/4$). Then

$$\begin{aligned} E[a\xi + \xi^2] &= \int_0^{+\infty} \mathcal{M}\{a\xi + \xi^2 \ge r\}\mathrm{d}r - \int_{y_0}^0 \mathcal{M}\{a\xi + \xi^2 \le r\}\mathrm{d}r \\ &= \int_0^{+\infty} \mathcal{M}\{(\xi \le x_1) \cup (\xi \ge x_2)\}\mathrm{d}r \\ &\quad - \int_{y_0}^0 \mathcal{M}\{(\xi \ge x_1) \cap (\xi \le x_2)\}\mathrm{d}r. \end{aligned} \tag{1.21}$$

Since

$$\begin{aligned} \mathcal{M}\{\xi \le x_2\} &= \mathcal{M}\{((\xi \ge x_1) \cap (\xi \le x_2)) \cup (\xi \le x_1)\} \\ &\le \mathcal{M}\{(\xi \ge x_1) \cap (\xi \le x_2)\} + \mathcal{M}\{\xi \le x_1\}, \end{aligned}$$

we have

$$\mathcal{M}\{(\xi \ge x_1) \cap (\xi \le x_2)\} \ge \mathcal{M}\{\xi \le x_2\} - \mathcal{M}\{\xi \le x_1\} = \Phi(x_2) - \Phi(x_1).$$

Notice that

$$\mathcal{M}\{(\xi \le x_1) \cup (\xi \ge x_2)\} \le \mathcal{M}\{\xi \le x_1\} + \mathcal{M}\{\xi \ge x_2\} = \Phi(x_1) + 1 - \Phi(x_2).$$

Hence, it follows from (1.21) that

$$
\begin{aligned}
E[a\xi+\xi^2] &< \int_0^{+\infty}\Phi(x_1)\mathrm{d}r+\int_0^{+\infty}(1-\Phi(x_2))\mathrm{d}r-\int_{y_0}^{0}(\Phi(x_2)-\Phi(x_1))\mathrm{d}r\\
&=\int_0^{+\infty}\frac{1}{1+\exp\left(-\frac{\pi x_1}{\sqrt{3}\sigma}\right)}\mathrm{d}r+\int_0^{+\infty}\frac{1}{1+\exp\left(\frac{\pi x_2}{\sqrt{3}\sigma}\right)}\mathrm{d}r\\
&\quad-\int_{y_0}^{0}\frac{1}{1+\exp\left(-\frac{\pi x_2}{\sqrt{3}\sigma}\right)}\mathrm{d}r+\int_{y_0}^{0}\frac{1}{1+\exp\left(-\frac{\pi x_1}{\sqrt{3}\sigma}\right)}\mathrm{d}r\\
&=\int_{-a}^{-\infty}\frac{a+2x}{1+\exp\left(-\frac{\pi x}{\sqrt{3}\sigma}\right)}\mathrm{d}x+\int_0^{+\infty}\frac{a+2x}{1+\exp\left(\frac{\pi x}{\sqrt{3}\sigma}\right)}\mathrm{d}x\\
&\quad-\int_{-a/2}^{0}\frac{a+2x}{1+\exp\left(-\frac{\pi x}{\sqrt{3}\sigma}\right)}\mathrm{d}x+\int_{-a/2}^{-a}\frac{a+2x}{1+\exp\left(-\frac{\pi x}{\sqrt{3}\sigma}\right)}\mathrm{d}x\\
&=\int_a^{+\infty}\frac{a-2x}{1+\exp\left(\frac{\pi x}{\sqrt{3}\sigma}\right)}(-\mathrm{d}x)+\int_0^{+\infty}\frac{a+2x}{1+\exp\left(\frac{\pi x}{\sqrt{3}\sigma}\right)}\mathrm{d}x\\
&\quad-\int_{a/2}^{0}\frac{a-2x}{1+\exp\left(\frac{\pi x}{\sqrt{3}\sigma}\right)}(-\mathrm{d}x)+\int_{a/2}^{a}\frac{a-2x}{1+\exp\left(-\frac{\pi x}{\sqrt{3}\sigma}\right)}(-\mathrm{d}x)\\
&=a\int_0^{a}\frac{1}{1+\exp\left(\frac{\pi x}{\sqrt{3}\sigma}\right)}\mathrm{d}x+2\int_a^{+\infty}\frac{x}{1+\exp\left(\frac{\pi x}{\sqrt{3}\sigma}\right)}\mathrm{d}x\\
&\quad+2\int_0^{+\infty}\frac{x}{1+\exp\left(\frac{\pi x}{\sqrt{3}\sigma}\right)}\mathrm{d}x-\int_0^{a}\frac{a-2x}{1+\exp\left(\frac{\pi x}{\sqrt{3}\sigma}\right)}\mathrm{d}x\\
&=4\int_0^{+\infty}\frac{x}{1+\exp\left(\frac{\pi x}{\sqrt{3}\sigma}\right)}\mathrm{d}x\\
&=\sigma^2.
\end{aligned}
\tag{1.22}
$$

On the other hand, since

$$\mathcal{M}\{(\xi\le x_1)\cup(\xi\ge x_2)\}\ge\mathcal{M}\{\xi\ge x_2\}=1-\Phi(x_2),$$

and

$$\mathcal{M}\{(\xi\ge x_1)\cap(\xi\le x_2)\}\le\mathcal{M}\{\xi\le x_2\}=\Phi(x_2),$$

it follows from (1.21) that

$$
\begin{aligned}
E[a\xi+\xi^2] &\ge \int_0^{+\infty}(1-\Phi(x_2))\mathrm{d}r-\int_{y_0}^{0}\Phi(x_2)\mathrm{d}r\\
&=\int_0^{+\infty}\frac{1}{1+\exp\left(\frac{\pi x_2}{\sqrt{3}\sigma}\right)}\mathrm{d}r-\int_{y_0}^{0}\frac{1}{1+\exp\left(-\frac{\pi x_2}{\sqrt{3}\sigma}\right)}\mathrm{d}r\\
&=\int_0^{+\infty}\frac{a+2x}{1+\exp\left(\frac{\pi x}{\sqrt{3}\sigma}\right)}\mathrm{d}x-\int_{-a/2}^{0}\frac{a+2x}{1+\exp\left(-\frac{\pi x}{\sqrt{3}\sigma}\right)}\mathrm{d}x\\
&=2\int_0^{+\infty}\frac{x}{1+\exp\left(\frac{\pi x}{\sqrt{3}\sigma}\right)}\mathrm{d}x+2\int_0^{a/2}\frac{x}{1+\exp\left(\frac{\pi x}{\sqrt{3}\sigma}\right)}\mathrm{d}x\\
&\quad+a\int_{a/2}^{+\infty}\frac{1}{1+\exp\left(\frac{\pi x}{\sqrt{3}\sigma}\right)}\mathrm{d}x\\
&=\frac{\sigma^2}{2}+\frac{6\sigma^2}{\pi^2}\int_0^{\frac{\pi a}{2\sqrt{3}\sigma}}\frac{z}{1+e^z}\mathrm{d}z+\frac{\sqrt{3}a\sigma}{\pi}\int_{\frac{\pi a}{2\sqrt{3}\sigma}}^{+\infty}\frac{1}{1+e^z}\mathrm{d}z\\
&\ge\frac{\sigma^2}{2}.
\end{aligned}
\tag{1.23}
$$

Combining (1.22) and (1.23) yields the conclusion. The theorem is completed.

Given an increasing function $\Phi(x)$ whose values are in $[0, 1]$, Peng and Iwamura [6] introduced an uncertainty space $(R, \mathcal{B}, \mathcal{M})$ as follows. Let $\mathcal{B}$ be the Borel algebra over R. Let $\mathcal{C}$ be the collection of all intervals of the form $(-\infty, a]$, $(b, +\infty)$, $\emptyset$ and R. The uncertain measure $\mathcal{M}$ is provided in such a way: first,

$$\mathcal{M}\{(-\infty,a]\}=\Phi(a),\quad \mathcal{M}\{(b,+\infty)\}=1-\Phi(b),\quad \mathcal{M}\{\emptyset\}=0,\quad \mathcal{M}\{R\}=1.$$

Second, for any $B \in \mathcal{B}$, there exists a sequence $\{A_i\}$ in $\mathcal{C}$ such that

$$B\subset\bigcup_{i=1}^{\infty}A_i.$$

Thus

$$
\mathcal{M}\{B\}=\begin{cases}
\inf\limits_{B\subset\cup A_i}\sum\limits_{i=1}^{\infty}\mathcal{M}\{A_i\}, & \text{if } \inf\limits_{B\subset\cup A_i}\sum\limits_{i=1}^{\infty}\mathcal{M}\{A_i\}<0.5\\
1-\inf\limits_{B^c\subset\cup A_i}\sum\limits_{i=1}^{\infty}\mathcal{M}\{A_i\}, & \text{if } \inf\limits_{B^c\subset\cup A_i}\sum\limits_{i=1}^{\infty}\mathcal{M}\{A_i\}<0.5\\
0.5, & \text{otherwise.}
\end{cases}
\tag{1.24}
$$

The uncertain variable defined by $\xi(\gamma)=\gamma$ from the uncertainty space $(R, \mathcal{B}, \mathcal{M})$ to R has the uncertainty distribution Φ.

Note that for monotone increasing function $\Phi(x)$ except $\Phi(x) \equiv 0$ and $\Phi(x) \equiv 1$, there may be multiple uncertain variables whose uncertainty distributions are just $\Phi(x)$. However, for any one ξ among them, the uncertain measure of the event $\{\xi \in B\}$ for Borel set B may not be analytically expressed by $\Phi(x)$. For any two ξ and η among them, the uncertain measure of $\{\xi \in B\}$ may differ from that of $\{\eta \in B\}$. These facts result in inconvenience of use in practice. Which one among them should we choose for reasonable and convenient use? Let us consider the uncertain variable ξ defined by $\xi(\gamma) = \gamma$ on the uncertainty space $(R, \mathcal{B}, \mathcal{M})$ with the uncertainty distribution $\Phi(x)$, where the uncertain measure $\mathcal{M}$ is defined by (1.24), and another uncertain variable ξ_1 on the uncertainty space $(\Gamma_1, \mathcal{L}_1, \mathcal{M}_1)$. For each $A \in \mathcal{C}$, we have $\mathcal{M}\{\xi \in A\} = \mathcal{M}_1\{\xi_1 \in A\}$. For any Borel set $B \subset R$, if $B \subset \cup_{i=1}^{\infty} A_i$ with $\sum_{i=1}^{\infty} \mathcal{M}\{A_i\} < 0.5$, then

$$\begin{aligned}\mathcal{M}_1\{\xi_1 \in B\} &\le \mathcal{M}_1\left\{\bigcup_{i=1}^{\infty}\{\xi_1 \in A_i\}\right\} \le \sum_{i=1}^{\infty} \mathcal{M}_1\{\xi_1 \in A_i\} \\ &= \sum_{i=1}^{\infty} \mathcal{M}\{\xi \in A_i\} < 0.5;\end{aligned}$$

if $B^c \subset \cup_{i=1}^{\infty} A_i$ with $\sum_{i=1}^{\infty} \mathcal{M}\{A_i\} < 0.5$, then

$$\begin{aligned}\mathcal{M}_1\{\xi_1 \in B\} &= 1 - \mathcal{M}_1\{\xi_1 \in B^c\} \ge 1 - \mathcal{M}_1\left\{\bigcup_{i=1}^{\infty}\{\xi_1 \in A_i\}\right\} \\ &\ge 1 - \sum_{i=1}^{\infty} \mathcal{M}_1\{\xi_1 \in A_i\} = 1 - \sum_{i=1}^{\infty} \mathcal{M}\{\xi \in A_i\} > 0.5.\end{aligned}$$

Thus

$$\mathcal{M}_1\{\xi_1 \in B\} \le \mathcal{M}\{\xi \in B\} = \inf_{B \subset \cup A_i} \sum_{i=1}^{\infty} \mathcal{M}\{A_i\} < 0.5$$

if $\inf_{B \subset \cup A_i} \sum_{i=1}^{\infty} \mathcal{M}\{A_i\} < 0.5$ and

$$\mathcal{M}_1\{\xi_1 \in B\} \ge \mathcal{M}\{\xi \in B\} = 1 - \inf_{B^c \subset \cup A_i} \sum_{i=1}^{\infty} \mathcal{M}\{A_i\} > 0.5$$

if $\inf_{B^c \subset \cup A_i} \sum_{i=1}^{\infty} \mathcal{M}\{A_i\} < 0.5$.

In other cases, $\mathcal{M}\{\xi \in B\} = 0.5$. Therefore, the uncertain measure of $\{\xi \in B\}$ is closer to 0.5 than that of $\{\xi_1 \in B\}$. Based on the maximum uncertainty principle [4], we adopt uncertain variable ξ defined on $(R, \mathcal{B}, \mathcal{M})$ for use in our discussion if only the uncertainty distribution is provided.

Definition 1.9 ([9]) An uncertain variable ξ with distribution $\Phi(x)$ is an ordinary uncertain variable if it is from the uncertainty space $(R, \mathcal{B}, \mathcal{M})$ to R defined by $\xi(\gamma) = \gamma$, where $\mathcal{B}$ is the Borel algebra over R and $\mathcal{M}$ is defined by (1.24).

Let $\Phi(x)$ be continuous. For uncertain measure $\mathcal{M}$ defined by (1.24), we know that $\mathcal{M}\{(-\infty, a)\} = \Phi(a)$ and $\mathcal{M}\{[b, +\infty)\} = 1 - \Phi(b)$.

Definition 1.10 ([9]) An uncertain vector $\boldsymbol{\xi} = (\xi_1, \xi_2, \ldots, \xi_n)$ is ordinary if every uncertain variable ξ_i is ordinary for $i = 1, 2, \ldots, n$.

1.2.1 Distribution of Function of Uncertain Variable

Let us discuss the distribution of $f(\xi)$ for an ordinary uncertain variable ξ or an ordinary uncertain vector. Assume $\mathcal{C}$ is the collection of all intervals of the form $(-\infty, a]$, $(b, +\infty)$, $\emptyset$ and R. Each element A_i emerging in the sequel is in $\mathcal{C}$.

Theorem 1.9 ([9]) *(i) Let ξ be an ordinary uncertain variable with the continuous distribution $\Phi(x)$ and $f(x)$ a Borel function. Then the distribution of the uncertain variable $f(\xi)$ is*

$$\Psi(x) = \mathcal{M}\{f(\xi) \le x\} = \begin{cases} \inf\limits_{\{f(\xi)\le x\}\subset \cup A_i} \sum\limits_{i=1}^{\infty} \mathcal{M}\{A_i\}, & \text{if } \inf\limits_{\{f(\xi)\le x\}\subset \cup A_i} \sum\limits_{i=1}^{\infty} \mathcal{M}\{A_i\} < 0.5 \\ 1 - \inf\limits_{\{f(\xi)> x\}\subset \cup A_i} \sum\limits_{i=1}^{\infty} \mathcal{M}\{A_i\}, & \text{if } \inf\limits_{\{f(\xi)> x\}\subset \cup A_i} \sum\limits_{i=1}^{\infty} \mathcal{M}\{A_i\} < 0.5 \\ 0.5, & \text{otherwise.} \end{cases} \tag{1.25}$$

(ii) Let $f : R^n \to R$ be a Borel function, and $\boldsymbol{\xi} = (\xi_1, \xi_2, \ldots, \xi_n)$ be an ordinary uncertain vector. Then the distribution of the uncertain variable $f(\boldsymbol{\xi})$ is

$$\begin{aligned} \Psi(x) &= \mathcal{M}\{f(\xi_1, \xi_2, \ldots, \xi_n) \le x\} \\ &= \mathcal{M}\{(\xi_1, \xi_2, \ldots, \xi_n) \in f^{-1}(-\infty, x)\} \\ &= \begin{cases} \sup\limits_{\Lambda_1\times\Lambda_2\times\cdots\times\Lambda_n\subset\Lambda} \min\limits_{1\le k\le n} \mathcal{M}_k\{\Lambda_k\} \\ \quad \text{if } \sup\limits_{\Lambda_1\times\Lambda_2\times\cdots\times\Lambda_n\subset\Lambda} \min\limits_{1\le k\le n} \mathcal{M}_k\{\Lambda_k\} > 0.5, \\ 1 - \sup\limits_{\Lambda_1\times\Lambda_2\times\cdots\times\Lambda_n\subset\Lambda^c} \min\limits_{1\le k\le n} \mathcal{M}_k\{\Lambda_k\} \\ \quad \text{if } \sup\limits_{\Lambda_1\times\Lambda_2\times\cdots\times\Lambda_n\subset\Lambda^c} \min\limits_{1\le k\le n} \mathcal{M}_k\{\Lambda_k\} > 0.5, \\ 0.5, \quad \text{otherwise} \end{cases} \end{aligned} \tag{1.26}$$

where $\Lambda = f^{-1}(-\infty, x)$, and each $\mathcal{M}_k\{\Lambda_k\}$ is derived from (1.24).

Proof The conclusions follow directly from (1.24) and (1.5), respectively.

Theorem 1.10 ([9]) *Let ξ be an ordinary uncertain variable with the continuous distribution $\Phi(x)$. For real numbers b and c, denote*

$$x_1 = \frac{-b - \sqrt{b^2 - 4(c-x)}}{2}, \quad x_2 = \frac{-b + \sqrt{b^2 - 4(c-x)}}{2}$$

for $x \geq c - b^2/4$. Then the distribution of the uncertain variable $\xi^2 + b\xi + c$ is

$$\Psi(x) = \begin{cases} 0, & \text{if } x < c - \dfrac{b^2}{4} \\ \Phi(x_2) \wedge (1 - \Phi(x_1)), & \text{if } \Phi(x_2) \wedge (1 - \Phi(x_1)) < 0.5 \\ \Phi(x_2) - \Phi(x_1), & \text{if } \Phi(x_2) - \Phi(x_1) > 0.5 \\ 0.5, & \text{otherwise.} \end{cases} \tag{1.27}$$

Proof For $x < c - b^2/4$, we have

$$\Psi(x) = \mathcal{M}\{\xi^2 + b\xi + c \leq x\} = \mathcal{M}\{\emptyset\} = 0.$$

Let $x \geq c - b^2/4$ in the sequel. Then

$$\Psi(x) = \mathcal{M}\{\xi^2 + b\xi + c \leq x\} = \mathcal{M}\{x_1 \leq \xi \leq x_2\} = \mathcal{M}\{[x_1, x_2]\}.$$

The conclusion will be proved by (1.24). Since $[x_1, x_2] \subset (-\infty, x_2]$ and $[x_1, x_2] \subset [x_1, +\infty)$, and $\mathcal{M}\{(-\infty, x_2]\} = \Phi(x_2)$ and $\mathcal{M}\{[x_1, +\infty)\} = 1 - \Phi(x_1)$, we have

$$\Psi(x) = \Phi(x_2) \wedge (1 - \Phi(x_1)) \text{ if } \Phi(x_2) \wedge (1 - \Phi(x_1)) < 0.5.$$

Since $[x_1, x_2]^c = (-\infty, x_1) \cup (x_2, +\infty)$, we have

$$\Psi(x) = 1 - (\Phi(x_1) + 1 - \Phi(x_2)) = \Phi(x_2) - \Phi(x_1)$$

if $\mathcal{M}\{(-\infty, x_1)\} + \mathcal{M}\{(x_2, +\infty)\} = \Phi(x_1) + 1 - \Phi(x_2) < 0.5$, or $\Phi(x_2) - \Phi(x_1) > 0.5$. Otherwise $\Psi(x) = 0.5$. The proof of the theorem is completed.

1.2.2 Expected Value of Function of Uncertain Variable

If the expected value of uncertain variable ξ with uncertainty distribution $\Phi(x)$ exists, then

$$E[\xi] = \int_0^{+\infty} (1 - \Phi(x))dx - \int_{-\infty}^0 \Phi(x)dx;$$

or

$$E[\xi] = \int_0^1 \Phi^{-1}(\alpha)d\alpha$$

provided that $\Phi^{-1}(\alpha)$ exists and unique for each $\alpha \in (0, 1)$. Thus, if we obtain the uncertainty distribution $\Psi(x)$ of $f(\xi)$, the expected value of $f(\xi)$ is easily derived from

$$E[f(\xi)] = \int_0^{+\infty} (1 - \Psi(x))dx - \int_{-\infty}^0 \Psi(x)dx. \tag{1.28}$$

For a monotone function $f(x)$, Theorem 1.6 gives a formula to compute the expected value of $f(\xi)$ with the uncertainty distribution $\Phi(x)$ of ξ. However, we may generally not present a formula to compute the expected value of $f(\xi)$ with $\Phi(x)$ for a nonmonotone function $f(x)$ because the uncertainty distribution $\Psi(x)$ of $f(\xi)$ may not be analytically expressed by $\Phi(x)$.

Now if we consider an ordinary uncertain variable ξ, the uncertainty distribution $\Psi(x)$ of $f(\xi)$ may be presented by (1.25), and then the expected value of $f(\xi)$ can be obtained by (1.28). Next, we will give some examples to show how to compute the expected value of $f(\xi)$ for an ordinary uncertain variable ξ no matter whether $f(x)$ is monotone.

Example 1.5 Let ξ be an ordinary linear uncertain variable $\mathcal{L}(a, b)$ with the distribution (also see Fig. 1.1)

$$\Phi(x) = \begin{cases} 0, & \text{if } x \le a \\ (x-a)/(b-a), & \text{if } a \le x \le b \\ 1, & \text{if } x \ge b. \end{cases}$$

The expected value of ξ is $e = (a+b)/2$. Now we consider the variance of ξ: $V[\xi] = E[(\xi - e)^2]$. Let the uncertainty distribution of $(\xi - e)^2$ be $\Psi(x)$. Let $x \ge 0$, and $x_1 = e - \sqrt{x}$, $x_2 = e + \sqrt{x}$. If $\sqrt{x} \ge (b-a)/2$, then $x_2 \ge b$ and $x_1 \le a$. Thus $\Psi(x) = \Phi(x_2) - \Phi(x_1) = 1$. If $\sqrt{x} \le (b-a)/2$, then $e \le x_2 \le b$ and $a \le x_1 \le e$. Thus $\Phi(x_2) \wedge (1 - \Phi(x_1)) > 0.5$. When $\Phi(x_2) - \Phi(x_1) = 2\sqrt{x}/(b-a) > 0.5$,

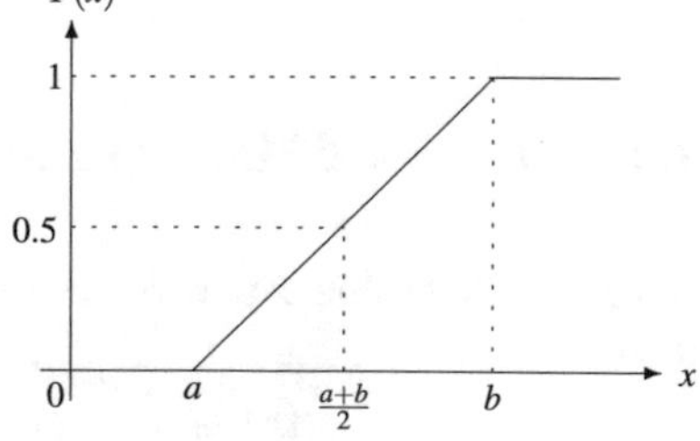

Fig. 1.1 Linear uncertainty distribution

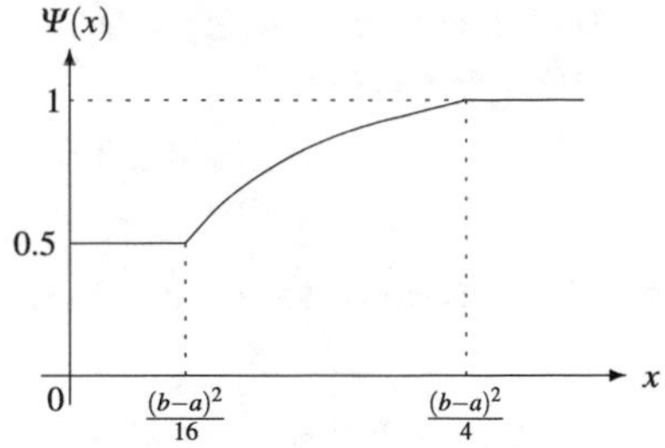

Fig. 1.2 Uncertainty distribution of $(\xi - e)^2$

that is, $\sqrt{x} > (b-a)/4$, $\Psi(x) = \Phi(x_2) - \Phi(x_1) = 2\sqrt{x}/(b-a)$. Hence, the uncertainty distribution of $(\xi - e)^2$ (also see Fig. 1.2) is

$$\Psi(x) = \begin{cases} 0, & \text{if } x < 0 \\ 0.5, & \text{if } 0 \le x \le (b-a)^2/16 \\ 2\sqrt{x}/(b-a), & \text{if } (b-a)^2/16 \le x \le (b-a)^2/4 \\ 1, & \text{if } x \ge (b-a)^2/4 \end{cases}$$

by (1.27).

The variance of ξ is

$$\begin{aligned} V[\xi] = E[(\xi - e)^2] &= \int_0^{+\infty} (1 - \Psi(x))\mathrm{d}x \\ &= \int_0^{(b-a)^2/16} 0.5\mathrm{d}x + \int_{(b-a)^2/16}^{(b-a)^2/4} \left(1 - \frac{2\sqrt{x}}{b-a}\right)\mathrm{d}x \\ &= \frac{7}{96}(b-a)^2. \end{aligned}$$

Example 1.6 Let ξ be an ordinary linear uncertain variable $\mathcal{L}(-1, 1)$ with the distribution

$$\Phi(x) = \begin{cases} 0, & \text{if } x \le -1 \\ (x+1)/2, & \text{if } -1 \le x \le 1 \\ 1, & \text{if } x \ge 1. \end{cases}$$

We will consider the expected value $E[\xi^2 + b\xi]$ for real number b. Let the uncertainty distribution of uncertain variable $\eta = \xi^2 + b\xi$ be $\Psi(x)$. For $x \ge -b^2/4$, denote

$$x_1 = \frac{-b - \sqrt{b^2 + 4x}}{2}, \quad x_2 = \frac{-b + \sqrt{b^2 + 4x}}{2}.$$

(I) If $b = 0$, then $E[\xi^2] = 7/24$ by Example 1.5.
(II) If $b \geq 2$, then

$$\Psi(x) = \begin{cases} 0, & \text{if } x < 1 - b \\ \Phi(x_2), & \text{if } x \geq 1 - b \end{cases}$$

by (1.27). Note that $x = x_2^2 + bx_2$. Thus

$$\begin{aligned} E[\xi^2 + b\xi] &= \int_0^{+\infty} (1 - \Psi(x))\mathrm{d}x - \int_{-\infty}^0 \Psi(x)\mathrm{d}x \\ &= \int_0^{1+b} (1 - \Phi(x_2))\mathrm{d}x - \int_{1-b}^0 \Phi(x_2)\mathrm{d}x \\ &= \int_0^1 \left(1 - \frac{y+1}{2}\right)(2y + b)\mathrm{d}y - \int_{-1}^0 \frac{y+1}{2}(2y + b)\mathrm{d}y \\ &= \frac{1}{3}. \end{aligned}$$

(III) If $1 \leq b < 2$, then

$$\Psi(x) = \begin{cases} 0, & \text{if } x < -b^2/4 \\ \Phi(x_2), & \text{if } x \geq -b^2/4. \end{cases}$$

Thus

$$\begin{aligned} E[\xi^2 + b\xi] &= \int_0^{+\infty} (1 - \Psi(x))\mathrm{d}x - \int_{-\infty}^0 \Psi(x)\mathrm{d}x \\ &= \int_0^{1+b} (1 - \Phi(x_2))\mathrm{d}x - \int_{-b^2/4}^0 \Phi(x_2)\mathrm{d}x \\ &= \int_0^1 \left(1 - \frac{y+1}{2}\right)(2y + b)\mathrm{d}y - \int_{-b/2}^0 \frac{y+1}{2}(2y + b)\mathrm{d}y \\ &= \frac{1}{48}(b^3 - 6b^2 + 12b + 8). \end{aligned}$$

(IV) If $0 < b < 1$, then

$$\Psi(x) = \begin{cases} 0, & \text{if } x < -b^2/4 \\ \Phi(x_2), & \text{if } -b^2/4 \leq x < 0 \\ 0.5, & \text{if } 0 \leq x \leq (1 - b^2)/4 \\ \Phi(x_2) - \Phi(x_1), & \text{if } x > (1 - b^2)/4. \end{cases}$$

Thus

$$
\begin{aligned}
E[\xi^2 + b\xi] &= \int_0^{+\infty} (1 - \Psi(x))\mathrm{d}x - \int_{-\infty}^{0} \Psi(x)\mathrm{d}x \\
&= \int_0^{(1-b^2)/4} \frac{1}{2}\mathrm{d}x + \int_{(1-b^2)/4}^{1-b} (1 - \Phi(x_2) + \Phi(x_1))\mathrm{d}x \\
&\quad + \int_{1-b}^{1+b} (1 - \Phi(x_2))\mathrm{d}x - \int_{-b^2/4}^{0} \Phi(x_2)\mathrm{d}x \\
&= \frac{1-b^2}{8} + \int_{(1-b)/2}^{1-b} \left(1 - \frac{y+1}{2} + \frac{-y-b+1}{2}\right)(2y+b)\mathrm{d}y \\
&\quad + \int_{1-b}^{1} \left(1 - \frac{y+1}{2}\right)(2y+b)\mathrm{d}y - \int_{-b/2}^{0} \frac{y+1}{2}(2y+b)\mathrm{d}y \\
&= \frac{1}{48}(b^3 + 12b^2 - 12b + 14).
\end{aligned}
$$

(V) If $b \le -2$, then

$$
\Psi(x) = \begin{cases} 0, & \text{if } x < 1+b \\ 1 - \Phi(x_1), & \text{if } x \ge 1+b. \end{cases}
$$

Also we have $E[\xi^2 + b\xi] = 1/3$.

(VI) If $-2 < b \le -1$, then

$$
\Psi(x) = \begin{cases} 0, & \text{if } x < -b^2/4 \\ 1 - \Phi(x_1), & \text{if } x \ge -b^2/4. \end{cases}
$$

Thus

$$
E[\xi^2 + b\xi] = \frac{1}{48}(-b^3 - 6b^2 - 12b + 8).
$$

(VII) If $-1 < b < 0$, then

$$
\Psi(x) = \begin{cases} 0, & \text{if } x < -b^2/4 \\ 1 - \Phi(x_1), & \text{if } -b^2/4 \le x < 0 \\ 0.5, & \text{if } 0 \le x \le (1-b^2)/4 \\ \Phi(x_2) - \Phi(x_1), & \text{if } x > (1-b^2)/4. \end{cases}
$$

Thus

$$
\begin{aligned}
E[\xi^2+b\xi] &= \int_0^{+\infty}(1-\Psi(x))\mathrm{d}x - \int_{-\infty}^0 \Psi(x)\mathrm{d}x \\
&= \int_0^{(1-b^2)/4}\frac{1}{2}\mathrm{d}x + \int_{(1-b^2)/4}^{1+b}(1-\Phi(x_2)+\Phi(x_1))\mathrm{d}x + \int_{1+b}^{1-b}\Phi(x_1)\mathrm{d}x \\
&\quad - \int_{-b^2/4}^{0}(1-\Phi(x_1))\mathrm{d}x \\
&= \frac{1-b^2}{8} + \int_{(1-b)/2}^{1}\left(1-\frac{y+1}{2}+\frac{-y-b+1}{2}\right)(2y+b)\mathrm{d}y \\
&\quad + \int_{(-1-b)/2}^{-1}\frac{y+1}{2}(2y+b)\mathrm{d}y - \int_{-b/2}^{0}\left(1-\frac{y+1}{2}\right)(2y+b)\mathrm{d}y \\
&= \frac{1}{48}(-b^3+12b^2+12b+14)
\end{aligned}
$$

1.3 Optimistic Value and Pessimistic Value

Definition 1.11 ([1]) Let ξ be an uncertain variable, and $\alpha \in (0, 1]$. Then $\xi_{\sup}(\alpha) = \sup\{r \mid \mathcal{M}\{\xi \geq r\} \geq \alpha\}$ is called the α-optimistic value to ξ; and $\xi_{\inf}(\alpha) = \inf\{r \mid \mathcal{M}\{\xi \leq r\} \geq \alpha\}$ is called the α-pessimistic value to ξ.

Example 1.7 Let ξ be a normal uncertain variable $\mathcal{N}(e, \sigma)(\sigma > 0)$. Then its α-optimistic value and α-pessimistic value are $\xi_{\sup}(\alpha) = e - \frac{\sqrt{3}\sigma}{\pi}\ln\frac{\alpha}{1-\alpha}$ and $\xi_{\inf}(\alpha) = e + \frac{\sqrt{3}\sigma}{\pi}\ln\frac{\alpha}{1-\alpha}$.

Theorem 1.11 ([1]) *Assume that ξ is an uncertain variable. Then we have*
(a) if $\lambda \geq 0$, then $(\lambda\xi)_{\sup}(\alpha) = \lambda\xi_{\sup}(\alpha)$, and $(\lambda\xi)_{\inf}(\alpha) = \lambda\xi_{\inf}(\alpha)$,
(b) if $\lambda < 0$, then $(\lambda\xi)_{\sup}(\alpha) = \lambda\xi_{\inf}(\alpha)$, and $(\lambda\xi)_{\inf}(\alpha) = \lambda\xi_{\sup}(\alpha)$.
(c) $(\xi+\eta)_{\sup}(\alpha) = \xi_{\sup}(\alpha) + \eta_{\sup}(\alpha)$ if ξ and η are independent.

Let us give an estimation for the α-optimistic value of $a\xi + b\xi^2$ if ξ is a normal uncertain variable ($\alpha \in (0, 1)$).

Theorem 1.12 ([10]) *Let ξ be a normal uncertain variable with expected value* 0 *and variance $\sigma^2(\sigma > 0)$, whose uncertainty distribution is*

$$
\Phi(x) = \left(1+\exp\left(\frac{-\pi x}{\sqrt{3}\sigma}\right)\right)^{-1}, \quad x \in R.
$$

Then for any real number a and any small enough $\varepsilon > 0$,

$$\left[a\xi + b\xi^2\right]_{\sup}(\alpha) \geq \frac{\sqrt{3}}{\pi}\ln\frac{1-\alpha}{\alpha}|a|\sigma + \left(\frac{\sqrt{3}}{\pi}\ln\frac{1-\alpha}{\alpha}\right)^2 b\sigma^2, \quad (1.29)$$

$$\left[a\xi + b\xi^2\right]_{\sup}(\alpha) \leq \frac{\sqrt{3}}{\pi}\ln\frac{1-\alpha+\varepsilon}{\alpha-\varepsilon}|a|\sigma + \left(\frac{\sqrt{3}}{\pi}\ln\frac{2-\varepsilon}{\varepsilon}\right)^2 b\sigma^2 \quad (1.30)$$

if $b > 0$; *and*

$$\left[a\xi + b\xi^2\right]_{\sup}(\alpha) \geq \frac{\sqrt{3}}{\pi}\ln\frac{1-\alpha-\varepsilon}{\alpha+\varepsilon}|a|\sigma + \left(\frac{\sqrt{3}}{\pi}\ln\frac{2-\varepsilon}{\varepsilon}\right)^2 b\sigma^2, \quad (1.31)$$

$$\left[a\xi + b\xi^2\right]_{\sup}(\alpha) \leq \frac{\sqrt{3}}{\pi}\ln\frac{1-\alpha}{\alpha}|a|\sigma + \left(\frac{\sqrt{3}}{\pi}\ln\frac{1-\alpha}{\alpha}\right)^2 b\sigma^2 \quad (1.32)$$

if $b < 0$; *and also*

$$\left[a\xi + b\xi^2\right]_{\sup}(\alpha) = \frac{\sqrt{3}}{\pi}\ln\frac{1-\alpha}{\alpha}|a|\sigma \quad (1.33)$$

if $b = 0$.

Proof (I) We first verify the conclusion under the case that $b > 0$. Let

$$x_1 = \frac{-a-\sqrt{a^2+4by}}{2b}, \quad x_2 = \frac{-a+\sqrt{a^2+4by}}{2b}$$

which are derived from the solutions of the equation $ax + bx^2 = y$ for any real number $y \geq y_0 = -\frac{a^2}{4b}$ (when $y < y_0$, $\mathcal{M}\left\{a\xi + b\xi^2 \geq y\right\} = 1$). If $a \geq 0$, we have

$$\begin{aligned}\mathcal{M}\left\{a\xi + b\xi^2 \geq y\right\} &= \mathcal{M}\{(\xi \leq x_1) \cup (\xi \geq x_2)\}\\ &\geq \mathcal{M}\{\xi \leq x_1\} \vee \mathcal{M}\{\xi \geq x_2\}\\ &= 1 - \Phi(x_2).\end{aligned}$$

Letting $1 - \Phi(x_2) = \alpha$, we get

$$\begin{aligned}y = ax_2 + bx_2^2 &= a\Phi(1-\alpha) + b\left[\Phi^{-1}(1-\alpha)\right]^2\\ &= a\frac{\sigma\sqrt{3}}{\pi}\ln\frac{1-\alpha}{\alpha} + b\frac{3\sigma^2}{\pi^2}\left(\ln\frac{1-\alpha}{\alpha}\right)^2.\end{aligned}$$

By the definition of α-optimistic value, we have

$$[a\xi + b\xi^2]_{\sup}(\alpha) \geq a\frac{\sigma\sqrt{3}}{\pi}\ln\frac{1-\alpha}{\alpha} + b\frac{3\sigma^2}{\pi^2}\left(\ln\frac{1-\alpha}{\alpha}\right)^2.$$

If $a < 0$, we have

$$\mathcal{M}\left\{a\xi + b\xi^2 \geq y\right\} \geq \mathcal{M}\{(\xi \leq x_1)\} \vee \mathcal{M}\{(\xi \geq x_2)\} = \Phi(x_1).$$

Letting $\Phi(x_1) = \alpha$, we have

$$[a\xi + b\xi^2]_{\sup}(\alpha) \geq -a\frac{\sigma\sqrt{3}}{\pi}\ln\frac{1-\alpha}{\alpha} + b\frac{3\sigma^2}{\pi^2}\left(\ln\frac{1-\alpha}{\alpha}\right)^2.$$

Hence, for any real number a, we obtain inequality (1.29).

On the other hand, for $\varepsilon > 0$ small enough, there exists a $d = d_\varepsilon > 0$ such that

$$\mathcal{M}\{\xi \leq -d\} = \mathcal{M}\{\xi \geq d\} = \frac{\varepsilon}{2}.$$

In fact, it follows from $\Phi(-d) = \mathcal{M}\{\xi \leq -d\} = \frac{\varepsilon}{2}$ that $d = \frac{\sqrt{3}\sigma}{\pi}\ln\frac{2-\varepsilon}{\varepsilon}$. Note that

$$\begin{aligned}&\{a\xi + b\xi^2 \geq y\}\\ &= \{a\xi + b\xi^2 \geq y,\ -d \leq \xi \leq d\} \cup \{a\xi + b\xi^2 \geq y,\ \xi < -d \text{ or } \xi > d\}.\end{aligned}$$

For each $\gamma \in \{\gamma \mid a\xi(\gamma) + b\xi(\gamma)^2 \geq y,\ -d \leq \xi(\gamma) \leq d\}$, we have

$$a\xi(\gamma) + b\xi(\gamma)^2 \leq a\xi(\gamma) + bd^2.$$

Then we get

$$\{a\xi + b\xi^2 \geq y\} \subseteq \{a\xi + bd^2 \geq y\} \cup \{\xi \leq -d\} \cup \{\xi \geq d\}.$$

So we have

$$\mathcal{M}\{a\xi + b\xi^2 \geq y\} \leq \mathcal{M}\{a\xi + bd^2 \geq y\} + \varepsilon.$$

Letting $\mathcal{M}\{a\xi + b\xi^2 \geq y\} \geq \alpha$, we have

$$\mathcal{M}\{a\xi + bd^2 \geq y\} + \varepsilon \geq \alpha, \quad \text{or} \quad \mathcal{M}\{a\xi + bd^2 \geq y\} \geq \alpha - \varepsilon. \tag{1.34}$$

It follows from inequality (1.34) and the definition of optimistic value that

$$y \leq (a\xi + bd^2)_{\sup}(\alpha - \varepsilon) = (a\xi)_{\sup}(\alpha - \varepsilon) + bd^2.$$

If $a \geq 0$, then

$$y \leq a\xi_{\sup}(\alpha - \varepsilon) + bd^2 = a\frac{\sigma\sqrt{3}}{\pi}\ln\frac{1-\alpha+\varepsilon}{\alpha-\varepsilon} + b\frac{3\sigma^2}{\pi^2}\left(\ln\frac{2-\varepsilon}{\varepsilon}\right)^2.$$

If $a < 0$, then

$$y \le a\xi_{\inf}(\alpha-\varepsilon)+bd^2 = -a\frac{\sigma\sqrt{3}}{\pi}\ln\frac{1-\alpha+\varepsilon}{\alpha-\varepsilon}+b\frac{3\sigma^2}{\pi^2}\left(\ln\frac{2-\varepsilon}{\varepsilon}\right)^2.$$

Therefore, inequality (1.30) holds.

(II) In the case of $b < 0$, we can prove the inequalities (1.31) and (1.32) by the similar method to the above process.

(III) When $b = 0$, if $a \ge 0$, $\left[a\xi + b\xi^2\right]_{\sup}(\alpha) = a\xi_{\sup}(\alpha) = a\sigma\frac{\sqrt{3}}{\pi}\ln\frac{1-\alpha}{\alpha}$; If $a < 0$, $\left[a\xi + b\xi^2\right]_{\sup}(\alpha) = a\xi_{\inf}(\alpha) = -a\sigma\frac{\sqrt{3}}{\pi}\ln\frac{1-\alpha}{\alpha}$. Thus, Eq. (1.33) is obtained. The theorem is proved.

Similarly, we can get an estimation for the α-pessimistic value of $a\xi + b\xi^2$ if ξ is a normal uncertain variable ($\alpha \in (0, 1)$).

Theorem 1.13 ([4]) *Let ξ be a normal uncertain variable with expected value 0 and variance $\sigma^2(\sigma > 0)$, whose uncertainty distribution is*

$$\Phi(x) = \left(1+\exp\left(\frac{-\pi x}{\sqrt{3}\sigma}\right)\right)^{-1}, \quad x \in R.$$

Then for any real number a and any small enough $\varepsilon > 0$,

$$\left[a\xi + b\xi^2\right]_{\inf}(\alpha) \ge \frac{\sqrt{3}}{\pi}\ln\frac{\alpha}{1-\alpha}|a|\sigma + \left(\frac{\sqrt{3}}{\pi}\ln\frac{\alpha}{1-\alpha}\right)^2 b\sigma^2, \tag{1.35}$$

$$\left[a\xi + b\xi^2\right]_{\inf}(\alpha) \le \frac{\sqrt{3}}{\pi}\ln\frac{\alpha+\varepsilon}{1-\alpha-\varepsilon}|a|\sigma + \left(\frac{\sqrt{3}}{\pi}\ln\frac{2-\varepsilon}{\varepsilon}\right)^2 b\sigma^2 \tag{1.36}$$

if $b > 0$; and

$$\left[a\xi + b\xi^2\right]_{\inf}(\alpha) \ge \frac{\sqrt{3}}{\pi}\ln\frac{\alpha-\varepsilon}{1-\alpha+\varepsilon}|a|\sigma + \left(\frac{\sqrt{3}}{\pi}\ln\frac{2-\varepsilon}{\varepsilon}\right)^2 b\sigma^2, \tag{1.37}$$

$$\left[a\xi + b\xi^2\right]_{\inf}(\alpha) \le \frac{\sqrt{3}}{\pi}\ln\frac{\alpha}{1-\alpha}|a|\sigma + \left(\frac{\sqrt{3}}{\pi}\ln\frac{\alpha}{1-\alpha}\right)^2 b\sigma^2 \tag{1.38}$$

if $b < 0$; and also

$$\left[a\xi + b\xi^2\right]_{\inf}(\alpha) = \frac{\sqrt{3}}{\pi}\ln\frac{\alpha}{1-\alpha}|a|\sigma \tag{1.39}$$

if $b = 0$.

Proof According to Theorem 1.11, we have

$$\left[a\xi + b\xi^2\right]_{\inf}(\alpha) = -\left[-a\xi - b\xi^2\right]_{\sup}(\alpha).$$

Then, via applying Theorem 1.12, the conclusions are easily proved.

1.4 Uncertain Simulation

It follows from Theorem 1.10 and the examples in the above section that the uncertainty distribution $\Psi(x)$ of $f(\boldsymbol{\xi})$ may be analytically expressed by (1.27) for a quadratic function $f(x)$. But $\Psi(x)$ may be hardly analytically expressed for other kinds of functions. Now we will introduce uncertain simulation approaches [9] for uncertainty distribution $\Psi(x)$, optimistic value $f_{\sup}$, and expected value $E[f(\boldsymbol{\xi})]$ of $f(\boldsymbol{\xi})$ based on (1.25) and (1.26).

(a) Let $\boldsymbol{\xi} = (\xi_1, \xi_2, \ldots, \xi_n)$ be an ordinary uncertain vector where ξ_i is an ordinary uncertain variable with continuous uncertainty distribution $\Phi_i(x)$ for $i = 1, 2, \ldots, n$, and $f : R^n \to R$ be a Borel function. We use Algorithm 1.1 to simulate the following uncertain measure:

$$L = \mathcal{M}\{f(\boldsymbol{\xi}) \le 0\}.$$

Algorithm 1.1 (Uncertain simulation for L)

Step 1. Set $m_1(i) = 0$ and $m_2(i) = 0$, $i = 1, 2, \ldots, n$.
Step 2. Randomly generate $\boldsymbol{u}_k = (\gamma_k^{(1)}, \gamma_k^{(2)}, \ldots, \gamma_k^{(n)})$ with $0 < \Phi_i(\gamma_k^{(i)}) < 1$, $i = 1, 2, \ldots, n$, $k = 1, 2, \ldots, N$.
Step 3. Rank $\gamma_k^{(i)}$ from small to large as $\gamma_1^{(i)} \le \gamma_2^{(i)} \le \cdots \le \gamma_N^{(i)}$, $i = 1, 2, \ldots, n$.
Step 4. From $k = 1$ to $k = N$, if $f(\boldsymbol{u}_k) \le 0$, $m_1(i) = m_1(i) + 1$, denote $x_{m_1(i)}^{(i)} = \gamma_k^{(i)}$; otherwise, $m_2(i) = m_2(i) + 1$, denote $y_{m_2(i)}^{(i)} = \gamma_k^{(i)}$, $i = 1, 2, \ldots, n$.
Step 5. Set $a^{(i)} = \Phi(x_{m_1(i)}^{(i)}) \wedge (1 - \Phi(x_1^{(i)})) \wedge (\Phi(x_1^{(i)}) + 1 - \Phi(x_2^{(i)})) \wedge \cdots \wedge (\Phi(x_{m_1(i)-1}^{(i)}) + 1 - \Phi(x_{m_1(i)}^{(i)}))$; $b^{(i)} = \Phi(y_{m_2(i)}^{(i)}) \wedge (1 - \Phi(y_1^{(i)})) \wedge (\Phi(y_1^{(i)}) + 1 - \Phi(y_2^{(i)})) \wedge \cdots \wedge (\Phi(y_{m_2(i)-1}^{(i)}) + 1 - \Phi(y_{m_2(i)}^{(i)}))$, $i = 1, 2, \ldots, n$.
Step 6. If $a^{(i)} < 0.5$, return $L_1^{(i)} = a^{(i)}$, $L_2^{(i)} = 1 - a^{(i)}$; if $b^{(i)} < 0.5$, return $L_1^{(i)} = 1 - b^{(i)}$, $L_2^{(i)} = b^{(i)}$; otherwise, return $L_1^{(i)} = 0.5$, $L_2^{(i)} = 0.5$, $i = 1, 2, \ldots, n$.
Step 7. If $a = L_1^{(1)} \wedge L_1^{(2)} \wedge \cdots \wedge L_1^{(n)} > 0.5$, then $L = a$; if $b = L_2^{(1)} \wedge L_2^{(2)} \wedge \cdots \wedge L_2^{(n)} > 0.5$, then $L = 1 - b$; otherwise, $L = 0.5$.

(b) Let $\boldsymbol{\xi} = (\xi_1, \xi_2, \ldots, \xi_n)$ be an ordinary uncertain vector where ξ_i is an ordinary uncertain variable with continuous uncertainty distribution $\Phi_i(x)$ for $i = 1, 2, \ldots, n$, and $f : R^n \to R$ be a Borel function. The Algorithm 1.2 is used to simulate the optimistic value:

$$f_{\sup} = \sup\{r \mid \mathcal{M}\{f(\boldsymbol{\xi}) \ge r\} \ge \alpha\}$$

where $\alpha \in (0, 1)$ is a predetermined confidence level.

Algorithm 1.2 (Uncertain simulation for $f_{\sup}$)

Step 1. Randomly generate $\boldsymbol{u}_k = (\gamma_k^{(1)}, \gamma_k^{(2)}, \ldots, \gamma_k^{(n)})$ with $0 < \Phi_i(\gamma_k^{(i)}) < 1$, $i = 1, 2, \ldots, n$, $k = 1, 2, \ldots, m$.
Step 2. Set $a = f(\boldsymbol{u}_1) \wedge f(\boldsymbol{u}_2) \wedge \cdots \wedge f(\boldsymbol{u}_m)$, $b = f(\boldsymbol{u}_1) \vee f(\boldsymbol{u}_2) \vee \cdots \vee f(\boldsymbol{u}_m)$.
Step 3. Set $r = (a + b)/2$.
Step 4. If $\mathcal{M}\{f(\boldsymbol{\xi}) \ge r\} \ge \alpha$, then $a \leftarrow r$.
Step 5. If $\mathcal{M}\{f(\boldsymbol{\xi}) \ge r\} < \alpha$, then $b \leftarrow r$.
Step 6. Repeat the third to fifth steps until $b - a < \epsilon$ for a sufficiently small number ϵ.
Step 7. $f_{\sup} = (a + b)/2$.

(c) Let $\boldsymbol{\xi} = (\xi_1, \xi_2, \ldots, \xi_n)$ be an ordinary uncertain vector where ξ_i is an ordinary uncertain variable with continuous uncertainty distribution $\Phi_i(x)$ for $i = 1, 2, \ldots, n$, and $f : R^n \to R$ be a Borel function. The expected value $E[f(\boldsymbol{\xi})]$ is approached by the Algorithm 1.3.

Algorithm 1.3 (Uncertain simulation for E)

Step 1. Set $E = 0$.
Step 2. Randomly generate $\boldsymbol{u}_k = (\gamma_k^{(1)}, \gamma_k^{(2)}, \ldots, \gamma_k^{(n)})$ with $0 < \Phi_i(\gamma_k^{(i)}) < 1$, $i = 1, 2, \ldots, n$, $k = 1, 2, \ldots, m$.
Step 3. Set $a = f(\boldsymbol{u}_1) \wedge f(\boldsymbol{u}_2) \wedge \cdots \wedge f(\boldsymbol{u}_m)$, $b = f(\boldsymbol{u}_1) \vee f(\boldsymbol{u}_2) \vee \cdots \vee f(\boldsymbol{u}_m)$.
Step 4. Randomly generate r from $[a, b]$.
Step 5. If $r \ge 0$, then $E \leftarrow E + \mathcal{M}\{f(\boldsymbol{\xi}) \ge r\}$.
Step 6. If $r < 0$, then $E \leftarrow E - \mathcal{M}\{f(\boldsymbol{\xi}) \le r\}$.
Step 7. Repeat the fourth to sixth steps for N times.
Step 8. $E[f(\boldsymbol{\xi})] = a \vee 0 + b \wedge 0 + E \cdot (b - a)/N$.

1.5 Uncertain Process

The study of uncertain process was started by Liu [11] in 2008 for modeling the evolution of uncertain phenomena.

Definition 1.12 ([11]) Let $(\Gamma, \mathcal{L}, \mathcal{M})$ be an uncertainty space and let T be a totally ordered set (e.g., time). An uncertain process is a function $X_t(\gamma)$ from $T \times (\Gamma, \mathcal{L}, \mathcal{M})$

to the set of real numbers such that $\{X_t \in B\}$ is an event for any Borel set B at each time t.

Remark 1.1 If X_t is an uncertain process, then X_t is an uncertain variable at each time t.

Example 1.8 Let a and b be real numbers with $a < b$. Assume X_t is a linear uncertain variable, i.e.,

$$X_t \sim \mathcal{L}(at, bt) \tag{1.40}$$

at each time t. Then X_t is an uncertain process.

Example 1.9 Let a, b, c be real numbers with $a < b < c$. Assume X_t is a zigzag uncertain variable, i.e.,

$$X_t \sim \mathcal{Z}(at, bt, ct) \tag{1.41}$$

at each time t. Then X_t is an uncertain process.

Example 1.10 Let e and σ be real numbers with $\sigma > 0$. Assume X_t is a normal uncertain variable, i.e.,

$$X_t \sim \mathcal{N}(et, \sigma t) \tag{1.42}$$

at each time t. Then X_t is an uncertain process.

Definition 1.13 ([11]) An uncertain process X_t is said to have independent increments if

$$X_{t_0},\ X_{t_1} - X_{t_0},\ X_{t_2} - X_{t_1},\ \ldots,\ X_{t_k} - X_{t_{k-1}} \tag{1.43}$$

are independent uncertain variables where t_0 is the initial time and $t_1, t_2, \ldots, t_k$ are any times with $t_0 < t_1 < \ldots < t_k$.

An uncertain process X_t is said to have *stationary increments* if its increments are identically distributed uncertain variables whenever the time intervals have the same length; i.e., for any given $t > 0$, the increments $X_{s+t} - X_s$ are identically distributed uncertain variables for all $s > 0$.

Definition 1.14 ([11]) An uncertain process is said to be a stationary independent increment process if it has not only stationary increments but also independent increments.

1.5.1 Liu Process

In 2009, Liu [5] investigated a type of stationary independent increment process whose increments are normal uncertain variables. Later, this process was named by the academic community as Liu process due to its importance and usefulness.

Definition 1.15 ([5]) An uncertain process C_t is said to be a canonical Liu process if

(i) $C_0 = 0$ and almost all sample paths are Lipschitz continuous,
(ii) C_t has stationary and independent increments,
(iii) every increment $C_{s+t} - C_s$ is a normal uncertain variable with expected value 0 and variance t^2.

1.5.2 Liu Integral

Definition 1.16 ([5]) Let X_t be an uncertain process and let C_t be a canonical Liu process. For any partition of closed interval $[a, b]$ with $a = t_1 < t_2 < \cdots < t_{k+1} = b$, the mesh is written as

$$\Delta = \max_{1 \le i \le k} |t_{i+1} - t_i|. \tag{1.44}$$

Then Liu integral of X_t with respect to C_t is defined as

$$\int_a^b X_t \mathrm{d}C_t = \lim_{\Delta \to 0} \sum_{i=1}^{k} X_{t_i} \cdot (C_{t_{i+1}} - C_{t_i}) \tag{1.45}$$

provided that the limit exists almost surely and is finite. In this case, the uncertain process X_t is said to be integrable.

Since X_t and C_t are uncertain variables at each time t, the limit in (1.45) is also an uncertain variable provided that the limit exists almost surely and is finite. Hence, an uncertain process X_t is integrable with respect to C_t if and only if the limit in (1.45) is an uncertain variable.

Theorem 1.14 ([5]) *Let $h(t, c)$ be a continuous differentiable function. Then $Z_t = h(t, C_t)$ is a Liu process and has an uncertain differential*

$$\mathrm{d}Z_t = \frac{\partial h}{\partial t}(t, C_t)\mathrm{d}t + \frac{\partial h}{\partial c}(t, C_t)\mathrm{d}C_t. \tag{1.46}$$

1.6 Uncertain Differential Equation

Definition 1.17 ([11]) Suppose C_t is a canonical Liu process, and f and g are two functions. Then

$$\mathrm{d}X_t = f(t, X_t)\mathrm{d}t + g(t, X_t)\mathrm{d}C_t \tag{1.47}$$

is called an uncertain differential equation. A solution is a Liu process X_t that satisfies (1.47) identically in t.

Remark 1.2 The uncertain differential equation (1.47) is equivalent to the uncertain integral equation

$$X_s = X_0 + \int_0^s f(t, X_t)\mathrm{d}t + \int_0^s g(t, X_t)\mathrm{d}C_t. \tag{1.48}$$

Theorem 1.15 *Let u_t and v_t be two integrable uncertain processes. Then the uncertain differential equation*

$$\mathrm{d}X_t = u_t\mathrm{d}t + v_t\mathrm{d}C_t \tag{1.49}$$

has a solution

$$X_t = X_0 + \int_0^t u_s\mathrm{d}s + \int_0^t v_s\mathrm{d}C_s. \tag{1.50}$$

Theorem 1.16 ([12], Existence and Uniqueness Theorem) *The uncertain differential equation*

$$\mathrm{d}X_t = f(t, X_t)\mathrm{d}t + g(t, X_t)\mathrm{d}C_t \tag{1.51}$$

has a unique solution if the coefficients $f(t, x)$ and $g(t, x)$ satisfy linear growth condition

$$|f(t, x)| + |g(t, x)| \le L(1 + |x|), \quad \forall x \in R, t \ge 0 \tag{1.52}$$

and Lipschitz condition

$$|f(t, x) - f(t, y)| + |g(t, x) - g(t, y)| \le L|x - y|, \quad \forall x, y \in R, t \ge 0 \tag{1.53}$$

for some constant L. Moreover, the solution is sample-continuous.

Definition 1.18 ([13]) Let α be a number with $0 < \alpha < 1$. An uncertain differential equation

$$\mathrm{d}X_t = f(t, X_t)\mathrm{d}t + g(t, X_t)\mathrm{d}C_t$$

is said to have an α-path X_t^α if it solves the corresponding ordinary differential equation

$$\mathrm{d}X_t^\alpha = f(t, X_t^\alpha)\mathrm{d}t + |g(t, X_t^\alpha)|\Phi^{-1}(\alpha)\mathrm{d}t$$

where $\Phi^{-1}(\alpha)$ is the inverse standard normal uncertain distribution, i.e.,

$$\Phi^{-1}(\alpha) = \frac{\sqrt{3}}{\pi} \ln \frac{\alpha}{1 - \alpha}.$$

Theorem 1.17 ([13]) *Let X_t and X_t^α be the solution and α-path of the uncertain differential equation*

$$\mathrm{d}X_t = f(t, X_t)\mathrm{d}t + g(t, X_t)\mathrm{d}C_t,$$

respectively. Then the solution X_t has an inverse uncertainty distribution

$$\Psi_t^{-1}(\alpha) = X_t^\alpha.$$

References

1. Liu B (2007) Uncertainty theory, 2nd edn. Springer, Berlin
2. Kahneman D, Tversky A (1979) Prospect theory: an analysis of decision under risk. Econometrica 47(4):263–292
3. Liu B (2012) Why is there a need for uncertainty theory? J Uncertain Syst 6(1):3–10
4. Liu B (2010) Uncertainty theory: a branch of mathematics for modeling human uncertainty. Springer, Berlin
5. Liu B (2009) Some research problems in uncertainty theory. J Uncertain Syst 3(1):3–10
6. Peng Z, Iwamura K (2010) A sufficient and necessary condition of uncertainty distribution. J Interdiscip Math 13(3):277–285
7. Liu Y, Ha M (2010) Expected value of function of uncertain variables. J Uncertain Syst 4(3):181–186
8. Zhu Y (2010) Uncertain optimal control with application to a portfolio selection model. Cybern Syst 41(7):535–547
9. Zhu Y (2012) Functions of uncertain variables and uncertain programming. J Uncertain Syst 6(4):278–288
10. Sheng L, Zhu Y (2013) Optimistic value model of uncertain optimal control. Int J Uncertain Fuzziness Knowl Based Syst 21(1):75–83
11. Liu B (2008) Fuzzy process, hybrid process and uncertain process. J Uncertain Syst 2(1):3–16
12. Chen X, Liu B (2010) Existence and uniqueness theorem for uncertain differential equations. Fuzzy Optim Decis Making 9(1):69–81
13. Yao K, Chen X (2013) A numerical method for solving uncertain differential equations. J Intell Fuzzy Syst 25(3):825–832

Chapter 2
Uncertain Expected Value Optimal Control

Uncertain optimal control problem is to choose the best decision such that some objective function related to an uncertain process driven by an uncertain differential equation is optimized. Because the objective function is an uncertain variable for any decision, we can not optimize it as a real function. A basic question is how to rank two different uncertain variables. In fact, there are many methods to do so but there is not a best one. These methods are established due to some criteria including, for example, expected value, optimistic value, pessimistic value, and uncertain measure [1]. In this chapter, we make use of the expected value-based method to optimize the uncertain objective function. That is, we assume that an uncertain variable is larger than the other if the expected value of it is larger than the expected value of the other.

2.1 Problem of Uncertain Optimal Control

Unless stated otherwise, we assume that C_t is a canonical Liu process. We consider the following uncertain expected value optimal control problem

$$J(0, x_0) \equiv \sup_{u_t \in U} E\left[\int_0^T f(s, u_s, X_s)\mathrm{d}s + G(T, X_T)\right] \tag{2.1}$$

subject to

$$\mathrm{d}X_s = \nu(s, u_s, X_s)\mathrm{d}s + \sigma(s, u_s, X_s)\mathrm{d}C_s \quad \text{and} \quad X_0 = x_0. \tag{2.2}$$

In the above problem, X_s is the state variable, u_s the decision variable (represents the function $u_s(s, X_s)$ of the time s and state X_s) with the value in U, f the objective function, and G the function of terminal reward. For a given u_s, X_s is provided by

Y. Zhu, *Uncertain Optimal Control*, Springer Uncertainty Research,
https://doi.org/10.1007/978-981-13-2134-4_2

the uncertain differential equation (2.2), where ν and σ are two functions of time s, u_s, and X_s. The function $J(0, x_0)$ is the expected optimal reward obtainable in $[0, T]$ with the initial condition that at time 0 we are in state x_0.

For any $0 < t < T$, $J(t, x)$ is the expected optimal reward obtainable in $[t, T]$ with the condition that at time t we are in state $X_t = x$. That is, we have

$$\begin{cases} J(t,x) \equiv \sup\limits_{u_t} E\left[\int_t^T f(s, u_s, X_s)\mathrm{d}s + G(T, X_T)\right] \\ \text{subject to} \\ \qquad \mathrm{d}X_s = \nu(s, u_s, X_s)\mathrm{d}s + \sigma(s, u_s, X_s)\mathrm{d}C_s \quad \text{and} \quad X_t = x. \end{cases} \tag{2.3}$$

2.2 Principle of Optimality

Now we present the following *principle of optimality* for uncertain optimal control.

Theorem 2.1 ([2]) *For any $(t, x) \in [0, T) \times R$, and $\Delta t > 0$ with $t + \Delta t < T$, we have*

$$J(t,x) = \sup_{u_t} E\left[f(t, u_t, X_t)\Delta t + J(t + \Delta t, x + \Delta X_t) + o(\Delta t)\right], \tag{2.4}$$

where $x + \Delta X_t = X_{t+\Delta t}$.

Proof We denote the right side of (2.4) by $\widetilde{J}(t, x)$. It follows from the definition of $J(t, x)$ that

$$\begin{aligned} J(t,x) \geq E\Bigg[&\int_t^{t+\Delta t} f(s, u_s|_{[t,t+\Delta t)}, X_s)\mathrm{d}s \\ &+ \int_{t+\Delta t}^T f(s, u_s|_{[t+\Delta t,T]}, X_s)\mathrm{d}s + G(T, X_T)\Bigg] \end{aligned} \tag{2.5}$$

for any u_t, where $u_s|_{[t,t+\Delta t)}$ and $u_s|_{[t+\Delta t,T]}$ are the values of decision variable u_t restricted on $[t, t + \Delta t)$ and $[t + \Delta t, T]$, respectively. Thus,

$$\begin{aligned} J(t,x) \geq E\big[&f(t, u_t, X_t)\Delta t + o(\Delta t) \\ &+ E\left[\int_{t+\Delta t}^T f(s, u_s|_{[t+\Delta t,T]}, X_s)\mathrm{d}s + G(T, X_T)\right]\Bigg]. \end{aligned} \tag{2.6}$$

Taking the supremum with respect to $u_s|_{[t+\Delta t,T]}$ first, and then $u_s|_{[t,t+\Delta t)}$ in (2.6), we get $J(t, x) \geq \widetilde{J}(t, x)$.

On the other hand, for all u_t, we have

$$\begin{aligned}
&E\left[\int_t^T f(s,u_s,X_s)\mathrm{d}s + G(T,X_T)\right]\\
&= E\left[\int_t^{t+\Delta t} f(s,u_s,X_s)\mathrm{d}s + E\left[\int_{t+\Delta t}^T f(s,u_s|_{[t+\Delta t,T]},X_s)\mathrm{d}s + G(T,X_T)\right]\right]\\
&\le E\left[f(t,u_t,X_t)\Delta t + o(\Delta t) + J(t+\Delta t, x+\Delta X_t)\right]\\
&\le \widetilde{J}(t,x).
\end{aligned}$$

Hence, $J(t,x) \le \widetilde{J}(t,x)$, and then $J(t,x) = \widetilde{J}(t,x)$. The theorem is proved.

Remark 2.1 It is easy to know that the principle of optimality is true for $x \in R^n$ under the multidimensional case.

2.3 Equation of Optimality

Consider the uncertain optimal control problem (2.3). Now let us give a fundamental result called *equation of optimality* in uncertain optimal control.

Theorem 2.2 (Equation of optimality, [2]) *Let $J(t,x)$ be twice differentiable on $[0,T]\times R$. Then we have*

$$-J_t(t,x) = \sup_{u_t}\left\{f(t,u_t,x) + J_x(t,x)\nu(t,u_t,x)\right\}, \tag{2.7}$$

where $J_t(t,x)$ and $J_x(t,x)$ are the partial derivatives of the function $J(t,x)$ in t and x, respectively.

Proof For any $\Delta t > 0$, by using Taylor series expansion, we get

$$\begin{aligned}
J(t+\Delta t, x+\Delta X_t) = {}& J(t,x) + J_t(t,x)\Delta t + J_x(t,x)\Delta X_t + \frac{1}{2}J_{tt}(t,x)\Delta t^2\\
&+ \frac{1}{2}J_{xx}(t,x)\Delta X_t{}^2 + J_{tx}(t,x)\Delta t\Delta X_t + o(\Delta t).
\end{aligned} \tag{2.8}$$

Substituting Eq. (2.8) into Eq. (2.4) yields

$$\begin{aligned}
0 = \sup_{u_t}\{&f(t,u_t,x)\Delta t + J_t(t,x)\Delta t + E[J_x(t,x)\Delta X_t + \frac{1}{2}J_{tt}(t,x)\Delta t^2\\
&+ \frac{1}{2}J_{xx}(t,x)\Delta X_t{}^2 + J_{tx}(t,x)\Delta t\Delta X_t] + o(\Delta t)\}.
\end{aligned} \tag{2.9}$$

Let ξ be an uncertain variable such that $\Delta X_t = \xi + \nu(t,u_t,x)\Delta t$. It follows from Eq. (2.9) that

$$
\begin{aligned}
0 &= \sup_{u_t}\{f(t,u_t,x)\Delta t + J_t(t,x)\Delta t + J_x(t,x)\nu(t,u_t,x)\Delta t + E[(J_x(t,x) \\
&\quad + J_{xx}(t,x)\nu(t,u_t,x)\Delta t + J_{tx}(t,x)\Delta t)\xi + \frac{1}{2}J_{xx}(t,x)\xi^2] + o(\Delta t)\} \\
&= \sup_{u_t}\{f(t,u_t,x)\Delta t + J_t(t,x)\Delta t + J_x(t,x)\nu(t,u_t,x)\Delta t \\
&\quad + E[a\xi + b\xi^2] + o(\Delta t)\}, \qquad (2.10)
\end{aligned}
$$

where $a \equiv J_x(t,x) + J_{xx}(t,x)\nu(t,u_t,x)\Delta t + J_{tx}(t,x)\Delta t$, and $b \equiv \frac{1}{2}J_{xx}(t,x)$. It follows from the uncertain differential equation, the constraint in (2.3), that $\xi = \Delta X_t - \nu(t,u_t,x)\Delta t$ is a normally distributed uncertain variable with expected value 0 and variance $\sigma^2(t,u_t,x)\Delta t^2$. If $b = 0$, then $E[a\xi + b\xi^2] = aE[\xi] = 0$. Otherwise, Theorem 1.8 implies that

$$
E[a\xi + b\xi^2] = bE\left[\frac{a}{b}\xi + \xi^2\right] = o(\Delta t). \qquad (2.11)
$$

Substituting Eq. (2.11) into Eq. (2.10) yields

$$
-J_t(t,x)\Delta t = \sup_{u_t}\{f(t,u_t,x)\Delta t + J_x(t,x)\nu(t,u_t,x)\Delta t + o(\Delta t)\}. \qquad (2.12)
$$

Dividing Eq. (2.12) by Δt, and letting $\Delta t \to 0$, we can obtain Eq. (2.7).

Remark 2.2 If the equation of optimality has solutions, then the optimal decision and optimal expected value of objective function are determined. If function f is convex in its arguments, then the equation will produce a minimum, and if f is concave in its arguments, then it will produce a maximum. We note that the boundary condition for the equation is $J(T, X_T) = E[G(T, X_T)]$.

Remark 2.3 We note that in the equation of optimality (Hamilton–Jacobi–Bellman equation) for stochastic optimal control, there is an extra term $\frac{1}{2}J_{xx}(t,x)\sigma^2(t,u_t,x)$.

2.4 Equation of Optimality for Multidimension Case

We now consider the optimal control model for multidimension case:

$$
J(t,\boldsymbol{x}) \equiv \sup_{\boldsymbol{u}_t \in U} E\left[\int_t^T f(s,\boldsymbol{u}_s,\boldsymbol{X}_s)\mathrm{d}s + G(T,\boldsymbol{X}_T)\right] \qquad (2.13)
$$

subject to

$$
\mathrm{d}\boldsymbol{X}_s = \boldsymbol{\nu}(s,\boldsymbol{u}_s,\boldsymbol{X}_s)\mathrm{d}s + \boldsymbol{\sigma}(s,\boldsymbol{u}_s,\boldsymbol{X}_s)\mathrm{d}\boldsymbol{C}_s \quad \text{and} \quad \boldsymbol{X}_t = \boldsymbol{x}. \qquad (2.14)
$$

In the above model, $\boldsymbol{X}_s$ is the state vector of dimension n with the initial condition that at time t we are in state $\boldsymbol{X}_t = \boldsymbol{x}$, $\boldsymbol{u}_s$ the decision vector of dimension r (represents the function $\boldsymbol{u}_s$ of time s and state $\boldsymbol{X}_s$) in a domain U, $f : [0, +\infty) \times R^r \times R^n \to R$ the objective function, and $G : [0, +\infty) \times R^n \to R$ the function of terminal reward. In addition, $\boldsymbol{v} : [0, +\infty) \times R^r \times R^n \to R^n$ is a column-vector function, $\boldsymbol{\sigma} : [0, +\infty) \times R^r \times R^n \to R^n \times R^k$ a matrix function, and $\boldsymbol{C}_s = (C_{s1}, C_{s2}, \ldots, C_{sk})^\tau$, where $C_{s1}, C_{s2}, \ldots, C_{sk}$ are independent canonical Liu processes. Note that $\boldsymbol{y}^\tau$ represents the transpose vector of the vector $\boldsymbol{y}$, and the final time $T > 0$ is fixed or free. We have the following equation of optimality.

Theorem 2.3 ([3]) *Let $J(t, \boldsymbol{x})$ be twice differentiable on $[0, T] \times R^n$. Then we have*

$$-J_t(t, \boldsymbol{x}) = \sup_{\boldsymbol{u}_t \in U} \left\{ f(t, \boldsymbol{u}_t, \boldsymbol{x}) + \boldsymbol{v}(t, \boldsymbol{u}_t, \boldsymbol{x})^\tau \nabla_{\boldsymbol{x}} J(t, \boldsymbol{x}) \right\}, \tag{2.15}$$

where $J_t(t, \boldsymbol{x})$ is the partial derivative of the function $J(t, \boldsymbol{x})$ in t, and $\nabla_{\boldsymbol{x}} J(t, \boldsymbol{x})$ is the gradient of $J(t, \boldsymbol{x})$ in $\boldsymbol{x}$.

Proof For Δt with $t + \Delta t \in [0, T]$, denote $\boldsymbol{X}_{t+\Delta t} = \boldsymbol{x} + \Delta \boldsymbol{X}_t$. By using Taylor series expansion, we have

$$\begin{aligned} J(t + \Delta t, \boldsymbol{x} + \Delta \boldsymbol{X}_t) = {} & J(t, \boldsymbol{x}) + J_t(t, \boldsymbol{x})\Delta t + \nabla_{\boldsymbol{x}} J(t, \boldsymbol{x})^\tau \Delta \boldsymbol{X}_t + \frac{1}{2} J_{tt}(t, \boldsymbol{x})\Delta t^2 \\ & + \frac{1}{2} \Delta \boldsymbol{X}_t{}^\tau \nabla_{\boldsymbol{xx}} J(t, \boldsymbol{x}) \Delta \boldsymbol{X}_t + \nabla_{\boldsymbol{x}} J_t(t, \boldsymbol{x})^\tau \Delta \boldsymbol{X}_t \Delta t \\ & + o(\Delta t) \end{aligned} \tag{2.16}$$

where $\nabla_{\boldsymbol{xx}} J(t, \boldsymbol{x})$ is the Hessian matrix of $J(t, \boldsymbol{x})$. Since

$$\Delta \boldsymbol{X}_t = \boldsymbol{v}(t, \boldsymbol{u}_t, \boldsymbol{X}_t)\Delta t + \boldsymbol{\sigma}(t, \boldsymbol{u}_t, \boldsymbol{X}_t)\Delta \boldsymbol{C}_t,$$

the expansion (2.16) may be rewritten as

$$\begin{aligned} & J(t + \Delta t, \boldsymbol{x} + \Delta \boldsymbol{X}_t) \\ = {} & J(t, \boldsymbol{x}) + J_t(t, \boldsymbol{x})\Delta t + \nabla_{\boldsymbol{x}} J(t, \boldsymbol{x})^\tau \boldsymbol{v}(t, \boldsymbol{u}_t, \boldsymbol{X}_t)\Delta t + \nabla_{\boldsymbol{x}} J(t, \boldsymbol{x})^\tau \boldsymbol{\sigma}(t, \boldsymbol{u}_t, \boldsymbol{X}_t)\Delta \boldsymbol{C}_t \\ & + \frac{1}{2} J_{tt}(t, \boldsymbol{x})\Delta t^2 + \frac{1}{2} \boldsymbol{v}(t, \boldsymbol{u}_t, \boldsymbol{X}_t)^\tau \nabla_{\boldsymbol{xx}} J(t, \boldsymbol{x}) \boldsymbol{v}(t, \boldsymbol{u}_t, \boldsymbol{X}_t)\Delta t^2 \\ & + \boldsymbol{v}(t, \boldsymbol{u}_t, \boldsymbol{X}_t)^\tau \nabla_{\boldsymbol{xx}} J(t, \boldsymbol{x}) \boldsymbol{\sigma}(t, \boldsymbol{u}_t, \boldsymbol{X}_t)\Delta \boldsymbol{C}_t \Delta t \\ & + \frac{1}{2} (\boldsymbol{\sigma}(t, \boldsymbol{u}_t, \boldsymbol{X}_t)\Delta \boldsymbol{C}_t)^\tau \nabla_{\boldsymbol{xx}} J(t, \boldsymbol{x}) \boldsymbol{\sigma}(t, \boldsymbol{u}_t, \boldsymbol{X}_t)\Delta \boldsymbol{C}_t + \nabla_{\boldsymbol{x}} J_t(t, \boldsymbol{x})^\tau \boldsymbol{v}(t, \boldsymbol{u}_t, \boldsymbol{X}_t)\Delta t^2 \\ & + \nabla_{\boldsymbol{x}} J_t(t, \boldsymbol{x})^\tau \boldsymbol{\sigma}(t, \boldsymbol{u}_t, \boldsymbol{X}_t)\Delta \boldsymbol{C}_t \Delta t + o(\Delta t) \\ = {} & J(t, \boldsymbol{x}) + J_t(t, \boldsymbol{x})\Delta t + \nabla_{\boldsymbol{x}} J(t, \boldsymbol{x})^\tau \boldsymbol{v}(t, \boldsymbol{u}_t, \boldsymbol{X}_t)\Delta t \\ & + \{\nabla_{\boldsymbol{x}} J(t, \boldsymbol{x})^\tau \boldsymbol{\sigma}(t, \boldsymbol{u}_t, \boldsymbol{X}_t) + \nabla_{\boldsymbol{x}} J_t(t, \boldsymbol{x})^\tau \boldsymbol{\sigma}(t, \boldsymbol{u}_t, \boldsymbol{X}_t)\Delta t \\ & + \boldsymbol{v}(t, \boldsymbol{u}_t, \boldsymbol{X}_t)^\tau \nabla_{\boldsymbol{xx}} J(t, \boldsymbol{x}) \boldsymbol{\sigma}(t, \boldsymbol{u}_t, \boldsymbol{X}_t)\Delta t\}\Delta \boldsymbol{C}_t \\ & + \frac{1}{2} \Delta \boldsymbol{C}_t{}^\tau \boldsymbol{\sigma}(t, \boldsymbol{u}_t, \boldsymbol{X}_t)^\tau \nabla_{\boldsymbol{xx}} J(t, \boldsymbol{x}) \boldsymbol{\sigma}(t, \boldsymbol{u}_t, \boldsymbol{X}_t)\Delta \boldsymbol{C}_t + o(\Delta t). \end{aligned} \tag{2.17}$$

Denote

$$
\begin{aligned}
\boldsymbol{a} &= \nabla_{\boldsymbol{x}} J(t,\boldsymbol{x})^{\tau}\boldsymbol{\sigma}(t,\boldsymbol{u}_t,\boldsymbol{X}_t) + \nabla_{\boldsymbol{x}} J_t(t,\boldsymbol{x})^{\tau}\boldsymbol{\sigma}(t,\boldsymbol{u}_t,\boldsymbol{X}_t)\Delta t \\
&\quad + \boldsymbol{\nu}(t,\boldsymbol{u}_t,\boldsymbol{X}_t)^{\tau}\nabla_{\boldsymbol{xx}} J(t,\boldsymbol{x})\boldsymbol{\sigma}(t,\boldsymbol{u}_t,\boldsymbol{X}_t)\Delta t, \\
\boldsymbol{B} &= \frac{1}{2}\boldsymbol{\sigma}(t,\boldsymbol{u}_t,\boldsymbol{X}_t)^{\tau}\nabla_{\boldsymbol{xx}} J(t,\boldsymbol{x})\boldsymbol{\sigma}(t,\boldsymbol{u}_t,\boldsymbol{X}_t).
\end{aligned}
$$

Hence, Eq. (2.17) may be simply expressed as

$$
\begin{aligned}
J(t+\Delta t,\boldsymbol{x}+\Delta\boldsymbol{X}_t) = J(t,\boldsymbol{x}) + J_t(t,\boldsymbol{x})\Delta t + \nabla_{\boldsymbol{x}} J(t,\boldsymbol{x})^{\tau}\boldsymbol{\nu}(t,\boldsymbol{u}_t,\boldsymbol{X}_t)\Delta t \\
+ \boldsymbol{a}\Delta\boldsymbol{C}_t + \Delta\boldsymbol{C}_t{}^{\tau}\boldsymbol{B}\Delta\boldsymbol{C}_t + o(\Delta t).
\end{aligned}
$$

It follows from the principle of optimality that

$$
J(t,\boldsymbol{x}) = \sup_{\boldsymbol{u}_t\in U} E[f(t,\boldsymbol{u}_t,\boldsymbol{x})\Delta t + J(t+\Delta t,\boldsymbol{x}+\Delta\boldsymbol{X}_t) + o(\Delta t)].
$$

Thus,

$$
\begin{aligned}
J(t,\boldsymbol{x}) = \sup_{\boldsymbol{u}_t\in U} \{ & f(t,\boldsymbol{u}_t,\boldsymbol{x})\Delta t + J(t,\boldsymbol{x}) + J_t(t,\boldsymbol{x})\Delta t + \nabla_{\boldsymbol{x}} J(t,\boldsymbol{x})^{\tau}\boldsymbol{\nu}(t,\boldsymbol{u}_t,\boldsymbol{X}_t)\Delta t \\
& + E[\boldsymbol{a}\Delta\boldsymbol{C}_t + \Delta\boldsymbol{C}_t{}^{\tau}\boldsymbol{B}\Delta\boldsymbol{C}_t]\} + o(\Delta t). \qquad (2.18)
\end{aligned}
$$

Let $\boldsymbol{a} = (a_1, a_2, \ldots, a_k)$, $\boldsymbol{B} = (b_{ij})_{k\times k}$. We have

$$
\boldsymbol{a}\Delta\boldsymbol{C}_t + \Delta\boldsymbol{C}_t{}^{\tau}\boldsymbol{B}\Delta\boldsymbol{C}_t = \sum_{i=1}^{k} a_i\Delta C_{ti} + \sum_{i=1}^{k}\sum_{j=1}^{k} b_{ij}\Delta C_{ti}\Delta C_{tj}.
$$

Since

$$
|b_{ij}\Delta C_{ti}\Delta C_{tj}| \le \frac{1}{2}|b_{ij}|(\Delta C_{ti}{}^2 + \Delta C_{tj}{}^2),
$$

we have

$$
\begin{aligned}
\sum_{i=1}^{k}\left\{ a_i\Delta C_{ti} - \left(\sum_{j=1}^{k}|b_{ij}|\right)\Delta C_{ti}{}^2 \right\} &\le \boldsymbol{a}\Delta\boldsymbol{C}_t + \Delta\boldsymbol{C}_t{}^{\tau}\boldsymbol{B}\Delta\boldsymbol{C}_t \\
&\le \sum_{i=1}^{k}\left\{ a_i\Delta C_{ti} + \left(\sum_{j=1}^{k}|b_{ij}|\right)\Delta C_{ti}{}^2 \right\}.
\end{aligned}
$$

It follows from the independence of $C_{t1}, C_{t2}, \ldots, C_{tk}$ that

$$\sum_{i=1}^{k} E\left[a_i \Delta C_{ti} - \left(\sum_{j=1}^{k} |b_{ij}|\right) \Delta C_{ti}{}^2\right] \le E[\boldsymbol{a}\Delta \boldsymbol{C}_t + \Delta \boldsymbol{C}_t{}^{\tau} \boldsymbol{B} \Delta \boldsymbol{C}_t]$$

$$\le \sum_{i=1}^{k} E\left[a_i \Delta C_{ti} + \left(\sum_{j=1}^{k} |b_{ij}|\right) \Delta C_{ti}{}^2\right].$$

It follows from Theorem 1.8 that

$$E\left[a_i \Delta C_{ti} - \left(\sum_{j=1}^{k} |b_{ij}|\right) \Delta C_{ti}{}^2\right] = o(\Delta t),$$

and

$$E\left[a_i \Delta C_{ti} + \left(\sum_{j=1}^{k} |b_{ij}|\right) \Delta C_{ti}{}^2\right] = o(\Delta t).$$

Hence, $E[\boldsymbol{a}\Delta \boldsymbol{C}_t + \Delta \boldsymbol{C}_t{}^{\tau} \boldsymbol{B} \Delta \boldsymbol{C}_t] = o(\Delta t)$. Therefore, Eq. (2.15) directly follows from Eq. (2.18). The theorem is proved.

2.5 Uncertain Linear Quadratic Model

We consider a kind of special optimal control model with a quadratic objective function subject to a linear uncertain differential equation.

$$\begin{cases} J(0, x) \equiv \min\limits_{u_t} E\left\{\displaystyle\int_0^T [\alpha_1(t)X_t^2 + \alpha_2(t)u_t^2 + \alpha_3(t)X_t u_t \right. \\ \qquad\qquad \left. + \alpha_4(t)X_t + \alpha_5(t)u_t + \alpha_6(t)]\mathrm{d}t + S_T X_T^2\right\} \\ \text{subject to} \\ \quad \mathrm{d}X_t = [\beta_1(t)X_t + \beta_2(t)u_t + \beta_3(t)]\,\mathrm{d}t + [\Delta_1(t)X_t + \Delta_2(t)u_t + \Delta_3(t)]\,\mathrm{d}C_t \\ \quad X_0 = x_0, \end{cases} \tag{2.19}$$

where x_0 denotes the initial state, $\alpha_i(t)$ $(i = 1, 2, \ldots, 6)$, $\beta_j(t)$, and $\Delta_j(t)$ $(j = 1, 2, 3)$ are all the functions of time t. The aim to discuss this model is to find an optimal control u_t^* which is a function of time t and state X_t. For any $0 < t < T$, use $J(t, x)$ to denote the optimal value obtainable in $[t, T]$ with the condition that at time t we are in state $X_t = x$.

Theorem 2.4 *Assume that $J(t, x)$ is a twice differentiable function on $[0, T] \times R$. Let $\alpha_i(t)$ $(i = 1, 2, \ldots, 6)$, $\beta_j(t)$, $\Delta_j(t)$ $(j = 1, 2, 3)$ and $\alpha_2^{-1}(t)$ be continuous* *bounded functions of t, and $\alpha_1(t) \geq 0$, $\alpha_2(t) > 0$. A necessary and sufficient condition that u_t^* is an optimal control for (2.19) is that*

$$u_t^* = -\frac{\alpha_3(t)x + \alpha_5(t) + \beta_2(t)\left[P(t)x + Q(t)\right]}{2\alpha_2(t)}, \tag{2.20}$$

where x is the state of the state variable X_t at time t obtained by applying the optimal control u_t^*, the function $P(t)$ satisfies the following Riccati differential equation and boundary condition

$$\begin{cases} \dfrac{\mathrm{d}P(t)}{\mathrm{d}t} = \dfrac{[\beta_2(t)]^2}{2\alpha_2(t)}P^2(t) + \left\{\dfrac{\alpha_3(t)\beta_2(t)}{\alpha_2(t)} - 2\beta_1(t)\right\}P(t) \\ \qquad + \dfrac{\alpha_3^2(t)}{2\alpha_2(t)} - 2\alpha_1(t) \\ P(T) = 2S_T, \end{cases} \tag{2.21}$$

and the function $Q(t)$ satisfies the following differential equation and boundary condition

$$\begin{cases} \dfrac{\mathrm{d}Q(t)}{\mathrm{d}t} = \left\{\dfrac{\alpha_3(t)\beta_2(t)}{2\alpha_2(t)} + \dfrac{[\beta_2(t)]^2 P(t)}{2\alpha_2(t)} - \beta_1(t)\right\}Q(t) \\ \qquad + \left\{\dfrac{\alpha_5(t)\beta_2(t)}{2\alpha_2(t)} - \beta_3(t)\right\}P(t) + \dfrac{\alpha_3(t)\alpha_5(t)}{2\alpha_2(t)} - \alpha_4(t) \\ Q(T) = 0. \end{cases} \tag{2.22}$$

The optimal value is

$$J(0, x_0) = \frac{1}{2}P(0)x_0^2 + Q(0)x_0 + R(0),$$

where

$$R(0) = \int_0^T \left\{\frac{[\beta_2(s)]^2}{4\alpha_2(s)}Q^2(s) + \left[\frac{\alpha_5(s)\beta_2(s)}{2\alpha_2(s)} - \beta_3(s)\right]Q(s) + \frac{\alpha_5^2(s)}{4\alpha_2(s)} - \alpha_6(s)\right\}\mathrm{d}s. \tag{2.23}$$

Proof The necessity will be proved first. It follows from the equation of optimality (2.7) that

$$\begin{aligned} -J_t &= \min_u \left\{\alpha_1(t)x^2 + \alpha_2(t)u^2 + \alpha_3(t)xu + \alpha_4(t)x + \alpha_5(t)u + \alpha_6(t)\right. \\ &\quad \left. + \left[\beta_1(t)x + \beta_2(t)u + \beta_3(t)\right]J_x\right\} \\ &= \min_u L(u), \end{aligned} \tag{2.24}$$

where $L(u)$ represents the term in the braces. The optimal u satisfies

$$\frac{\partial L(u)}{\partial u} = 2\alpha_2(t)u + \alpha_3(t)x + \alpha_5(t) + \beta_2(t)J_x = 0.$$

Since

$$\frac{\partial^2 L(u)}{\partial u^2} = 2\alpha_2(t) > 0,$$

we know that

$$u_t^* = -\frac{\alpha_3(t)x + \alpha_5(t) + [\beta_2(t) + r\Delta_2(t)]\,J_x}{2\alpha_2(t)} \tag{2.25}$$

is the minimum point of $L(u)$. By Eq. (2.24), we have

$$\begin{aligned}&\frac{\partial J}{\partial t} + \alpha_1(t)x^2 + \alpha_2(t)u_t^{*2} + \alpha_3(t)xu_t^* + \alpha_4(t)x + \alpha_5(t)u_t^* + \alpha_6(t)\\&+ \left\{\left[\beta_1(t)x + \beta_2(t)u_t^* + \beta_3(t)\right]\right\} J_x\\&= 0.\end{aligned} \tag{2.26}$$

Taking derivative in both sides of (2.26) with respect to x yields that

$$\begin{aligned}&\frac{\partial^2 J}{\partial x \partial t} + 2\alpha_1(t)x + 2\alpha_2(t)u_t^*\frac{\partial u_t^*}{\partial x} + \alpha_3(t)u_t^* + \alpha_3(t)x\frac{\partial u_t^*}{\partial x} + \alpha_4(t) + \alpha_5(t)\frac{\partial u_t^*}{\partial x}\\&+ \left\{\left[\beta_1(t)x + \beta_2(t)u_t^* + \beta_3(t)\right]\right\}\frac{\partial^2 J}{\partial x^2} + \left[\beta_1(t) + \beta_2(t)\frac{\partial u_t^*}{\partial x}\right] J_x\\&= 0,\end{aligned}$$

or

$$\begin{aligned}&\frac{\partial^2 J}{\partial x \partial t} + 2\alpha_1(t)x + \alpha_3(t)u_t^* + \alpha_4(t) + \beta_1(t)J_x + \left[\beta_1(t)x + \beta_2(t)u_t^* + \beta_3(t)\right]\frac{\partial^2 J}{\partial x^2}\\&+ \left\{2\alpha_2(t)u_t^* + \alpha_3(t)x + \alpha_5(t) + \beta_2(t)J_x\right\}\frac{\partial u_t^*}{\partial x}\\&= 0.\end{aligned}$$

By (2.25), we get

$$\begin{aligned}&\frac{\partial^2 J}{\partial x \partial t} + 2\alpha_1(t)x + \alpha_3(t)u_t^* + \alpha_4(t) + \beta_1(t)J_x\\&+ \left[\beta_1(t)x + \beta_2(t)u_t^* + \beta_3(t)\right]\frac{\partial^2 J}{\partial x^2} = 0.\end{aligned}$$

Hence,

$$\frac{\partial^2 J}{\partial x \partial t} = -2\alpha_1(t)x - \alpha_3(t)u_t^* - \alpha_4(t) - \beta_1(t)J_x$$
$$- \left[\beta_1(t)x + \beta_2(t)u_t^* + \beta_3(t)\right]\frac{\partial^2 J}{\partial x^2}. \tag{2.27}$$

Let

$$\lambda(t) = J_x. \tag{2.28}$$

Since $J(T, x) = S_T x^2$, we conjecture that

$$J_x = \lambda(t) = P(t)x(t) + Q(t). \tag{2.29}$$

Taking derivative in both sides of (2.29) with respect to x, we have

$$\frac{\partial^2 J}{\partial x^2} = P(t). \tag{2.30}$$

Substituting (2.29) into (2.25) yields that

$$u_t^* = -\frac{\alpha_3(t)x + \alpha_5(t) + \beta_2(t)[P(t)x + Q(t)]}{2\alpha_2(t)}. \tag{2.31}$$

Taking derivative in both sides of (2.28) with respect to t yields that

$$\frac{\mathrm{d}\lambda(t)}{\mathrm{d}t} = \frac{\partial^2 J}{\partial x \partial t} + \frac{\partial^2 J}{\partial x^2} \cdot \frac{\mathrm{d}x}{\mathrm{d}t}. \tag{2.32}$$

Substituting (2.27), (2.29), and (2.30) into (2.32) yields that

$$\frac{\mathrm{d}\lambda(t)}{\mathrm{d}t} = -2\alpha_1(t)x - \alpha_3(t)u_t^* - \alpha_4(t) - \beta_1(t)[P(t)x + Q(t)]$$
$$+ P(t) \cdot \frac{\mathrm{d}x}{\mathrm{d}t} - \left[\beta_1(t)x + \beta_2(t)u_t^* + \beta_3(t)\right]P(t). \tag{2.33}$$

Substituting (2.31) into (2.33), we have

$$\frac{\mathrm{d}\lambda(t)}{\mathrm{d}t} = \left\{\frac{[\beta_2(t)]^2}{2\alpha_2(t)}P^2(t) + \left[\frac{\alpha_3(t)\,[\beta_2(t)]}{\alpha_2(t)} - 2\beta_1(t)\right]P(t) + \frac{\alpha_3^2(t)}{2\alpha_2(t)} - 2\alpha_1(t)\right\}x$$
$$+ \left\{\frac{\alpha_5(t)\beta_2(t)}{2\alpha_2(t)} - \beta_3(t)\right\}P(t) + \left\{\frac{[\beta_2(t)]^2}{2\alpha_2(t)}P(t) + \frac{\alpha_3(t)\beta_2(t)}{2\alpha_2(t)}\right.$$
$$\left. - \beta_1(t)\right\}Q(t) + \frac{\alpha_3(t)\alpha_5(t)}{2\alpha_2(t)} - \alpha_4(t) + P(t)\frac{\mathrm{d}x}{\mathrm{d}t}. \tag{2.34}$$

Taking derivative in both sides of (2.29) with respect to t yields that

$$\frac{\mathrm{d}\lambda(t)}{\mathrm{d}t} = \frac{\mathrm{d}P(t)}{\mathrm{d}t}x + P(t)\frac{\mathrm{d}x}{\mathrm{d}t} + \frac{\mathrm{d}Q(t)}{\mathrm{d}t}. \tag{2.35}$$

By (2.34) and (2.35), we get

$$\frac{\mathrm{d}P(t)}{\mathrm{d}t} = \frac{[\beta_2(t)]^2}{2\alpha_2(t)}P^2(t) + \left\{\frac{\alpha_3(t)\beta_2(t)}{\alpha_2(t)} - 2\beta_1(t)\right\} P(t) + \frac{\alpha_3^2(t)}{2\alpha_2(t)} - 2\alpha_1(t),$$

and

$$\begin{aligned}\frac{\mathrm{d}Q(t)}{\mathrm{d}t} &= \left\{\frac{\alpha_3(t)\beta_2(t)}{2\alpha_2(t)} + \frac{[\beta_2(t)]^2\,P(t)}{2\alpha_2(t)} - \beta_1(t)\right\} Q(t)\\ &+ \left\{\frac{\alpha_5(t)\beta_2(t)}{2\alpha_2(t)} - \beta_3(t)\right\} P(t) + \frac{\alpha_3(t)\alpha_5(t)}{2\alpha_2(t)} - \alpha_4(t).\end{aligned}$$

It follows from (2.28) and (2.29) that

$$\lambda(T) = 2S_T x(T) \quad \text{and} \quad \lambda(T) = P(T)x(T) + Q(T).$$

So we have

$$P(T) = 2S_T \quad \text{and} \quad Q(T) = 0.$$

Hence, $P(t)$ satisfies the Riccati differential equation and boundary condition (2.21), and the function $Q(t)$ satisfies the differential equation and boundary condition (2.22). By solving the above equations, the expressions of $P(t)$ and $Q(t)$ can be obtained, respectively. In other words, the optimal control u_t^* is provided for the linear quadratic model (2.19) by (2.20).

Next we will verify the sufficient condition of the theorem. Suppose that u_t^*, $P(t)$, $Q(t)$ satisfy (2.20), (2.21), (2.22), respectively. Now we prove that u_t^* is an optimal control for the linear quadratic model (2.19). By the equation of the optimality (2.7), we have

$$\begin{aligned}-\frac{\partial J}{\partial t} &= \min_u \left\{\alpha_1(t)x^2 + \alpha_2(t)u^2 + \alpha_3(t)xu + \alpha_4(t)x + \alpha_5(t)u + \alpha_6(t)\right.\\ &\quad \left. + \,[\beta_1(t)x + \beta_2(t)u + \beta_3(t)]\,J_x\right\}.\end{aligned}$$

So

$$\begin{aligned}&\frac{\partial J}{\partial t} + \min_u \left\{\alpha_1(t)x^2 + \alpha_2(t)u^2 + \alpha_3(t)xu + \alpha_4(t)x + \alpha_5(t)u + \alpha_6(t)\right.\\ &\left. + \,[\beta_1(t)x + \beta_2(t)u + \beta_3(t)]\,J_x\right\}\\ &= 0.\end{aligned} \tag{2.36}$$

We conjecture that

$$J(t,x)=\frac{1}{2}P(t)x^2+Q(t)x+R(t).$$

where $R(t)$ is provided by

$$R(t)=\int_t^T\left\{\frac{[\beta_2(s)]^2}{4\alpha_2(s)}Q^2(s)+\left[\frac{\alpha_5(s)\beta_2(s)}{2\alpha_2(s)}-\beta_3(s)\right]Q(s)\right.$$
$$\left.+\frac{\alpha_5^2(s)}{4\alpha_2(s)}-\alpha_6(s)\right\}\mathrm{d}s. \tag{2.37}$$

Then

$$\frac{\partial J}{\partial t}+\alpha_1(t)x^2+\alpha_2(t)u_t^{*2}+\alpha_3(t)xu_t^*+\alpha_4(t)x+\alpha_5(t)u_t^*+\alpha_6(t)$$
$$+\left[\beta_1(t)x+\beta_2(t)u_t^*+\beta_3(t)\right]J_x$$
$$=\frac{\partial J}{\partial t}+\alpha_1(t)x^2+\alpha_4(t)x+\alpha_6(t)+\alpha_2(t)u_t^{*2}+[\alpha_3(t)x+\alpha_5(t)]\,u_t^*$$
$$+\{[\beta_1(t)+r\Delta_1(t)]\,x+\beta_3(t)\}\,J_x+\beta_2(t)u_t^*J_x$$
$$=\frac{1}{2}\frac{\mathrm{d}P(t)}{\mathrm{d}t}x^2+\frac{\mathrm{d}Q(t)}{\mathrm{d}t}x+\frac{\mathrm{d}R(t)}{\mathrm{d}t}+\alpha_1(t)x^2+\alpha_4(t)x+\alpha_6(t)$$
$$+\alpha_2(t)\left\{-\frac{\alpha_3(t)x+\alpha_5(t)+\beta_2(t)\,[P(t)x+Q(t)]}{2\alpha_2(t)}\right\}^2$$
$$+[\alpha_3(t)x+\alpha_5(t)]\left\{-\frac{\alpha_3(t)x+\alpha_5(t)+\beta_2(t)\,[P(t)x+Q(t)]}{2\alpha_2(t)}\right\}$$
$$+\{\beta_1(t)x+\beta_3(t)\}\,[P(t)x+Q(t)]$$
$$+\beta_2(t)\left\{-\frac{\alpha_3(t)x+\alpha_5(t)+\beta_2(t)\,[P(t)x+Q(t)]}{2\alpha_2(t)}\right\}[P(t)x+Q(t)]$$
$$=\frac{1}{2}\left\{\frac{\mathrm{d}P(t)}{\mathrm{d}t}+2\alpha_1(t)-\frac{\alpha_3^2(t)}{2\alpha_2(t)}-\frac{[\beta_2(t)]^2}{2\alpha_2(t)}P^2(t)-\frac{\alpha_3(t)\beta_2(t)}{\alpha_2(t)}P(t)\right\}x^2$$
$$+\left\{\frac{\mathrm{d}Q(t)}{\mathrm{d}t}-\frac{\alpha_3(t)\beta_2(t)}{2\alpha_2(t)}Q(t)-P(t)\frac{[\beta_2(t)]^2}{2\alpha_2(t)}Q(t)+\beta_1(t)Q(t)+\beta_3(t)P(t)\right.$$
$$\left.-\frac{\alpha_5(t)\beta_2(t)}{2\alpha_2(t)}P(t)-\frac{\alpha_3(t)\alpha_5(t)}{2\alpha_2(t)}+\alpha_4(t)\right\}x+\frac{\mathrm{d}R(t)}{\mathrm{d}t}-\frac{\alpha_5^2(t)}{4\alpha_2(t)}+\alpha_6(t)$$
$$-\frac{[\beta_2(t)]^2}{4\alpha_2(t)}Q^2(t)-\frac{\alpha_5(t)\beta_2(t)}{2\alpha_2(t)}Q(t)+\beta_3(t)Q(t)$$
$$=0.$$

Therefore, we know that u_t^* is a solution of Eq. (2.36). Because objective function is convex, Eq. (2.36) produces a minimum. That is u_t^* is an optimal control. At the same time, we also get the optimal value

$$J(0, x_0) = \frac{1}{2}P(0)x_0^2 + Q(0)x_0 + R(0).$$

The theorem is proved.

2.6 Optimal Control Problem of the Singular Uncertain System

We consider the following continuous-time singular uncertain system

$$\begin{cases} F\mathrm{d}X_t = g(t)AX_t\mathrm{d}t + h(t)BX_t\mathrm{d}C_t, \quad t \geq 0 \\ X_0 = x_0. \end{cases} \tag{2.38}$$

where $X_t \in R^n$ is the state vector of the system, and $g(t), h(t) : [0, +\infty) \to (0, +\infty)$ are both bounded functions, and $A \in R^{n\times n}$, $B \in R^{n\times n}$ are known coefficient matrices associated with X_t. The F is a known (singular) matrix with $\mathrm{rank}(F) = q \leq n$, and $deg(det(zF - A)) = r$ where z is a complex variable. Notice that $det(zF - A)$ is the determinant of the matrix $zF - A$ and $deg(det(zF - A))$ is the degree of the polynomial $det(zF - A)$. The C_t is a canonical Liu process representing the noise of the system. For a matrix $A = [a_{ij}]_{n\times n}$ and a vector $X = (x_1, x_2, \ldots, x_n)^T$, we define

$$\|A\| = \sum_{i,j=1}^{n} | a_{ij} |, \|X\| = \sum_{i=1}^{n} | x_i | .$$

For the system (2.38), the matrices F and A play main roles. Notice that (F, A) is said to be regular if $det(zF - A)$ is not identically zero and (F, A) is said to be impulse-free if $deg(det(zF - A)) = rank(F)$.

Lemma 2.1 ([4]) *If (F, A) is regular, impulse-free and $rank[F, B] = rank(F) = r$, there exist a pair of nonsingular matrices $P \in R^{n\times n}$ and $Q \in R^{n\times n}$ for the triplet (F, A, B) such that the following conditions are satisfied:*

$$PFQ = \begin{bmatrix} I_r & 0 \\ 0 & 0 \end{bmatrix}, \quad PAQ = \begin{bmatrix} A_1 & 0 \\ 0 & I_{n-r} \end{bmatrix}, \quad PBQ = \begin{bmatrix} B_1 & B_2 \\ 0 & 0 \end{bmatrix}$$

where $A_1 \in R^{r\times r}$, $B_1 \in R^{r\times r}$, $B_2 \in R^{r\times n-r}$.

Lemma 2.2 ([5]) *System (2.38) has a unique solution if* (F, A) *is regular, impulse-free and* $rank[F, B] = rank F$. *Moreover, the solution is sample-continuous.*

Proof Let $\begin{bmatrix} X_{1,t} \\ X_{2,t} \end{bmatrix} = Q^{-1}X_t$, where $X_{1,t} \in R^r$ and $X_{2,t} \in R^{n-r}$. Then system (2.38) is equivalent to

$$\begin{cases} \mathrm{d}X_{1,t} = g(t)A_1X_{1,t}\mathrm{d}t + h(t)[B_1X_{1,t} + B_2X_{2,t}]\mathrm{d}C_t, \\ 0 = g(t)X_{2,t}dt, \end{cases}$$

or

$$\begin{cases} \mathrm{d}X_{1,t} = g(t)A_1X_{1,t}\mathrm{d}t + h(t)B_1X_{1,t}\mathrm{d}C_t, \\ 0 = X_{2,t}, \end{cases} \tag{2.39}$$

for all $t \geq 0$. By [6], the equation

$$\mathrm{d}X_{1,t} = g(t)A_1X_{1,t}\mathrm{d}t + h(t)B_1X_{1,t}\mathrm{d}C_t$$

has a unique solution $X_{1,t}$ on interval $[0, +\infty)$. Obviously, $X_t = Q\begin{bmatrix} X_{1,t} \\ X_{2,t} \end{bmatrix}$ for all $t \geq 0$, which is the unique solution to (2.38) on $[0, +\infty)$. Finally, for each $\gamma \in \Gamma$, according to the result in [6], we have

$$\|X_t(\gamma) - X_r(\gamma)\| = \|Q\int_r^t g(s)A_1X_{1,s}(\gamma)ds + Q\int_r^t h(s)B_1X_{1,s}(\gamma)\mathrm{d}C_s(\gamma)\| \to 0.$$

as $r \to t$. Thus, X_t is sample-continuous and this completes the proof.

Unless stated otherwise, it is always assumed that system (2.38) is regular and impulse-free. Then, under this assumption, we will introduce the following optimal control problem for an uncertain singular system:

$$\begin{cases} J(0, X_0) = \sup\limits_{u(s)\in U} E\left[\int_0^T f(s, u(s), X_s)\,ds + G(T, X_T)\right] \\ \text{subject to} \\ \qquad F\mathrm{d}X_s = g(s)\left[AX_s + Bu(s)\right]\mathrm{d}s + h(s)Du(s)\mathrm{d}C_s, \text{ and } X_0 = x_0. \end{cases}$$

In the above problem, $X_s \in R^n$ is the state vector, $u(s) \in U \subset R^m$ is the input vector, f is the objective function, and G is the function of terminal reward, $A \in R^{n\times n}$, $B \in R^{n\times m}$, $D \in R^{n\times m}$. For a given $u(s)$, X_s is defined by the uncertain differential equations, where $g(s), h(s) : [0, +\infty) \to (0, +\infty)$ are both bounded functions. The function $J(0, X_0)$ is the expected optimal value obtainable in $[0, T]$ with the initial state that at time 0 we are in state X_0.

For any $0 < t < T$, $J(t, X)$ is the expected optimal reward obtainable in $[t, T]$ with the condition that at time t we are in state $X_t = \boldsymbol{x}$. That is, we have

$$\begin{cases} J(t, \boldsymbol{x}) = \sup\limits_{u(s) \in U} E\left[\int_t^T f(s, u(s), X_s)\, ds + G(T, X_T)\right] \\ \text{subject to} \\ \qquad F dX_s = g(s)\left[AX_s + Bu(s)\right] ds + h(s) Du(s) dC_s, \text{ and } X_t = \boldsymbol{x}. \end{cases} \tag{2.40}$$

Now let us give the following equation of optimality.

Theorem 2.5 (Equation of Optimality, [5]) *The (F, A) is assumed to be regular and impulse-free, and $P_2 u_t = 0$. Let $J(t, X)$ be twice differentiable on $[0, T] \times R^n$ and $u(s)$ derivable on $[0, T]$. Then we get*

$$-J_t(t, \boldsymbol{x}) = \sup_{u(t) \in U} \left\{ f(t, u(t), \boldsymbol{x}) + \nabla_x J(t, \boldsymbol{x})^T p \right\}, \tag{2.41}$$

where $p = Q \begin{bmatrix} g(t)\left(A_1 X_1 + B_1 u(t)\right) \\ -B_2 \dot{u}(t) \end{bmatrix}$ *and* $P = \begin{bmatrix} P_1 \\ P_2 \end{bmatrix}$, $P_1 \in R^{r \times n}$, $P_2 \in R^{(n-r) \times n}$.

Proof Because (F, A) is regular and impulse-free, by Lemma 2.1 there exist invertible matrices P and Q such that

$$PFQ = \begin{bmatrix} I_r & 0 \\ 0 & 0 \end{bmatrix}, \quad PAQ = \begin{bmatrix} A_1 & 0 \\ 0 & I_{n-r} \end{bmatrix}, \quad PB = \begin{bmatrix} B_1 \\ B_2 \end{bmatrix},$$

and from $P_2 u_t = 0$ we get

$$PD = \begin{bmatrix} P_1 \\ P_2 \end{bmatrix} u_t = \begin{bmatrix} u_{t1} \\ 0 \end{bmatrix},$$

where $u_{t1} = P_1 u_t$. Let $X_s = Q \begin{bmatrix} X_{1,s} \\ X_{2,s} \end{bmatrix}$ for any $s \in [t, T]$ and especially at time t denote $X = Q \begin{bmatrix} X_1 \\ X_2 \end{bmatrix}$, so we are easy to obtain

$$\begin{cases} \mathrm{d}X_{1,s} = g(s)\left[A_1 X_{1,s} + B_1 u(s)\right] \mathrm{d}s + h(s) u_{t1} u(s) \mathrm{d}C_s, \\ 0 = g(s)\left[X_{2,s} + B_2 u(s)\right] ds \end{cases}$$

where $s \in [t, T]$. Thus at any time $s \in [t, T]$ we have

$$X_{2,s} = -B_2 u(s).$$

Letting $s = t$ and $s = t + \Delta t$, respectively, gets the following two equations

$$\begin{aligned} X_{2,t} &= -B_2 u(t) \\ X_{2,t+\Delta t} &= -B_2 u(t + \Delta t). \end{aligned}$$

Using the latter equation minus the former one, we obtain

$$\Delta X_{2,t} = -B_2\dot{u}(t)\Delta t + o(\Delta t),$$

where $u(t + \Delta t) = u(t) + \dot{u}(t)\Delta t + o(\Delta t)$, because $u(s)$ is derivable on $[t, T]$. Obviously we know

$$\Delta X_{1,t} = g(t)\left[A_1X_1 + B_1u(t)\right]\Delta t + h(t)u_{t1}u(t)\Delta C_t,$$

where $\Delta C_t \sim \mathcal{N}(0, \Delta t)$ which means ΔC_t is a normal uncertain variable with expected value 0 and variance Δt^2. Because $X_s = Q\begin{bmatrix} X_{1,s} \\ X_{2,s} \end{bmatrix}$, we obtain

$$\Delta X_t = Q\begin{bmatrix} g(t)[A_1X_1 + B_2u(t)] \\ -B_2\dot{u}(t) \end{bmatrix}\Delta t + h(t)Q_1u_{t1}u(t)\Delta C_t + o(\Delta t)$$

where $Q = \begin{bmatrix} Q_1 & Q_2 \end{bmatrix}$ and $Q_1 \in R^{n\times r}$, $Q_2 \in R^{n\times(n-r)}$. Now denote

$$\begin{aligned} p &= Q\begin{bmatrix} g(t)[A_1X_1 + B_2u(t)] \\ -B_2\dot{u}(t) \end{bmatrix}, \\ q &= h(t)Q_1u_{t1}u(t). \end{aligned}$$

Then we have

$$\Delta X_t = p\Delta t + q\Delta C_t + o(\Delta t).$$

By employing Taylor series expansion, we obtain

$$\begin{aligned} J(t + \Delta t, X + \Delta X_t) = {} & J(t, X) + J_t(t, X)\Delta t + \nabla_X J(t, X)^T\Delta X_t + \frac{1}{2}J_{tt}(t, X)\Delta t^2 \\ & + \nabla_X J_t(t, X)^T\Delta X_t\Delta t + \frac{1}{2}\Delta X_t^T\nabla_{XX}J(t, X)\Delta X_t \\ & + o(\Delta t). \end{aligned} \tag{2.42}$$

Substituting Eq. (2.42) into Eq. (2.4) yields

$$\begin{aligned} 0 = \sup_{u(t)}\Big\{ & f(X, u(t), t)\Delta t + J_t(t, X)\Delta t + E\Big[\nabla_X J(t, X)^T\Delta X_t + \nabla_X J_t(t, X)^T\Delta X_t\Delta t \\ & + \frac{1}{2}\Delta X_t^T\nabla_{XX}J(t, X)\Delta X_t\Big] + o(\Delta t)\Big\} \end{aligned} \tag{2.43}$$

Applying Theorem 1.8, we know

$$
\begin{aligned}
& E\left[\nabla_X J(t,X)^T \Delta X_t + \nabla_X J_t(t,X)^T \Delta X_t \Delta t + \frac{1}{2}\Delta X_t^T \nabla_X J_{XX}(t,X)\Delta X_t\right] \\
&= E\Big[\nabla_X J(t,X)^T (p\Delta t + q\Delta C_t + \text{o}(\Delta t)) + p\Delta t + \nabla_X J_t(t,X)^T (p\Delta t + q\Delta C_t \\
&\quad + \text{o}\,(\Delta t))\Delta t + \frac{1}{2}(p\Delta t + q\Delta C_t + \text{o}(\Delta t))^T \nabla_{XX} J(t,X)(p\Delta t + q\Delta C_t + \text{o}(\Delta t))\Big] \\
&= \nabla_X J(t,X)^T p\Delta t + E\Big[\left(\nabla_X J(t,X)q + \nabla_X J_t(t,X)q\Delta t + p^T \nabla_X J_{XX}(t,X)q\right)\Delta C_t \\
&\quad + \frac{1}{2}q^T \nabla_{XX} J(t,X)q\Delta C_t^2\Big] + \text{o}(\Delta t) \\
&= \nabla_X J(t,X)^T p\Delta t + E\left[a\Delta C_t + b\Delta C_t^2\right] + \text{o}(\Delta t) \\
&= \nabla_X J(t,X)^T p\Delta t + bE\left[\frac{a}{b}\Delta C_t + \Delta C_t^2\right] + \text{o}(\Delta t) \\
&= \nabla_X J(t,X)^T p\Delta t + \text{o}(\Delta t)
\end{aligned} \tag{2.44}
$$

where $a = \nabla_X J(t,X)q + \nabla_X J_t(t,X)q\Delta t + p^T \nabla_X J_{XX}(t,X)q$ and $b = \frac{1}{2}q^T \nabla_{XX} J(t,X)q$. Substituting Eq. (2.44) into (2.43), we obtain

$$
-J_t(t,X)\Delta t = \sup_{u(t)}\left[f(X,u(t),t)\Delta t + \nabla_X J(t,X)^T p\Delta t + \text{o}(\Delta t))\right]. \tag{2.45}
$$

Dividing Eq. (2.45) by Δt and letting $\Delta t \to 0$, we are able to get Eq. (2.41).

Remark 2.4 Note that when F is invertible, the uncertain singular system becomes uncertain normal system and the optimal control problem of the uncertain normal system [2] has been tackled in recent years.

Remark 2.5 The solutions of the presented model (2.40) may be obtained from settling the equation of optimality (2.41). The vector $p = Q\begin{bmatrix} g(t)\,(A_1X_1 + B_1u(t)) \\ -B_2\dot{u}(t)\end{bmatrix}$ is related to the function $\dot{u}(t)$ which is totally different from the optimal control problem of the uncertain normal system, and it will bring lots of matters in solving Eq. (2.41).

Example Consider the following problem:

$$
\begin{cases}
J(t,x_t) = \sup\limits_{u(t)\in U_{ad}} E\left[\displaystyle\int_t^T \alpha^\tau(s)X_s u(s)ds + \alpha^\tau(T)X_T\right] \\
\text{subject to} \\
\qquad F\mathrm{d}X_s = g(s)\left[AX_s + Bu(s)\right]\mathrm{d}s + h(s)Du(s)\mathrm{d}C_s, \text{ and } X_t = \boldsymbol{x}
\end{cases} \tag{2.46}
$$

where $X_s \in R^4$ is the state vector, $\alpha(s) \in R^4$ is the coefficient of X_s, $U_{ad} = [-1, 1]$, $\alpha^\tau(s) = [1, 1, 1, 2]e^{-s}$, $g(s) = 1$, $h(s) = s + 1$, and

$$F = \begin{bmatrix} 1 & 0 & 0 & 0 \\ 0 & 0 & 1 & 0 \\ 0 & 0 & 0 & 0 \\ 0 & 0 & 0 & 0 \end{bmatrix}, \ A = \begin{bmatrix} 0 & 1 & 0 & 0 \\ 1 & 0 & 0 & 0 \\ -1 & 0 & 0 & 1 \\ 0 & 1 & 1 & 1 \end{bmatrix}, \ B = \begin{bmatrix} 1 \\ 0 \\ -1 \\ 1 \end{bmatrix}, \ D = \begin{bmatrix} 1 \\ -1 \\ 0 \\ 0 \end{bmatrix}.$$

Through calculating, we know

$$det(zF - A) = det \begin{bmatrix} z & -1 & 0 & 0 \\ -1 & 0 & z & 0 \\ 1 & 0 & 0 & -1 \\ 0 & -1 & -1 & -1 \end{bmatrix} = z^2 + z + 1.$$

Obviously, $det(zF - A)$ is not identically zero and $deg(det(zF - A)) = rank(F)$, namely, the given system is regular and impulse-free. By using Lemma 2.1, we obtain two invertible matrices

$$P = \begin{bmatrix} 1 & 0 & 1 & -1 \\ 0 & 1 & 0 & 0 \\ 0 & 0 & -1 & 1 \\ 0 & 0 & 1 & 0 \end{bmatrix}, \ Q = \begin{bmatrix} 1 & 0 & 0 & 0 \\ -1 & -1 & 1 & 0 \\ 0 & 1 & 0 & 0 \\ 1 & 0 & 0 & 1 \end{bmatrix},$$

such that

$$PFQ = \begin{bmatrix} 1 & 0 & 0 & 0 \\ 0 & 1 & 0 & 0 \\ 0 & 0 & 0 & 0 \\ 0 & 0 & 0 & 0 \end{bmatrix}, \ PAQ = \begin{bmatrix} -1 & -1 & 0 & 0 \\ 1 & 0 & 0 & 0 \\ 0 & 0 & 1 & 0 \\ 0 & 0 & 0 & 1 \end{bmatrix}, \ PB = \begin{bmatrix} -1 \\ 0 \\ 2 \\ -1 \end{bmatrix}, \ PD = \begin{bmatrix} 1 \\ -1 \\ 0 \\ 0 \end{bmatrix}.$$

Easily, we can see

$$A_1 = \begin{bmatrix} -1 & -1 \\ 1 & 0 \end{bmatrix}, \ B_1 = \begin{bmatrix} -1 \\ 0 \end{bmatrix}, \ B_2 = \begin{bmatrix} 2 \\ -1 \end{bmatrix}, \ P_2 u_t = \begin{bmatrix} 0 \\ 0 \end{bmatrix}$$

where $P_2 = \begin{bmatrix} 0 & 0 & -1 & 1 \\ 0 & 0 & 1 & 0 \end{bmatrix}$. Denote $\boldsymbol{x} = [x_1, x_2, x_3, x_4]^\tau$, and we assume that $x_1 + x_3 = 0$. Because

$$Q^{-1} = \begin{bmatrix} 1 & 0 & 0 & 0 \\ 0 & 0 & 1 & 0 \\ 1 & 1 & 1 & 1 \\ -1 & 0 & 0 & 1 \end{bmatrix},$$

and $\begin{bmatrix} \boldsymbol{x}_1 \\ \boldsymbol{x}_2 \end{bmatrix} = Q^{-1}\boldsymbol{x}$, we obtain $\boldsymbol{x}_1 = [x_1, x_3]^T$. Combining these results and Eq. (2.41), we know

$$p = Q\begin{bmatrix} g(t)\,(A_1 X_1 + B_1 u(t)) \\ -B_2 \dot{u}(t) \end{bmatrix}$$
$$= \begin{bmatrix} -(x_1 + x_3) - u(t) \\ x_3 + u(t) - 2\dot{u}(t) \\ x_1 \\ -(x_1 + x_3) - u(t) + \dot{u}(t) \end{bmatrix}$$
$$= \begin{bmatrix} -u(t) \\ x_3 + u(t) - 2\dot{u}(t) \\ x_1 \\ -u(t) + \dot{u}(t) \end{bmatrix}.$$

We conjecture that $J(t, \boldsymbol{x}) = k\alpha^T(t)\boldsymbol{x} - k\alpha^\tau(T)E[X_T] + \alpha^\tau(T)E[X_T]$. Then

$$J_t(t, \boldsymbol{x}) = -k\alpha^T(t)\boldsymbol{x},\ \nabla_{\boldsymbol{x}} J(t, \boldsymbol{x}) = k\alpha(t),$$

and

$$\begin{aligned} &\alpha^\tau(t)\boldsymbol{x}u(t) + \nabla_{\boldsymbol{x}} J(t, \boldsymbol{x})^\tau p \\ &= (x_1 + x_2 + x_3 + 2x_4)e^{-t}u(t) + k\,[-u(t) + (x_3 + u(t) - 2\dot{u}(t)) + x_1 \\ &\quad + 2(-u(t) + \dot{u}(t))]\,e^{-t} \\ &= [(x_1 + x_2 + x_3 + 2x_4) - 2k]\,e^{-t}u(t). \end{aligned}$$

Applying Eq. (2.41), we get

$$\begin{aligned} k(x_1 + x_2 + x_3 + 2x_4)e^{-t} &= \sup_{u(t)\in[-1,1]} [(x_1 + x_2 + x_3 + 2x_4) - 2k]\,e^{-t}u(t) \\ &= e^{-t} \cdot \sup_{u(t)\in[-1,1]} [(x_1 + x_2 + x_3 + 2x_4) - 2k]\,u(t) \\ &= e^{-t} \mid (x_1 + x_2 + x_3 + 2x_4) - 2k \mid . \end{aligned} \tag{2.47}$$

Dividing Eq. (2.47) by e^{-t}, we obtain

$$k(x_1 + x_2 + x_3 + 2x_4) = \mid (x_1 + x_2 + x_3 + 2x_4) - 2k \mid, \tag{2.48}$$

and

$$k^2(x_1 + x_2 + x_3 + 2x_4)^2 = [(x_1 + x_2 + x_3 + 2x_4) - 2k]^2,$$

namely

$$(a^2 - 4)k^2 + 4ak - a^2 = 0,$$

where $a = x_1 + x_2 + x_3 + 2x_4$. According to Eq. (2.48), the symbols of k and a must keep coincidence, so we know

$$k = \begin{cases} \dfrac{a}{4}, & \text{if } a = \pm 2 \\ 0, & \text{if } a = 0 \\ \dfrac{-a}{a-2}, & \text{if } a < -2 \text{ or } 0 < a < 2 \\ \dfrac{a}{a+2}, & \text{if } -2 < a < 0 \text{ or } a > 2. \end{cases}$$

Thus the optimal control is

$$u^*(t) = sign(a - 2k).$$

References

1. Liu B (2009) Theory and practice of uncertain programming, 2nd edn. Springer, Berlin
2. Zhu Y (2010) Uncertain optimal control with application to a portfolio selection model. Cybern Syst 41(7):535–547
3. Xu X, Zhu Y (2012) Uncertain bang-bang control for continuous time model. Cybern Syst Int J 43(6):515–527
4. Dai L (1989) Singular control systems. Springer, Berlin
5. Shu Y, Zhu Y (2017) Stability and optimal control for uncertain continuous-time singular systems. Eur J Control 34:16–23
6. Ji X, Zhou J (2015) Multi-dimensional uncertain differential equation: existence and uniqueness of solution. Fuzzy Optim Decis Mak 14(4):477–491

Chapter 3
Optimistic Value-Based Uncertain Optimal Control

Expected value is the weighted average of uncertain variables in the sense of uncertain measure. However, in some cases, we need to take other characters of uncertain variables into account. For instance, if the student test scores presented two levels of differentiation phenomenon, and the difference between higher performance and lower performance is too large, then average grade may not be considered only. In this case, critical value (optimistic value or pessimistic value) of test scores may be discussed. We may investigate the problem such as which point the lowest of the 95% test scores is up to.

Different from the expected value optimal control problems, in this chapter, we will introduce another kind of uncertain optimal control problems, namely optimal control problems, for uncertain differential systems based on optimistic value criterion.

3.1 Optimistic Value Model

Assume that $\boldsymbol{C}_t = (C_{t1}, C_{t2}, \ldots, C_{tk})^\tau$, where $C_{t1}, C_{t2}, \ldots, C_{tk}$ are independent canonical Liu processes. For any $0 < t < T$, and confidence level $\alpha \in (0, 1)$, we introduce an uncertain optimistic value optimal control problem for multidimensional case as follows [1].

$$\begin{cases} J(t, \boldsymbol{x}) \equiv \sup\limits_{\boldsymbol{u}_t \in U} F_{\sup}(\alpha) \\ \text{subject to} \\ \quad \mathrm{d}\boldsymbol{X}_s = \boldsymbol{\mu}(s, \boldsymbol{u}_s, \boldsymbol{X}_s)\mathrm{d}s + \boldsymbol{\sigma}(s, \boldsymbol{u}_s, \boldsymbol{X}_s)\mathrm{d}\boldsymbol{C}_s \quad \text{and} \quad \boldsymbol{X}_t = \boldsymbol{x} \end{cases} \tag{3.1}$$

where $F = \int_t^T f(s, \boldsymbol{u}_s, \boldsymbol{X}_s)\mathrm{d}s + G(T, \boldsymbol{X}_T)$, and $F_{\sup}(\alpha) = \sup\left\{\overline{F} | \mathcal{M}\left\{F \geq \overline{F}\right\} \geq \alpha\right\}$ which denotes the α-optimistic value to F. The vector $\mathbf{X}_s$ is a state vector of dimension n, $\boldsymbol{u}_s$ is a control vector of dimension r subject to a constraint set U.

Y. Zhu, *Uncertain Optimal Control*, Springer Uncertainty Research,
https://doi.org/10.1007/978-981-13-2134-4_3

The function $f : [0, T] \times R^r \times R^n \to R$ is an objective function, and $G : [0, T] \times R^n \to R$ is a function of terminal reward. In addition, $\boldsymbol{\mu} : [0, T] \times R^r \times R^n \to R^n$ is a vector-value function, and $\boldsymbol{\sigma} : [0, T] \times R^r \times R^n \to R^n \times R^k$ is a matrix-value function. All functions mentioned are continuous. We first present the following principle of optimality.

Theorem 3.1 ([1]) *For any $(t, \boldsymbol{x}) \in [0, T) \times R^n$, and $\Delta t > 0$ with $t + \Delta t < T$, we have*

$$J(t, \boldsymbol{x}) = \sup_{\boldsymbol{u}_t \in U} \left\{ f(t, \boldsymbol{u}_t, \boldsymbol{x})\Delta t + J(t + \Delta t, \boldsymbol{x} + \Delta \boldsymbol{X}_t) + o(\Delta t) \right\}, \tag{3.2}$$

where $\boldsymbol{x} + \Delta \boldsymbol{X}_t = \boldsymbol{X}_{t+\Delta t}$.

Proof We denote the right side of (3.2) by $\tilde{J}(t, \boldsymbol{x})$. For arbitrary $\boldsymbol{u}_t \in U$, it follows from the definition of $J(t, \boldsymbol{x})$ that

$$\begin{aligned} J(t, \boldsymbol{x}) \geq \Bigg[&\int_t^{t+\Delta t} f(s, \boldsymbol{u}_s|_{[t,t+\Delta t)}, \boldsymbol{X}_s)\mathrm{d}s \\ &+ \int_{t+\Delta t}^{T} f(s, \boldsymbol{u}_s|_{[t+\Delta t,T]}, \boldsymbol{X}_s)\mathrm{d}s + G(T, \boldsymbol{X}_T) \Bigg]_{\sup} (\alpha), \end{aligned}$$

where $\boldsymbol{u}_s|_{[t,t+\Delta t)}$ and $\boldsymbol{u}_s|_{[t+\Delta t,T]}$ are control vector $\boldsymbol{u}_s$ restricted on $[t, t + \Delta t)$ and $[t + \Delta t, T]$, respectively. Since for any $\Delta t > 0$,

$$\int_t^{t+\Delta t} f(s, \boldsymbol{u}_s|_{[t,t+\Delta t)}, \boldsymbol{X}_s)\mathrm{d}s = f(t, \boldsymbol{u}_t, \boldsymbol{x})\Delta t + o(\Delta t),$$

we have

$$\begin{aligned} J(t, \boldsymbol{x}) \geq\ & f(t, \boldsymbol{u}_t, \boldsymbol{x})\Delta t + o(\Delta t) \\ &+ \left[\int_{t+\Delta t}^{T} f(s, \boldsymbol{u}_s|_{[t+\Delta t,T]}, \boldsymbol{X}_s)\mathrm{d}s + G(T, \boldsymbol{X}_T) \right]_{\sup} (\alpha). \end{aligned} \tag{3.3}$$

Taking the supremum with respect to $\boldsymbol{u}_s|_{[t+\Delta t,T]}$ in (3.3), we get $J(t, \boldsymbol{x}) \geq \tilde{J}(t, \boldsymbol{x})$.

On the other hand, for all $\boldsymbol{u}_t$, we have

$$\begin{aligned} &\left[\int_t^{T} f(s, \boldsymbol{u}_s, \boldsymbol{X}_s)\mathrm{d}s + G(T, \boldsymbol{X}_T) \right]_{\sup} (\alpha) \\ &= f(t, \boldsymbol{u}_t, \boldsymbol{x})\Delta t + o(\Delta t) + \left[\int_{t+\Delta t}^{T} f(s, \boldsymbol{u}_t|_{[t+\Delta t,T]}, \boldsymbol{X}_s)\mathrm{d}s + G(T, \boldsymbol{X}_T) \right]_{\sup} (\alpha) \\ &\leq f(t, \boldsymbol{u}_t, \boldsymbol{x})\Delta t + o(\Delta t) + J(t + \Delta t, \boldsymbol{x} + \Delta \boldsymbol{X}_t) \\ &\leq \tilde{J}(t, \boldsymbol{x}). \end{aligned}$$

Hence, $J(t, \boldsymbol{x}) \leq \tilde{J}(t, \boldsymbol{x})$, and then $J(t, \boldsymbol{x}) = \tilde{J}(t, \boldsymbol{x})$. Theorem 3.1 is proved.

3.2 Equation of Optimality

Consider the uncertain optimal control problem (3.1). Now let us give an equation of optimality in optimistic value model.

Theorem 3.2 ([1]) *Let $J(t, \boldsymbol{x})$ be twice differentiable on $[0, T] \times R^n$. Then we have*

$$- J_t(t, \boldsymbol{x}) = \sup_{\boldsymbol{u} \in U} \left\{ f(t, \boldsymbol{u}_t, \boldsymbol{x}) + \nabla_{\boldsymbol{x}} J(t, \boldsymbol{x})^\tau \boldsymbol{\mu}(t, \boldsymbol{u}_t, \boldsymbol{x}) + \frac{\sqrt{3}}{\pi} \ln \frac{1-\alpha}{\alpha} \|\nabla_{\boldsymbol{x}} J(t, \boldsymbol{x})^\tau \boldsymbol{\sigma}(t, \boldsymbol{x}, \boldsymbol{u})\|_1 \right\} \tag{3.4}$$

where $J_t(t, \boldsymbol{x})$ is the partial derivative of the function $J(t, \boldsymbol{x})$ in t, $\nabla_{\boldsymbol{x}} J(t, \boldsymbol{x})$ is the gradient of $J(t, \boldsymbol{x})$ in $\boldsymbol{x}$, and $\| \cdot \|_1$ is the 1-norm for vectors, that is, $\|p\|_1 = \sum_{i=1}^{n} |p_i|$ for $p = (p_1, p_2, \ldots, p_n)$.

Proof By Taylor expansion, we get

$$\begin{aligned} J(t + \Delta t, \boldsymbol{x} + \Delta \boldsymbol{X}_t) = {} & J(t, \boldsymbol{x}) + J_t(t, \boldsymbol{x})\Delta t + \nabla_{\boldsymbol{x}} J(t, \boldsymbol{x})^\tau \Delta \boldsymbol{X}_t \\ & + \frac{1}{2} J_{tt}(t, \boldsymbol{x})\Delta t^2 + \frac{1}{2} \Delta \boldsymbol{X}_t^\tau \nabla_{\boldsymbol{xx}} J(t, \boldsymbol{x}) \Delta \boldsymbol{X}_t \\ & + \nabla_{\boldsymbol{x}} J_t(t, \boldsymbol{x})^\tau \Delta \boldsymbol{X}_t \Delta t + o(\Delta t) \end{aligned} \tag{3.5}$$

where $\nabla_{\boldsymbol{xx}} J(t, \boldsymbol{x})$ is the Hessian matrix of $J(t, \boldsymbol{x})$. Substituting Eq. (3.5) into Eq. (3.2) yields that

$$\begin{aligned} 0 = \sup_{\boldsymbol{u} \in U} \Big\{ & f(t, \boldsymbol{u}_t, \boldsymbol{x})\Delta t + J_t(t, \boldsymbol{x})\Delta t + \Big[\nabla_{\boldsymbol{x}} J(t, \boldsymbol{x})^\tau \Delta \boldsymbol{X}_t \\ & + \frac{1}{2} \Delta \boldsymbol{X}_t^\tau \nabla_{\boldsymbol{xx}} J(t, \boldsymbol{x}) \Delta \boldsymbol{X}_t + \nabla_{\boldsymbol{x}} J_t(t, \boldsymbol{x})^\tau \Delta \boldsymbol{X}_t \Delta t \Big]_{\sup} (\alpha) \\ & + o(\Delta t) \Big\}. \end{aligned} \tag{3.6}$$

Note that $\Delta \boldsymbol{X}_t = \boldsymbol{\mu}(t, \boldsymbol{u}_t, \boldsymbol{x})\Delta t + \boldsymbol{\sigma}(t, \boldsymbol{u}_t, \boldsymbol{x})\Delta \boldsymbol{C}_t$. It follows from (3.6) that

$$\begin{aligned} 0 = \sup_{\boldsymbol{u} \in U} \Big\{ & f(t, \boldsymbol{u}_t, \boldsymbol{x})\Delta t + J_t(t, \boldsymbol{x})\Delta t + \nabla_{\boldsymbol{x}} J(t, \boldsymbol{x})^\tau \boldsymbol{\mu}(t, \boldsymbol{u}_t, \boldsymbol{x})\Delta t \\ & + [\boldsymbol{a} \Delta \boldsymbol{C}_t + \Delta \boldsymbol{C}_t^\tau \boldsymbol{B} \Delta \boldsymbol{C}_t]_{\sup}(\alpha) + o(\Delta t) \Big\}, \end{aligned} \tag{3.7}$$

where

$$\begin{aligned} \boldsymbol{a} &= \nabla_{\boldsymbol{x}} J(t, \boldsymbol{x})^\tau \boldsymbol{\sigma}(t, \boldsymbol{u}_t, \boldsymbol{x}) + \nabla_{\boldsymbol{x}} J_t(t, \boldsymbol{x})^\tau \boldsymbol{\sigma}(t, \boldsymbol{u}_t, \boldsymbol{x})\Delta t \\ &\quad + \boldsymbol{\mu}(t, \boldsymbol{u}_t, \boldsymbol{x})^\tau \nabla_{\boldsymbol{xx}} J(t, \boldsymbol{x}) \boldsymbol{\sigma}(t, \boldsymbol{u}_t, \boldsymbol{x})\Delta t, \\ \boldsymbol{B} &= \frac{1}{2} \boldsymbol{\sigma}(t, \boldsymbol{u}_t, \boldsymbol{x})^\tau \nabla_{\boldsymbol{xx}} J(t, \boldsymbol{x}) \boldsymbol{\sigma}(t, \boldsymbol{u}_t, \boldsymbol{x}). \end{aligned}$$

Let $\boldsymbol{a} = (a_1, a_2, \ldots, a_k)$, $\boldsymbol{B} = (b_{ij})_{k\times k}$. Then we have

$$\boldsymbol{a}\Delta\boldsymbol{C}_t + \Delta\boldsymbol{C}_t^\tau \boldsymbol{B}\Delta\boldsymbol{C}_t = \sum_{i=1}^k a_i \Delta C_{ti} + \sum_{i=1}^k \sum_{j=1}^k b_{ij}\Delta C_{ti}\Delta C_{tj}.$$

Since $|b_{ij}\Delta C_{ti}\Delta C_{tj}| \le \frac{1}{2}|b_{ij}|(\Delta C_{ti}{}^2 + \Delta C_{tj}{}^2)$, we have

$$\sum_{i=1}^k \left\{a_i\Delta C_{ti} - \left(\sum_{j=1}^k |b_{ij}|\right)\Delta C_{ti}{}^2\right\} \le \boldsymbol{a}\Delta\boldsymbol{C}_t + \Delta\boldsymbol{C}_t^\tau \boldsymbol{B}\Delta\boldsymbol{C}_t$$
$$\le \sum_{i=1}^k \left\{a_i\Delta C_{ti} + \left(\sum_{j=1}^k |b_{ij}|\right)\Delta C_{ti}{}^2\right\}.$$

Because of the independence of $C_{t1}, C_{t2}, \ldots, C_{tk}$, we have

$$\sum_{i=1}^k \left[a_i\Delta C_{ti} - \left(\sum_{j=1}^k |b_{ij}|\right)\Delta C_{ti}{}^2\right]_{\sup}(\alpha) \le [\boldsymbol{a}\Delta\boldsymbol{C}_t + \Delta\boldsymbol{C}_t^\tau \boldsymbol{B}\Delta\boldsymbol{C}_t]_{\sup}(\alpha)$$
$$\le \sum_{i=1}^k \left[a_i\Delta C_{ti} + \left(\sum_{j=1}^k |b_{ij}|\right)\Delta C_{ti}{}^2\right]_{\sup}(\alpha).$$

It follows from Theorem 1.12 that for any small enough $\varepsilon > 0$, we have

$$[\boldsymbol{a}\Delta\boldsymbol{C}_t + \Delta\boldsymbol{C}_t^\tau \boldsymbol{B}\Delta\boldsymbol{C}_t]_{\sup}(\alpha) \le \frac{\sqrt{3}}{\pi}\ln\frac{1-\alpha+\varepsilon}{\alpha-\varepsilon}\cdot\Delta t\cdot\sum_{i=1}^k |a_i|$$
$$+\left(\frac{\sqrt{3}}{\pi}\ln\frac{2-\varepsilon}{\varepsilon}\right)^2\cdot\Delta t^2\cdot\sum_{i=1}^k\sum_{j=1}^k |b_{ij}|, \quad (3.8)$$

and

$$[\boldsymbol{a}\Delta\boldsymbol{C}_t + \Delta\boldsymbol{C}_t^\tau \boldsymbol{B}\Delta\boldsymbol{C}_t]_{\sup}(\alpha) \ge \frac{\sqrt{3}}{\pi}\ln\frac{1-\alpha-\varepsilon}{\alpha+\varepsilon}\cdot\Delta t\cdot\sum_{i=1}^k |a_i|$$
$$-\left(\frac{\sqrt{3}}{\pi}\ln\frac{2-\varepsilon}{\varepsilon}\right)^2\cdot\Delta t^2\cdot\sum_{i=1}^k\sum_{j=1}^k |b_{ij}|. \quad (3.9)$$

By Eq. (3.7) and inequality (3.8), for $\Delta t > 0$, there exists a control $\boldsymbol{u}_t \equiv \boldsymbol{u}_{\varepsilon,\Delta t}$ such that

$$
\begin{aligned}
-\varepsilon\Delta t \le &\ \{f(t, \boldsymbol{u}_t, \boldsymbol{x})\Delta t + J_t(t, \boldsymbol{x})\Delta t + \nabla_{\boldsymbol{x}} J(t, \boldsymbol{x})^\tau \boldsymbol{\mu}(t, \boldsymbol{u}_t, \boldsymbol{x})\Delta t \\
&+ [\boldsymbol{a}\Delta \boldsymbol{C}_t + \Delta \boldsymbol{C}_t^\tau \boldsymbol{B}\Delta \boldsymbol{C}_t]_{\sup}(\alpha) + o(\Delta t)\} \\
\le &\ f(t, \boldsymbol{u}_t, \boldsymbol{x})\Delta t + J_t(t, \boldsymbol{x})\Delta t + \nabla_{\boldsymbol{x}} J(t, \boldsymbol{x})^\tau \boldsymbol{\mu}(t, \boldsymbol{u}_t, \boldsymbol{x})\Delta t \\
&+ \frac{\sqrt{3}}{\pi} \ln \frac{1-\alpha+\varepsilon}{\alpha-\varepsilon} \cdot \Delta t \cdot \sum_{i=1}^{k} |a_i| \\
&+ \left(\frac{\sqrt{3}}{\pi} \ln \frac{2-\varepsilon}{\varepsilon}\right)^2 \cdot \Delta t^2 \cdot \sum_{i=1}^{k}\sum_{j=1}^{k} |b_{ij}| + o(\Delta t).
\end{aligned}
$$

Dividing both sides of the above inequality by Δt, we get

$$
\begin{aligned}
-\varepsilon \le &\ f(t, \boldsymbol{u}_t, \boldsymbol{x}) + J_t(t, \boldsymbol{x}) + \nabla_{\boldsymbol{x}} J(t, \boldsymbol{x})^\tau \boldsymbol{\mu}(t, \boldsymbol{u}_t, \boldsymbol{x}) \\
&+ \frac{\sqrt{3}}{\pi} \ln \frac{1-\alpha+\varepsilon}{\alpha-\varepsilon} \|\nabla_{\boldsymbol{x}} J(t, \boldsymbol{x})^\tau \boldsymbol{\sigma}(t, \boldsymbol{u}_t, \boldsymbol{x})\|_1 + h_1(\varepsilon, \Delta t) + h_2(\Delta t) \\
\le &\ J_t(t, \boldsymbol{x}) + \sup_{\boldsymbol{u}\in U}\{f(t, \boldsymbol{u}_t, \boldsymbol{x}) + \nabla_{\boldsymbol{x}} J(t, \boldsymbol{x})^\tau \boldsymbol{\mu}(t, \boldsymbol{u}_t, \boldsymbol{x}) \\
&+ \left.\frac{\sqrt{3}}{\pi} \ln \frac{1-\alpha+\varepsilon}{\alpha-\varepsilon} \|\nabla_{\boldsymbol{x}} J(t, \boldsymbol{x})^\tau \boldsymbol{\sigma}(t, \boldsymbol{u}_t, \boldsymbol{x})\|_1 \right\} + h_1(\varepsilon, \Delta t) + h_2(\Delta t)
\end{aligned}
$$

since

$$
\sum_{i=1}^{k} |a_i| \to \|\nabla_{\boldsymbol{x}} J(t, \boldsymbol{x})^\tau \boldsymbol{\sigma}(t, \boldsymbol{u}_t, \boldsymbol{x})\|_1
$$

as $\Delta t \to 0$, where $h_1(\varepsilon, \Delta t) \to 0$ and $h_2(\Delta t) \to 0$ as $\Delta t \to 0$. Letting $\Delta t \to 0$, and then $\varepsilon \to 0$ results in

$$
\begin{aligned}
0 \le &\ J_t(t, \boldsymbol{x}) + \sup_{\boldsymbol{u}\in U}\{f(t, \boldsymbol{u}_t, \boldsymbol{x}) + \nabla_{\boldsymbol{x}} J(t, \boldsymbol{x})^\tau \boldsymbol{\mu}(t, \boldsymbol{u}_t, \boldsymbol{x}) \\
&+ \left.\frac{\sqrt{3}}{\pi} \ln \frac{1-\alpha}{\alpha} \|\nabla_{\boldsymbol{x}} J(t, \boldsymbol{x})^\tau \boldsymbol{\sigma}(t, \boldsymbol{u}_t, \boldsymbol{x})\|_1 \right\}.
\end{aligned}
\tag{3.10}
$$

On the other hand, by Eq. (3.7) and inequality (3.9), applying the similar method, we can obtain

$$
\begin{aligned}
0 \ge &\ J_t(t, \boldsymbol{x}) + \sup_{\boldsymbol{u}\in U}\{f(t, \boldsymbol{u}_t, \boldsymbol{x}) + \nabla_{\boldsymbol{x}} J(t, \boldsymbol{x})^\tau \boldsymbol{\mu}(t, \boldsymbol{u}_t, \boldsymbol{x}) \\
&+ \left.\frac{\sqrt{3}}{\pi} \ln \frac{1-\alpha}{\alpha} \|\nabla_{\boldsymbol{x}} J(t, \boldsymbol{x})^\tau \boldsymbol{\sigma}(t, \boldsymbol{u}_t, \boldsymbol{x})\|_1 \right\}.
\end{aligned}
\tag{3.11}
$$

Combining (3.10) and (3.11), we obtain the Eq. (3.4). The theorem is proved.

Remark 3.1 The solutions of the proposed model (3.1) may be derived from solving the equation of optimality (3.4).

Remark 3.2 Note that in the case of stochastic optimal control, we cannot obtain the similar conclusion to (3.4) due to the difficulty of calculating optimistic value of the variables with the form of $a\eta + b\eta^2$, where η is a normally distributed random variable, while random normal distribution function has no analytic expression.

Remark 3.3 Particularly, for one-dimensional case, the equation of optimality has a simple form:

$$
\begin{aligned}
-J_t(t,x) = \sup_{u_t \in U} \Big\{ & f(t,u_t,x) + J_x(t,x)\mu(t,u_t,x) \\
& + \frac{\sqrt{3}}{\pi} \ln \frac{1-\alpha}{\alpha} \left| J_x(t,x)\sigma(t,u_t,x) \right| \Big\}. \qquad (3.12)
\end{aligned}
$$

3.3 Uncertain Optimal Control Model with Hurwicz Criterion

Grounded on uncertain measure, the optimistic value criterion and pessimistic value criterion of uncertain variables have been introduced for handling optimization problems in uncertain environments. Applying the optimistic value criterion to consider the objectives is essentially a maximum approach, which maximizes the uncertain return. This approach suggests that the decision maker who is attracted by high payoffs to take some adventures. As opposed to the optimistic value criterion, using the pessimistic value criterion for uncertain decision system is essentially a maximin approach, which the underlying philosophy is based on selecting the alternative that provides the least bad uncertain return. It suggests the decision maker who is in pursuit of cautious that there is at least a known minimum payoff in the event of an unfavourable outcome.

The Hurwicz criterion can also be called optimism coefficient method, designed by economics professor Leonid Hurwicz [2] in 1951. It is a complex decision-making criterion attempting to find the intermediate area between the extremes posed by the optimistic and pessimistic criteria. Instead of assuming totally optimistic or pessimistic, Hurwicz criterion incorporates a measure of both by assigning a certain percentage weight to optimism and the balance to pessimism. With the Hurwicz criterion, the decision maker first should subjectively select a coefficient ρ denoting the optimism degree, note that $0 \le \rho \le 1$. Simultaneously, $1 - \rho$ represents a measure of the decision maker's pessimism. For every decision alternative, let the maximum return be multiplied by the coefficient of optimism ρ, and the minimum return be multiplied by the coefficient $1 - \rho$, then sum the results obtained. After computing each alternative's weighted average return, select the alternative with the best return as the chosen decision. Particularly, by changing the coefficient ρ, the Hurwicz criterion becomes various criteria. If $\rho = 1$, it reduces the Hurwicz criterion to the optimistic value criterion; if $\rho = 0$, the criterion is the pessimistic value criterion.

Assume that $\boldsymbol{C}_t = (C_{t1}, C_{t2}, \ldots, C_{tk})^\tau$, where $C_{t1}, C_{t2}, \ldots, C_{tk}$ are independent canonical processes. A selected coefficient $\rho \in (0, 1)$ denoting the optimism degree, and predetermined confidence level $\alpha \in (0, 1)$. For any $0 < t < T$, we present an uncertain optimal control model with Hurwicz criterion for multidimensional case as follows [3].

$$\begin{cases} J(t, \boldsymbol{x}) \equiv \sup\limits_{\boldsymbol{u}_t \in U} H_\alpha^\rho = \sup\limits_{\boldsymbol{u}_t \in U} \left\{\rho F_{\sup}(\alpha) + (1-\rho) F_{\inf}(\alpha)\right\} \\ \text{subject to} \\ \quad \mathrm{d}\boldsymbol{X}_s = \boldsymbol{\mu}(s, \boldsymbol{u}_s, \boldsymbol{X}_s)\mathrm{d}s + \boldsymbol{\sigma}(s, \boldsymbol{u}_s, \boldsymbol{X}_s)\mathrm{d}\boldsymbol{C}_s \quad \text{and} \quad \boldsymbol{X}_t = \boldsymbol{x} \end{cases} \tag{3.13}$$

where $F = \int_t^T f(s, \boldsymbol{u}_s, \boldsymbol{X}_s)ds + G(T, \boldsymbol{X}_T)$, and $F_{\sup}(\alpha) = \sup\left\{\overline{F} | \mathcal{M}\left\{F \geq \overline{F}\right\} \geq \alpha\right\}$ which denotes the α-optimistic value to F, $F_{\inf}(\alpha) = \inf\left\{\overline{F} | \mathcal{M}\left\{F \leq \overline{F}\right\} \geq \alpha\right\}$ reflects the α-pessimistic value to F. The vector $\mathbf{X}_s$ is the state vector of dimension n, $\boldsymbol{u}$ is a control vector of dimension r subject to a constraint set U. The function $f : [0, T] \times R^r \times R^n \to R$ is the objective function, and $G : [0, T] \times R^n \to R$ is the function of terminal reward. In addition, $\boldsymbol{\mu} : [0, T] \times R^r \times R^n \to R^n$ is a vector-value function, and $\boldsymbol{\sigma} : [0, T] \times R^r \times R^n \to R^n \times R^k$ is a matrix-value function.

For the purpose of solving the proposed model, now we present the following principle of optimality and equation of optimality.

Theorem 3.3 ([3]) *For any $(t, \boldsymbol{x}) \in [0, T) \times R^n$, and $\Delta t > 0$ with $t + \Delta t < T$, we have*

$$J(t, \boldsymbol{x}) = \sup_{\boldsymbol{u}_t \in U} \left\{f(t, \boldsymbol{u}_t, \boldsymbol{x})\Delta t + J(t + \Delta t, \boldsymbol{x} + \Delta \boldsymbol{X}_t) + o(\Delta t)\right\}, \tag{3.14}$$

where $\boldsymbol{x} + \Delta \boldsymbol{X}_t = \boldsymbol{X}_{t+\Delta t}$.

Proof The proof is similar to that of Theorem 3.1.

Theorem 3.4 ([3]) *Suppose $J(t, \boldsymbol{x}) \in C^2([0, T] \times R^n)$. Then we have*

$$\begin{aligned} -J_t(t, \boldsymbol{x}) = & \sup_{\boldsymbol{u}_t \in U} \left\{f(t, \boldsymbol{u}_t, \boldsymbol{x}) + \nabla_{\boldsymbol{x}} J(t, \boldsymbol{x})^\tau \boldsymbol{b}(t, \boldsymbol{u}_t, \boldsymbol{x}) \right. \\ & \left. + (2\rho - 1)\left(\frac{\sqrt{3}}{\pi} \ln \frac{1-\alpha}{\alpha}\right) \|\nabla_{\boldsymbol{x}} J(t, \boldsymbol{x})^\tau \boldsymbol{\sigma}(t, \boldsymbol{u}_t, \boldsymbol{x})\|_1 \right\} \end{aligned} \tag{3.15}$$

where $J_t(t, \boldsymbol{x})$ is the partial derivative of the function $J(t, \boldsymbol{x})$ in t, $\nabla_{\boldsymbol{x}} J(t, \boldsymbol{x})$ is the gradient of $J(t, \boldsymbol{x})$ in $\boldsymbol{x}$, and $\|\cdot\|_1$ is the 1-norm for vectors, that is, $\|p\|_1 = \sum_{i=1}^n |p_i|$ for $p = (p_1, p_2, \ldots, p_n)$.

Proof By using Taylor expansion, we get

$$\begin{aligned} &J(t+\Delta t, \boldsymbol{x}+\Delta \boldsymbol{X}_t) \\ &\quad = J(t,\boldsymbol{x}) + J_t(t,\boldsymbol{x})\Delta t + \nabla_{\boldsymbol{x}} J(t,\boldsymbol{x})^\tau \Delta \boldsymbol{X}_t + \frac{1}{2} J_{tt}(t,\boldsymbol{x})\Delta t^2 \\ &\qquad + \frac{1}{2}\Delta \boldsymbol{X}_t^\tau \nabla_{\boldsymbol{xx}} J(t,\boldsymbol{x})\Delta \boldsymbol{X}_t + \nabla_{\boldsymbol{x}} J_t(t,\boldsymbol{x})^\tau \Delta t \Delta \boldsymbol{X}_t + o(\Delta t) \end{aligned} \tag{3.16}$$

where $\nabla_{\boldsymbol{xx}} J(t,\boldsymbol{x})$ is the Hessian matrix of $J(t,\boldsymbol{x})$ in $\boldsymbol{x}$. Note that $\Delta \boldsymbol{X}_t = \boldsymbol{b}(t,\boldsymbol{u}_t,\boldsymbol{x})\Delta t + \boldsymbol{\sigma}(t,\boldsymbol{u}_t,\boldsymbol{x})\Delta \boldsymbol{C}_t$. Substituting Eq. (3.16) into Eq. (3.14) and simplifying the resulting expression yields that

$$\begin{aligned} 0 = \sup_{\boldsymbol{u}_t \in U} \{ & f(t,\boldsymbol{u}_t,\boldsymbol{x})\Delta t + J_t(t,\boldsymbol{x})\Delta t + \nabla_{\boldsymbol{x}} J(t,\boldsymbol{x})^\tau \boldsymbol{b}(t,\boldsymbol{u}_t,\boldsymbol{x})\Delta t \\ & + H_\alpha^\rho[\boldsymbol{a}\Delta \boldsymbol{C}_t + \Delta \boldsymbol{C}_t^\tau \boldsymbol{B} \Delta \boldsymbol{C}_t] + o(\Delta t) \}, \end{aligned} \tag{3.17}$$

where

$$\begin{aligned} \boldsymbol{a} &= \nabla_{\boldsymbol{x}} J(t,\boldsymbol{x})^\tau \boldsymbol{\sigma}(t,\boldsymbol{u}_t,\boldsymbol{x}) + \nabla_{\boldsymbol{x}} J_t(t,\boldsymbol{x})^\tau \boldsymbol{\sigma}(t,\boldsymbol{u}_t,\boldsymbol{x})\Delta t \\ &\quad + \boldsymbol{b}(t,\boldsymbol{u}_t,\boldsymbol{x})^\tau \nabla_{\boldsymbol{xx}} J(t,\boldsymbol{x})\boldsymbol{\sigma}(t,\boldsymbol{u}_t,\boldsymbol{x})\Delta t, \\ \boldsymbol{B} &= \frac{1}{2}\boldsymbol{\sigma}(t,\boldsymbol{u}_t,\boldsymbol{x})^\tau \nabla_{\boldsymbol{xx}} J(t,\boldsymbol{x})\boldsymbol{\sigma}(t,\boldsymbol{u}_t,\boldsymbol{x}). \end{aligned}$$

Let $\boldsymbol{a} = (a_1, a_2, \ldots, a_k)$, $\boldsymbol{B} = (b_{ij})_{k\times k}$. We have

$$\boldsymbol{a}\Delta \boldsymbol{C}_t + \Delta \boldsymbol{C}_t^\tau \boldsymbol{B} \Delta \boldsymbol{C}_t = \sum_{i=1}^k a_i \Delta C_{ti} + \sum_{i=1}^k \sum_{j=1}^k b_{ij} \Delta C_{ti} \Delta C_{tj}.$$

Since $|b_{ij}\Delta C_{ti}\Delta C_{tj}| \le \frac{1}{2}|b_{ij}|(\Delta C_{ti}{}^2 + \Delta C_{tj}{}^2)$, we have

$$\begin{aligned} \sum_{i=1}^k \left\{ a_i \Delta C_{ti} - \left(\sum_{j=1}^k |b_{ij}|\right)\Delta C_{ti}{}^2 \right\} &\le \boldsymbol{a}\Delta \boldsymbol{C}_t + \Delta \boldsymbol{C}_t^\tau \boldsymbol{B} \Delta \boldsymbol{C}_t \\ &\le \sum_{i=1}^k \left\{ a_i \Delta C_{ti} + \left(\sum_{j=1}^k |b_{ij}|\right)\Delta C_{ti}{}^2 \right\}. \end{aligned}$$

Because of the independence of $C_{t1}, C_{t2}, \ldots, C_{tk}$, we have

$$\begin{aligned} \sum_{i=1}^k H_\alpha^\rho\left[a_i \Delta C_{ti} - \left(\sum_{j=1}^k |b_{ij}|\right)\Delta C_{ti}{}^2 \right] &\le H_\alpha^\rho[\boldsymbol{a}\Delta \boldsymbol{C}_t + \Delta \boldsymbol{C}_t^\tau \boldsymbol{B} \Delta \boldsymbol{C}_t] \\ &\le \sum_{i=1}^k H_\alpha^\rho\left[a_i \Delta C_{ti} + \left(\sum_{j=1}^k |b_{ij}|\right)\Delta C_{ti}{}^2 \right]. \end{aligned}$$

By Eq. (3.17), for $\Delta t > 0$ and any small enough $\varepsilon > 0$, there exists a control $\boldsymbol{u}_t \equiv \boldsymbol{u}_{\varepsilon,\Delta t}$ such that

$$\begin{aligned}-\varepsilon\Delta t \le \{&f(t,\boldsymbol{u}_t,\boldsymbol{x})\Delta t + J_t(t,\boldsymbol{x})\Delta t + \nabla_{\boldsymbol{x}} J(t,\boldsymbol{x})^\tau \boldsymbol{\mu}(t,\boldsymbol{u}_t,\boldsymbol{x})\Delta t\\ &+ H_\alpha^\rho[\boldsymbol{a}\Delta \boldsymbol{C}_t + \Delta \boldsymbol{C}_t^\tau \boldsymbol{B}\Delta \boldsymbol{C}_t] + o(\Delta t)\}.\end{aligned}$$

Applying Theorems 1.12 and 1.13, we have

$$\begin{aligned}-\varepsilon\Delta t \le\ & f(t,\boldsymbol{u}_t,\boldsymbol{x})\Delta t + J_t(t,\boldsymbol{x})\Delta t + \nabla_{\boldsymbol{x}} J(t,\boldsymbol{x})^\tau \boldsymbol{\mu}(t,\boldsymbol{u}_t,\boldsymbol{x})\Delta t\\ &+(2\rho-1)\left\{\frac{\sqrt{3}}{\pi}\ln\frac{1-\alpha+\varepsilon}{\alpha-\varepsilon}\cdot\Delta t\cdot\sum_{i=1}^{k}|a_i|\right.\\ &\left.+\left(\frac{\sqrt{3}}{\pi}\ln\frac{2-\varepsilon}{\varepsilon}\right)^2\cdot\Delta t^2\cdot\sum_{i=1}^{k}\sum_{j=1}^{k}|b_{ij}|\right\}+o(\Delta t).\end{aligned}$$

Dividing both sides of the above inequality by Δt, and taking the supremum with respect to u_t, we get

$$\begin{aligned}-\varepsilon \le\ & J_t(t,\boldsymbol{x}) + \sup_{\boldsymbol{u}\in U}\{f(t,\boldsymbol{u}_t,\boldsymbol{x}) + \nabla_{\boldsymbol{x}} J(t,\boldsymbol{x})^\tau \boldsymbol{\mu}(t,\boldsymbol{u}_t,\boldsymbol{x})\\ &\left.+(2\rho-1)\frac{\sqrt{3}}{\pi}\ln\frac{1-\alpha+\varepsilon}{\alpha-\varepsilon}\|\nabla_{\boldsymbol{x}} J(t,\boldsymbol{x})^\tau\boldsymbol{\sigma}(t,\boldsymbol{u}_t,\boldsymbol{x})\|_1\right\}+h_1(\varepsilon,\Delta t)+h_2(\Delta t)\end{aligned}$$

since

$$\sum_{i=1}^{k}|a_i| \to \|\nabla_{\boldsymbol{x}} J(t,\boldsymbol{x})^\tau\boldsymbol{\sigma}(t,\boldsymbol{u}_t,\boldsymbol{x})\|_1$$

as $\Delta t \to 0$; where $h_1(\varepsilon,\Delta t) \to 0$ and $h_2(\Delta t) \to 0$ as $\Delta t \to 0$. Letting $\Delta t \to 0$, and then $\varepsilon \to 0$ results in

$$\begin{aligned}0 \le\ & J_t(t,\boldsymbol{x}) + \sup_{\boldsymbol{u}_t\in U}\{f(t,\boldsymbol{u}_t,\boldsymbol{x}) + \nabla_{\boldsymbol{x}} J(t,\boldsymbol{x})^\tau \boldsymbol{\mu}(t,\boldsymbol{u}_t,\boldsymbol{x})\\ &\left.+\ (2\rho-1)\frac{\sqrt{3}}{\pi}\ln\frac{1-\alpha}{\alpha}\|\nabla_{\boldsymbol{x}} J(t,\boldsymbol{x})^\tau\boldsymbol{\sigma}(t,\boldsymbol{u}_t,\boldsymbol{x})\|_1\right\}.\end{aligned}\tag{3.18}$$

On the other hand, by Theorems 1.12 and 1.13 again and applying the similar process, we can obtain

$$\begin{aligned}0 \ge\ & J_t(t,\boldsymbol{x}) + \sup_{\boldsymbol{u}\in U}\{f(t,\boldsymbol{u}_t,\boldsymbol{x}) + \nabla_{\boldsymbol{x}} J(t,\boldsymbol{x})^\tau \boldsymbol{b}(t,\boldsymbol{u}_t,\boldsymbol{x})\\ &\left.+(2\rho-1)\frac{\sqrt{3}}{\pi}\ln\frac{1-\alpha}{\alpha}\|\nabla_{\boldsymbol{x}} J(t,\boldsymbol{x})^\tau\boldsymbol{\sigma}(t,\boldsymbol{u}_t,\boldsymbol{x})\|_1\right\}.\end{aligned}\tag{3.19}$$

Combining (3.18) and (3.19), we obtain the Eq. (3.15). The theorem is proved.

Remark 3.4 If we consider a discounted infinite horizon optimal control problem, we assume that the objective function f, drift $\boldsymbol{\mu}$ and diffusion $\boldsymbol{\sigma}$ are independent of time. Thus, we replace $f(s, \boldsymbol{u}_s, \boldsymbol{X}_s)$, $\boldsymbol{b}(s, \boldsymbol{u}_s, \boldsymbol{X}_s)$, and $\boldsymbol{\sigma}(s, \boldsymbol{u}_s, \boldsymbol{X}_s)$ by $f(\boldsymbol{u}_s, \boldsymbol{X}_s)$, $\boldsymbol{\mu}(\boldsymbol{u}_s, \boldsymbol{X}_s)$ and $\boldsymbol{\sigma}(\boldsymbol{u}_s, \boldsymbol{X}_s)$, respectively. The problem is stated as follows:

$$\begin{cases} J(\boldsymbol{x}) \equiv \sup\limits_{\boldsymbol{u}\in U} H_\alpha^\rho \left[\displaystyle\int_t^\infty e^{-\gamma s} f(\boldsymbol{u}_s, \boldsymbol{X}_s) ds \right] \\ \text{subject to} \\ \quad \mathrm{d}\boldsymbol{X}_s = \boldsymbol{\mu}(\boldsymbol{u}_s, \boldsymbol{X}_s)ds + \boldsymbol{\sigma}(\boldsymbol{u}_s, \boldsymbol{X}_s)\mathrm{d}\boldsymbol{C}_s \quad \text{and} \quad \boldsymbol{X}_t = \boldsymbol{x}. \end{cases} \tag{3.20}$$

At time 0, the present value of the objective is given by $e^{-\gamma t} J(\boldsymbol{x})$. Using the relations from Eq. (3.15), we obtain the present value by

$$\begin{aligned} \gamma J(\boldsymbol{x}) = & \sup_{\boldsymbol{u}_t \in U} \{ f(\boldsymbol{x}, \boldsymbol{u}) + \nabla_{\boldsymbol{x}} J(\boldsymbol{x})^\tau \boldsymbol{\mu}(\boldsymbol{x}, \boldsymbol{u}) \\ & + (2\rho - 1) \left(\frac{\sqrt{3}}{\pi} \ln \frac{1-\alpha}{\alpha} \right) \| \nabla_{\boldsymbol{x}} J(\boldsymbol{x})^\tau \boldsymbol{\sigma}(\boldsymbol{x}, \boldsymbol{u}) \|_1 \Bigg\}. \end{aligned} \tag{3.21}$$

Example 3.1 Consider the following optimization problem comes from the Vidale-Wolfe advertising model [4] in uncertain environments:

$$\begin{cases} J(0, x_0) \equiv \max\limits_{u \in U} H_\alpha^\rho \left[\displaystyle\int_0^\infty e^{-\gamma t} (\delta X_t - u^2) \mathrm{d}t \right] \\ \text{subject to} \\ \quad \mathrm{d}X_t = [ru\sqrt{1 - X_t} - kX_t]\mathrm{d}t + \sigma(X_t)\mathrm{d}C_t, \end{cases}$$

where $X_t \in [0, 1]$ is the fraction of market potential, $u \geq 0$ denotes the rate of advertising effort, $r > 0$, $k > 0$, σ is a small diffusion coefficient, $\sigma \geq 0$, γ is a discount factor. In this case, we have $F = \int_0^\infty e^{-\gamma t} (\delta X_t - u^2) \mathrm{d}t$. Applying Eq. (3.15), we obtain

$$\begin{aligned} \gamma J &= \max_u \left\{ (\delta x - u^2) + (ru\sqrt{1-x} - kx) J_x + (2\rho - 1) \frac{\sqrt{3}}{\pi} \ln \frac{1-\alpha}{\alpha} |J_x| \sigma \right\} \\ &= \max_u L(u) \end{aligned} \tag{3.22}$$

where $L(u)$ denotes the term in the braces. Setting $\mathrm{d}L(u)/\mathrm{d}u = 0$, we obtain the necessary condition for optimality

$$u = \frac{r\sqrt{1-x}}{2} J_x(t, x).$$

Substituting the equality into Eq. (3.22), we have

$$\gamma J = \delta x + \frac{r^2(1-x)}{4} J_x^2 - kxJ_x + (2\rho - 1)\frac{\sqrt{3}}{\pi}\ln\frac{1-\alpha}{\alpha}\sigma\,|J_x| \tag{3.23}$$

We conjecture that $J(t,x) = Px + Q\ (P > 0)$. This gives $J_x = P$. Using the expression in Eq. (3.23), we have the following condition for optimality

$$\left(4\gamma P + r^2P^2 - 4\delta + 4kP\right)x + 4\gamma Q - r^2P^2 - 4P(2\rho-1)\frac{\sqrt{3}}{\pi}\ln\frac{1-\alpha}{\alpha}\sigma = 0,$$

or

$$4\gamma P + r^2P^2 - 4\delta + 4kP = 0, \text{ and } 4\gamma Q - r^2P^2 - 4P(2\rho-1)\frac{\sqrt{3}}{\pi}\ln\frac{1-\alpha}{\alpha}\sigma = 0.$$

The solution is given by

$$P = \frac{-2(\gamma+k) + 2\sqrt{(\gamma+k)^2 + r^2\delta}}{r^2} \text{ and } Q = \frac{r^2P^2 + 4P(2\rho-1)\frac{\sqrt{3}}{\pi}\ln\frac{1-\alpha}{\alpha}\sigma}{\gamma}.$$

The optimal decision is determined by $u^* = \frac{rP\sqrt{1-x}}{2}$.

3.4 Uncertain Linear Quadratic Model Under Optimistic Value Criterion

We discuss an optimal control problem of uncertain linear quadratic model under optimistic value criterion. The problem is of the form:

$$\begin{cases} J(0, \boldsymbol{x}_0) = \inf\limits_{\boldsymbol{u}_s}\left\{\displaystyle\int_0^T \left(\boldsymbol{X}_s^\tau Q(s)\boldsymbol{X}_s + \boldsymbol{u}_s^\tau R(s)\boldsymbol{u}_s\right)\mathrm{d}s + \boldsymbol{X}_T^\tau S_T \boldsymbol{X}_T\right\}_{\sup}(\alpha) \\ \text{subject to} \\ \quad \mathrm{d}\boldsymbol{X}_s = (A(s)\boldsymbol{X}_s + B(s)\boldsymbol{u}_s)\mathrm{d}s + M(s)\boldsymbol{X}_s\mathrm{d}C_s \\ \quad \boldsymbol{X}_0 = \boldsymbol{x}_0, \end{cases} \tag{3.24}$$

where $\boldsymbol{X}_s$ is a state vector of dimension n, $\boldsymbol{u}_s$ is a decision vector of dimension r, S_T is a symmetric matrix and $\boldsymbol{x}_s \in [a,b]^n$, where $\boldsymbol{x}_s$ represents the state of $\boldsymbol{X}_s$ at time s. The matrices $Q(s)$, $R(s)$, S_T, $A(s)$, $B(s)$, and $M(s)$ are appropriate size matrix functions, where $Q(s)$ is a symmetric nonnegative definite matrix and $R(s)$ is a symmetric positive definite matrix.

For any $0 < t < T$, we use $\boldsymbol{x}$ to denote the state of $\boldsymbol{X}_s$ at time t and $J(t,\boldsymbol{x})$ to denote the optimal value obtainable in $[t,T]$. First, we shall make the following two assumptions: (i) the elements of $Q(s)$, $R(s)$, $A(s)$, $B(s)$, $M(s)$, and $R^{-1}(s)$ are continuous and bounded functions on $[0, T]$; (ii) the optimal value $J(t, \boldsymbol{x})$ is a twice differentiable function on $[0, T] \times [a, b]^n$. Then, applying the equation of optimality (3.4), we obtain

$$\inf_{\boldsymbol{u}_t} \Big\{ \boldsymbol{x}^\tau Q(t)\boldsymbol{x} + \boldsymbol{u}_t^\tau R(t)\boldsymbol{u}_t + \nabla_{\boldsymbol{x}} J(t, \boldsymbol{x})^\tau (A(t)\boldsymbol{x} + B(t)\boldsymbol{u}_t) + \frac{\sqrt{3}}{\pi} \ln \frac{1-\alpha}{\alpha} |\nabla_{\boldsymbol{x}} J(t, \boldsymbol{x})^\tau M(t)\boldsymbol{x}| + J_t(t, \boldsymbol{x}) \Big\} = 0. \quad (3.25)$$

Theorem 3.5 ([5]) *A necessary and sufficient condition that $\boldsymbol{u}_t^*$ be an optimal control for model (3.24) is that*

$$\boldsymbol{u}_t^* = -\frac{1}{2} R^{-1}(t) B^\tau(t) P(t)\boldsymbol{x}, \quad (3.26)$$

where the function $P(t)$ satisfies the following Riccati differential equation

$$\frac{\mathrm{d}P(t)}{\mathrm{d}t} = \begin{cases} -2Q(t) - A^\tau(t)P(t) - P(t)A(t) \\ \quad -\frac{\sqrt{3}}{\pi} \ln \frac{1-\alpha}{\alpha} P(t)M(t) - \frac{\sqrt{3}}{\pi} \ln \frac{1-\alpha}{\alpha} M^\tau(t)P(t) \\ \quad +\frac{1}{2} P(t)B(t)R^{-1}(t)B^\tau(t)P(t) \quad \text{if } (t, \boldsymbol{x}) \in \Omega_1, \\ -2Q(t) - A^\tau(t)P(t) - P(t)A(t) \\ \quad +\frac{\sqrt{3}}{\pi} \ln \frac{1-\alpha}{\alpha} P(t)M(t) + \frac{\sqrt{3}}{\pi} \ln \frac{1-\alpha}{\alpha} M^\tau(t)P(t) \\ \quad +\frac{1}{2} P(t)B(t)R^{-1}(t)B^\tau(t)P(t) \quad \text{if } (t, \boldsymbol{x}) \in \Omega_2 \end{cases} \quad (3.27)$$

and boundary condition $P(T) = 2S_T$, where

$$\Omega_1 = \left\{ (t, \boldsymbol{x}) \mid \boldsymbol{x}^\tau P(t)M(t)\boldsymbol{x} \geq 0, (t, \boldsymbol{x}) \in [0, T] \times [a, b]^n \right\},$$
$$\Omega_2 = \left\{ (t, \boldsymbol{x}) \mid \boldsymbol{x}^\tau P(t)M(t)\boldsymbol{x} < 0, (t, \boldsymbol{x}) \in [0, T] \times [a, b]^n \right\}.$$

The optimal value of model (3.24) is

$$J(0, \boldsymbol{x}_0) = \frac{1}{2} \boldsymbol{x}_0^\tau P(0)\boldsymbol{x}_0. \quad (3.28)$$

Proof Denote

$$\psi(\boldsymbol{u}_t) = \boldsymbol{x}^\tau Q(t)\boldsymbol{x} + \boldsymbol{u}_t^\tau R(t)\boldsymbol{u}_t + \nabla_{\boldsymbol{x}} J(t, \boldsymbol{x})^\tau (A(t)\boldsymbol{x} + B(t)\boldsymbol{u}_t) + \frac{\sqrt{3}}{\pi} \ln \frac{1-\alpha}{\alpha} |\nabla_{\boldsymbol{x}} J(t, \boldsymbol{x})^\tau M(t)\boldsymbol{x}| + J_t(t, \boldsymbol{x}). \quad (3.29)$$

First, we verify the necessity. Since $J(T, \boldsymbol{X}_T) = \boldsymbol{x}_T^\tau S_T \boldsymbol{x}_T$, we conjecture that

$$\nabla_{\boldsymbol{x}} J(t, \boldsymbol{x}) = P(t)\boldsymbol{x}$$

with the boundary condition $P(T) = 2S_T$. Setting $\dfrac{\partial \psi(\boldsymbol{u}_t)}{\partial \boldsymbol{u}_t} = 0$, we have

$$\boldsymbol{u}_t = -\frac{1}{2} R^{-1}(t) B^{\tau}(t) P(t) \boldsymbol{x}. \tag{3.30}$$

Because $\dfrac{\partial^2 \psi(\boldsymbol{u}_t)}{\partial \boldsymbol{u}_t^2} = 2R(t) > 0$, $\boldsymbol{u}_t$ is the optimal control of model (3.24), i.e.,

$$\boldsymbol{u}_t^* = -\frac{1}{2} R^{-1}(t) B^{\tau}(t) P(t) \boldsymbol{x}. \tag{3.31}$$

If $(t, \boldsymbol{x}) \in \Omega_1$, taking the derivative of $\psi(\boldsymbol{u}_t^*)$ with respect to $\boldsymbol{x}$, we have

$$\Bigg(2Q(t) + A^{\tau}(t)P(t) + P(t)A(t) + \frac{\sqrt{3}}{\pi} \ln \frac{1-\alpha}{\alpha} P(t)M(t)$$
$$+ \frac{\sqrt{3}}{\pi} \ln \frac{1-\alpha}{\alpha} M^{\tau}(t)P(t) - \frac{1}{2} P(t)B(t)R^{-1}(t)B^{\tau}(t)P(t) + \frac{\mathrm{d}P(t)}{\mathrm{d}t}\Bigg) \boldsymbol{x} = 0.$$

That is,

$$\frac{\mathrm{d}P(t)}{\mathrm{d}t} = -2Q(t) - A^{\tau}(t)P(t) - P(t)A(t) - \frac{\sqrt{3}}{\pi} \ln \frac{1-\alpha}{\alpha} P(t)M(t)$$
$$- \frac{\sqrt{3}}{\pi} \ln \frac{1-\alpha}{\alpha} M^{\tau}(t)P(t) + \frac{1}{2} P(t)B(t)R^{-1}(t)B^{\tau}(t)P(t).$$

If $(t, \boldsymbol{x}) \in \Omega_2$, by the same method, we obtain

$$\frac{\mathrm{d}P(t)}{\mathrm{d}t} = -2Q(t) - A^{\tau}(t)P(t) - P(t)A(t) + \frac{\sqrt{3}}{\pi} \ln \frac{1-\alpha}{\alpha} P(t)M(t)$$
$$+ \frac{\sqrt{3}}{\pi} \ln \frac{1-\alpha}{\alpha} M^{\tau}(t)P(t) + \frac{1}{2} P(t)B(t)R^{-1}(t)B^{\tau}(t)P(t).$$

Hence, the solution $P(t)$ is a symmetric matrix. Because $\nabla_{\boldsymbol{x}} J(t, \boldsymbol{x}) = P(t)\boldsymbol{x}$ and $J(T, \boldsymbol{X}_T) = \boldsymbol{x}_T^{\tau} S_T \boldsymbol{x}_T$, we have $J(t, \boldsymbol{x}) = \frac{1}{2}\boldsymbol{x}^{\tau} P(t)\boldsymbol{x}$. Then, the optimal value $J(0, \boldsymbol{x}_0)$ is

$$J(0, \boldsymbol{x}_0) = \frac{1}{2} \boldsymbol{x}_0^{\tau} P(0) \boldsymbol{x}_0. \tag{3.32}$$

Then, we prove the sufficient condition. Because $J(T, \boldsymbol{X}_T) = \boldsymbol{x}_T^{\tau} S_T \boldsymbol{x}_T$, we assume that $J(t, \boldsymbol{x}) = \frac{1}{2}\boldsymbol{x}^{\tau} P(t)\boldsymbol{x}$, where $P(t)$ satisfies the Riccati differential equation (3.27) with the boundary condition $P(T) = 2S_T$. Substituting Eqs. (3.26) and (3.27) into $\psi(\boldsymbol{u}_t)$, we have $\psi(\boldsymbol{u}_t^*) = 0$. Because the objective function of model

(3.24) is convex, there must be an optimal control solution. Hence, $\boldsymbol{u}_t^*$ is the optimal control and $J(t, \boldsymbol{x}) = \frac{1}{2}\boldsymbol{x}^\tau P(t)\boldsymbol{x}$. Furthermore, the optimal value $J(0, \boldsymbol{x}_0)$ is

$$J(0, \boldsymbol{x}_0) = \frac{1}{2}\boldsymbol{x}_0^\tau P(0)\boldsymbol{x}_0. \tag{3.33}$$

The theorem is proved.

Remark 3.5 We know that there is yet no simple and effective method to solve the Riccati differential equation with absolute value function. In order to obtain the solution of $P(t)$, we need to make a judgment about the sign of $\boldsymbol{x}^\tau P(t)M(t)\boldsymbol{x}$. The procedure is as follows. First, we assume that $\boldsymbol{x}^\tau P(t)M(t)\boldsymbol{x} \geq 0$ or $\boldsymbol{x}^\tau P(t)M(t)\boldsymbol{x} < 0$ and use the four-order Runge-Kutta method to solve the numerical solution of $P(t)$. Then we check whether the result is consistent with the assumption. If they are consistent, the numerical solution of $P(t)$ is serviceable and we can use Theorem 3.5 to obtain the optimal control. If they are both inconsistent, then we can not solve the optimal control problem in this case. Moreover, if we can verify the positive or negative definiteness of $P(t)M(t)$, then the Theorem 3.5 can be used immediately. Hence, here we only consider the reconcilable cases.

3.5 Optimistic Value Optimal Control for Singular System

Consider the following optimal control problem for an continuous-time singular uncertain system:

$$\begin{cases} J(0, X_0) = \sup\limits_{u_s \in U} \left\{ \int_0^T f(s, u_s, X_s) \mathrm{d}s + G(T, X_T) \right\}_{\sup} (\alpha) \\ \text{subject to} \\ \qquad F\mathrm{d}X_s = [AX_s + Bu(s)]\,\mathrm{d}s + Du(s)\mathrm{d}C_s, \text{ and } X_0 = x_0, \end{cases}$$

where $X_s \in R^n$ is the state vector, $u_s \in U \subset R^m$ is the input variable, f is the objective function, and G is the function of terminal reward. For a given u_s, X_s is defined by the uncertain differential equations. The function $J(0, X_0)$ is the expected optimal value obtainable in $[0, T]$ with the initial state that at time 0 we are in state x_0.

For any $0 < t < T$, $J(t, X)$ is the expected optimal reward obtainable in $[t, T]$ with the condition that at time t we are in state $X_t = x$. That is, we have

$$\begin{cases} J(t, X) = \sup\limits_{u_s \in U} \left[\int_t^T f(s, u_s, X_s)\,\mathrm{d}s + G(X_T, T) \right]_{\sup} (\alpha) \\ \text{subject to} \\ \qquad F\mathrm{d}X_s = [AX_s + Bu_s]\,\mathrm{d}s + Du_s\mathrm{d}C_s, \text{ and } X_t = x. \end{cases} \tag{3.34}$$

If (F, A) is regular and impulse-free and $rank(F) = r$, by Lemma 2.1 there exist invertible matrices P and Q such that

$$PFQ = \begin{bmatrix} I_r & 0 \\ 0 & 0 \end{bmatrix}, \quad PAQ = \begin{bmatrix} A_1 & 0 \\ 0 & I_{n-r} \end{bmatrix}, \quad PB = \begin{bmatrix} B_1 \\ B_2 \end{bmatrix}.$$

We have the following equation of optimality.

Theorem 3.6 ([6]) *The (F, A) is assumed to be regular and impulse-free, and $P_2 D = 0$. Let $J(t, x)$ be twice differentiable on $[0, T] \times R^n$ and u_s derivable on $[0, T]$. Then, we have*

$$-J_t(t, x) = \sup_{u_t \in U} \left\{ f(t, u_t, x) + \nabla_x J(t, x)^\tau p + \frac{\sqrt{3}}{\pi} \ln \frac{1-\alpha}{\alpha} |\nabla_x J(t, x)^\tau q| \right\} \tag{3.35}$$

where $p = Q \begin{bmatrix} A_1 x_1 + B_1 u_t \\ -B_2 \dot{u}_t \end{bmatrix}$, $q = Q_1 D_1 u_t$, *and* $Q = \begin{bmatrix} Q_1 & Q_2 \end{bmatrix}$, $Q_1 \in R^{n \times r}$, $Q_2 \in R^{n \times (n-r)}$, $P = \begin{bmatrix} P_1 \\ P_2 \end{bmatrix}$, $D_1 = P_1 D$, $P_1 \in R^{r \times n}$, $P_2 \in R^{(n-r) \times n}$, $x = Q \begin{bmatrix} x_1 \\ x_2 \end{bmatrix}$, $x_1 \in R^r$, $x_2 \in R^{n-r}$.

Proof It follows from $P_2 D = 0$ that

$$PD = \begin{bmatrix} P_1 \\ P_2 \end{bmatrix} D = \begin{bmatrix} D_1 \\ 0 \end{bmatrix}.$$

Let $X_s = Q \begin{bmatrix} X_{1,s} \\ X_{2,s} \end{bmatrix}$ for any $s \in [t, T]$ and especially at time t, $x = Q \begin{bmatrix} x_1 \\ x_2 \end{bmatrix}$. So we are easy to obtain

$$\begin{cases} \mathrm{d}X_{1,s} = \left[A_1 X_{1,s} + B_1 u_s \right] \mathrm{d}s + D_1 u_s \mathrm{d}C_s, \\ 0 = \left[X_{2,s} + B_2 u_s \right] \mathrm{d}s, \end{cases}$$

where $s \in [t, T]$. Since at any time $s \in [t, T]$ we have

$$X_{2,s} = -B_2 u_s$$

Let $s = t$ and $s = t + \Delta t$, respectively. We get the following two equations:

$$X_2 = -B_2 u_t$$
$$X_{2,t+\Delta t} = -B_2 u_{t+\Delta t}$$

Using the latter equation minus the former one, we obtain

$$\Delta X_{2,t} = -B_2 \dot{u}(t) \Delta t + o(\Delta t),$$

where $u_{t+\Delta t} = u_t + \dot{u}_t \Delta t + o(\Delta t)$, because u_s is derivable on $[t, T]$. Obviously we know

$$\Delta X_{1,t} = [A_1 X_1 + B_1 u_t]\,\Delta t + D_1 u_t \Delta C_t,$$

where $\Delta C_t \sim \mathcal{N}(0, \Delta t^2)$ which means ΔC_t is a normally uncertain variable with expected value 0 and variance Δt^2. We have

$$\Delta X_t = Q \begin{bmatrix} A_1 X_1 + B_2 u_t \\ -B_2 \dot{u}(t) \end{bmatrix} \Delta t + Q_1 D_1 u_t \Delta C_t + o(\Delta t).$$

Now denote

$$p = Q \begin{bmatrix} A_1 X_1 + B_2 u_t \\ -B_2 \dot{u}(t) \end{bmatrix}, q = Q_1 D_1 u_t.$$

Then we have

$$\Delta X_t = p\Delta t + q\Delta C_t + o(\Delta t).$$

By employing Taylor series expansion, we obtain

$$\begin{aligned} & J(t+\Delta t, x+\Delta X_t) \\ = & J(t,x) + J_t(t,x)\Delta t + \nabla_x J(t,x)^\tau \Delta X_t + \frac{1}{2} J_{tt}(t,x)\Delta t^2 \\ & + \nabla_x J_t(t,x)^\tau \Delta X_t \Delta t + \frac{1}{2} \Delta X_t^\tau \nabla_{xx} J(t,x) \Delta X_t + o(\Delta t). \end{aligned} \quad (3.36)$$

Substituting Eq. (3.36) into Eq. (3.2) yields

$$\begin{aligned} 0 = \sup_{u_t} \Big\{ & f(t,u_t,x)\Delta t + J_t(t,x)\Delta t + \Big[\nabla_x J(t,x)^\tau \Delta X_t + \nabla_x J_t(t,x)^\tau \Delta X_t \Delta t \\ & + \frac{1}{2} \Delta X_t^\tau \nabla_{xx} J(t,x) \Delta X_t \Big]_{\sup}(\alpha) + o(\Delta t) \Big\}. \end{aligned} \quad (3.37)$$

Then, we know

$$\begin{aligned} & \Big[\nabla_x J(t,x)^\tau \Delta X_t + \nabla_x J(t,x)^\tau \Delta X_t \Delta t + \frac{1}{2} \Delta X_t^\tau \nabla_{xx} J(t,x) \Delta X_t \Big]_{\sup}(\alpha) \\ = & \Big[\nabla_x J(t,x)^\tau (p\Delta t + q\Delta C_t + o(\Delta t)) + p\Delta t + \nabla_x J(t,x)^\tau (p\Delta t + q\Delta C_t + o(\Delta t))\Delta t \\ & + \frac{1}{2} (p\Delta t + q\Delta C_t + o(\Delta t))^T \nabla_{xx} J(t,x) (p\Delta t + q\Delta C_t + o(\Delta t)) \Big]_{\sup}(\alpha) \\ = & \nabla_x J(t,x)^\tau p\Delta t + \Big[\Big(\nabla_x J(t,x) q + \nabla_x J_t(t,x) q\Delta t + p^T \nabla_{xx} J(t,x) q\Delta t \Big) \Delta C_t \\ & + \frac{1}{2} q^T \nabla_{xx} J(t,x) q \Delta C_t^2 \Big]_{\sup}(\alpha) + o(\Delta t) \\ = & \nabla_x J(t,x)^\tau p\Delta t + \Big[a\Delta C_t + b\Delta C_t^2 \Big]_{\sup}(\alpha) + o(\Delta t), \end{aligned} \quad (3.38)$$

where $a = \nabla_x J(t,x)q + \nabla_x J_t(t,x)q\Delta t + p^\tau \nabla_{xx} J(t,x)q\Delta t$, and $b = \frac{1}{2}q^\tau \nabla_{xx} J(t,x)q$. Substituting Eq. (3.38) into (3.37) results in

$$0 = \sup_{u_t} \Big\{ f(t,u_t,x)\Delta t + J_t(t,x)\Delta t + \nabla_x J(t,x)^T p\Delta t + \left[a\Delta C_t + b\Delta C_t^2\right]_{\sup}(\alpha) + o(\Delta t) \Big\}. \tag{3.39}$$

Obviously, we have

$$a\Delta C_t - \mid b \mid \Delta C_t^2 \le a\Delta C_t + b\Delta C_t^2 \le a\Delta C_t + \mid b \mid \Delta C_t^2. \tag{3.40}$$

Applying Theorem 1.12 that for any small enough $\varepsilon > 0$, we get

$$\left[a\Delta C_t + \mid b \mid \Delta C_t^2\right]_{\sup}(\alpha) \le \frac{\sqrt{3}}{\pi}\ln\frac{1-\alpha+\varepsilon}{\alpha-\varepsilon} \mid a \mid \Delta t + \left(\frac{\sqrt{3}}{\pi}\ln\frac{2-\varepsilon}{\varepsilon}\right)^2 \mid b \mid \Delta t^2, \tag{3.41}$$

$$\left[a\Delta C_t - \mid b \mid \Delta C_t^2\right]_{\sup}(\alpha) \ge \frac{\sqrt{3}}{\pi}\ln\frac{1-\alpha-\varepsilon}{\alpha+\varepsilon} \mid a \mid \Delta t - \left(\frac{\sqrt{3}}{\pi}\ln\frac{2-\varepsilon}{\varepsilon}\right)^2 \mid b \mid \Delta t^2. \tag{3.42}$$

Combining inequalities (3.40), (3.41), and (3.42), we obtain

$$\left[a\Delta C_t + b\Delta C_t^2\right]_{\sup}(\alpha) \le \frac{\sqrt{3}}{\pi}\ln\frac{1-\alpha+\varepsilon}{\alpha-\varepsilon} \mid a \mid \Delta t + \left(\frac{\sqrt{3}}{\pi}\ln\frac{2-\varepsilon}{\varepsilon}\right)^2 \mid b \mid \Delta t^2, \tag{3.43}$$

$$\left[a\Delta C_t + b\Delta C_t^2\right]_{\sup}(\alpha) \ge \frac{\sqrt{3}}{\pi}\ln\frac{1-\alpha-\varepsilon}{\alpha+\varepsilon} \mid a \mid \Delta t - \left(\frac{\sqrt{3}}{\pi}\ln\frac{2-\varepsilon}{\varepsilon}\right)^2 \mid b \mid \Delta t^2. \tag{3.44}$$

According to Eq. (3.39) and inequality (3.43), for $\Delta t > 0$, there exists a control u_t such that

$$
\begin{aligned}
-\varepsilon\Delta t \le & \{f(t,u_t,x)\Delta t + J_t(t,x)\Delta t + \nabla_x J(t,x)^\tau p\Delta t \\
& + \left[a\Delta C_t + b\Delta C_t^2\right]_{\sup}(\alpha) + o(\Delta t)\Big\} \\
\le & f(t,u_t,x)\Delta t + J_t(t,x)\Delta t + \nabla_x J(t,x)^\tau p\Delta t + \frac{\sqrt{3}}{\pi}\ln\frac{1-\alpha+\varepsilon}{\alpha-\varepsilon}|a|\Delta t \\
& + \left(\frac{\sqrt{3}}{\pi}\ln\frac{2-\varepsilon}{\varepsilon}\right)^2 |b|\Delta t^2 + o(\Delta t).
\end{aligned}
$$

Dividing both sides of this inequality by Δt, we have

$$
\begin{aligned}
-\varepsilon \le & f(t,u_t,x) + J_t(t,x) + \nabla_x J(t,x)^\tau p + \frac{\sqrt{3}}{\pi}\ln\frac{1-\alpha+\varepsilon}{\alpha-\varepsilon} \mid a \mid \\
& + \left(\frac{\sqrt{3}}{\pi}\ln\frac{2-\varepsilon}{\varepsilon}\right)^2 \mid b \mid \Delta t + \frac{o(\Delta t)}{\Delta t} \\
\le & J_t(t,x) + \sup_{u_t}\left\{f(t,u_t,x) + \nabla_x J(t,x)^\tau p + \frac{\sqrt{3}}{\pi}\ln\frac{1-\alpha+\varepsilon}{\alpha-\varepsilon} \mid a \mid\right\} \\
& + \left(\frac{\sqrt{3}}{\pi}\ln\frac{2-\varepsilon}{\varepsilon}\right)^2 \mid b \mid \Delta t + \frac{o(\Delta t)}{\Delta t}.
\end{aligned}
$$

Since $|a| \to |\nabla_x J(t,x)^\tau q|$ as $\Delta t \to 0$, letting $\Delta t \to 0$ and then $\varepsilon \to 0$, it is easy to know

$$
0 \le J_t(t,x) + \sup_{u_t}\left\{f(t,u_t,x) + \nabla_x J(t,x)^\tau p + \frac{\sqrt{3}}{\pi}\ln\frac{1-\alpha}{\alpha}|\nabla_x J(t,x)^\tau q|\right\}. \quad (3.45)
$$

On the other hand, according Eq. (3.39) and inequality (3.44), using the similar approach, we are able to obtain

$$
0 \ge J_t(t,X) + \sup_{u_t}\left\{f(t,u_t,x) + \nabla_x J(t,x)^\tau p + \frac{\sqrt{3}}{\pi}\ln\frac{1-\alpha}{\alpha}|\nabla_x J(t,x)^\tau q|\right\}. \quad (3.46)
$$

By inequalities (3.45) and (3.46), we get the Eq. (3.35). This completes the proof.

Remark 3.6 The solutions of the presented model (3.34) may be obtained from settling the equation of optimality (3.35). The vector $p = Q\begin{bmatrix} g(t)\,(A_1X_1 + B_1u_t) \\ -B_2\dot{u}(t) \end{bmatrix}$ is related to the function $\dot{u}(t)$ which is totally different from the optimal control problem of the uncertain normal system, and it will bring lots of matters in solving equation (3.35). In some special cases, this equation of optimality may be settled to get analytical solution such as the following example. Otherwise, we have to employ numerical methods to obtain the solution approximately.

3.5.1 Example

Consider the following problem:

$$\begin{cases} J(t, X) = \sup_{u_t \in U_{ad}} \left[\int_t^{+\infty} \rho^\tau(s) X_s u_s \mathrm{d}s \right]_{\sup} (\alpha) \\ \text{subject to} \\ \qquad F \mathrm{d}X_s = [AX_s + Bu_s]\,\mathrm{d}s + Du_s \mathrm{d}C_s, \text{ and } X_t = x. \end{cases} \tag{3.47}$$

where $X_s \in R^3$ is the state vector, $U_{ad} = [-1, 1]$, and

$$F = \begin{bmatrix} 1 & 0 & -1 \\ 0 & 0 & 1 \\ 0 & 0 & 0 \end{bmatrix}, \quad A = \begin{bmatrix} 1 & -1 & 0 \\ 1 & 0 & 1 \\ -1 & 1 & -1 \end{bmatrix}, \quad B = \begin{bmatrix} 1 \\ 2 \\ -1 \end{bmatrix}, \quad D = \begin{bmatrix} \frac{2}{3} \\ -\frac{1}{3} \\ 0 \end{bmatrix},$$

and

$$\rho^\tau(s) = [1, 0, -2]e^{-s}.$$

Through calculating, we know

$$det(zF - A) = det \begin{bmatrix} z-1 & 1 & -1 \\ -1 & 0 & z-1 \\ 1 & -1 & 1 \end{bmatrix} = (z-1)^2.$$

Obviously, $det(zE - A)$ is not identically zero and $deg(det(zF - A)) = rank(F)$, namely, (F, A) is regular and impulse free. By using Lemma 2.1, through deduction we obtain two invertible matrices P and Q:

$$P = \begin{bmatrix} 0 & \frac{1}{4} & 0 \\ 1 & 1 & 1 \\ 0 & 0 & -1 \end{bmatrix}, \quad Q = \begin{bmatrix} 0 & 1 & 0 \\ 4 & 1 & -1 \\ 4 & 0 & 0 \end{bmatrix}$$

such that

$$PFQ = \begin{bmatrix} 1 & 0 & 0 \\ 0 & 1 & 0 \\ 0 & 0 & 0 \end{bmatrix}, \quad PAQ = \begin{bmatrix} 1 & \frac{1}{4} & 0 \\ 0 & 1 & 0 \\ 0 & 0 & 1 \end{bmatrix}, \quad PB = \begin{bmatrix} \frac{1}{2} \\ 2 \\ 1 \end{bmatrix}, \quad PD = \begin{bmatrix} -\frac{1}{12} \\ \frac{1}{3} \\ 0 \end{bmatrix}.$$

Easily, we can see

$$A_1 = \begin{bmatrix} 1 & \frac{1}{4} \\ 0 & 1 \end{bmatrix}, \quad B_1 = \begin{bmatrix} \frac{1}{2} \\ 2 \end{bmatrix}, \quad B_2 = 1, \quad D_1 = \begin{bmatrix} -\frac{1}{12} \\ \frac{1}{3} \end{bmatrix}, \quad P_2 D = \begin{bmatrix} 0 \\ 0 \end{bmatrix}, \quad Q_1 = \begin{bmatrix} 0 & 1 \\ 4 & 1 \\ 4 & 0 \end{bmatrix}$$

where $P_2 = \begin{bmatrix} 0 & 0 & -1 \end{bmatrix}$. Denote $x = [x_1, x_2, x_3]^\tau$ and assume that $x_1 + 2x_3 = 0$. Because

$$Q^{-1} = \begin{bmatrix} 0 & 0 & \frac{1}{4} \\ 1 & 0 & 0 \\ 1 & -1 & 1 \end{bmatrix},$$

and $[x_1, x_2]^\tau = Q^{-1}x$, we obtain $x_1 = [\frac{1}{4}x_3, x_1]^\tau$. Combining these results and Theorem 3.6, we know

$$p = Q\begin{bmatrix} A_1X_1 + B_1u_t \\ -B_2\dot{u}_t \end{bmatrix} = \begin{bmatrix} x_1 + 2u_t \\ 2x_1 + x_3 + 4u_t + \dot{u}_t \\ x_1 + x_3 + 2u_t \end{bmatrix},$$

$$q = Q_1D_1u_t = \begin{bmatrix} \frac{1}{3}u_t \\ 0 \\ -\frac{1}{3}u_t \end{bmatrix}.$$

We conjecture that $J(t, x) = k\rho^\tau(t)x$, and let $\alpha = 0.2$. Then

$$J_t(t, x) = -k\rho^\tau(t)x, \ \ \nabla_x J(t, x) = k\rho(t),$$

and

$$\begin{aligned} &\rho^\tau(t)xu(t) + \nabla_x J(t, x)^\tau p + \frac{\sqrt{3}}{\pi}\ln\frac{1-\alpha}{\alpha} \mid \nabla_x J(t, x)^\tau q \mid \\ &= (x_1 - 2x_3)e^{-t}u_t + k\left[x_1 + 2u_t - 2\left(x_1 + x_3 + 2u_t\right)\right]e^{-t} + \frac{\sqrt{3}}{\pi}\ln\frac{1-\alpha}{\alpha} \mid ku(t) \mid e^{-t} \\ &= (x_1 - 2x_3 - 6k)e^{-t}u_t + \frac{\sqrt{3}}{\pi}\ln 4|ku_t|e^{-t}. \end{aligned}$$

Applying Eq. (3.35), we get

$$\begin{aligned} k(x_1 - 2x_3)e^{-t} &= \sup_{u_t\in[-1,1]}\left[(x_1 - 2x_3 - 2k)e^{-t}u_t + \frac{\sqrt{3}}{\pi}\ln 4|ku(t)|e^{-t}\right] \\ &= e^{-t}\cdot \sup_{u_t\in[-1,1]}\left[(x_1 - 2x_3 - 2k)u_t + \frac{\sqrt{3}}{\pi}\ln 4|ku(t)|\right]. \end{aligned} \tag{3.48}$$

Dividing Eq. (3.48) by e^{-t}, we obtain

$$k(x_1 - 2x_3) = \sup_{u_t\in[-1,1]}\left[(x_1 - 2x_3 - 2k)u_t + \frac{\sqrt{3}}{\pi}\ln 4|ku_t|\right]. \tag{3.49}$$

If $ku_t \geq 0$, Eq. (3.49) turns to be

$$k(x_1 - 2x_3) = \sup_{u_t \in [-1,1]} \left[(x_1 + 2x_3 - 2k)u_t + \frac{\sqrt{3}}{\pi} k u_t \ln 4 \right] = \left| x_1 - 2x_3 - \left(2 - \frac{\sqrt{3}}{\pi} \ln 4 \right) k \right|, \tag{3.50}$$

and then

$$k^2(x_1 - 2x_3)^2 = \left[x_1 - 2x_3 - \left(2 - \frac{\sqrt{3}}{\pi} \ln 4 \right) k \right]^2,$$

namely

$$(a^2 - b_1^2)k^2 + 2ab_1 k - a^2 = 0,$$

where $a = x_1 - 2x_3$, and $b_1 = 2 - \frac{\sqrt{3}}{\pi} \ln 4 > 0$. Because $ku_t \geq 0$, and by Eq. (3.50) the symbols of k and a must keep coincident, we know

$$k = \begin{cases} \dfrac{a}{2b_1}, & \text{if } a = \pm b_1 \\ 0, & \text{if } a = 0 \\ \dfrac{-a}{a - b_1}, & \text{if } a < -b_1 \\ \dfrac{a}{a + b_1}, & \text{if } a > b_1. \end{cases}$$

The optimal control is

$$u_t^* = sign(a - b_1 k).$$

If $ku_t < 0$, Eq. (3.49) turns to be

$$k(x_1 - 2x_3) = \sup_{u_t \in [-1,1]} \left[(x_1 + 2x_3 - 2k)u_t - \frac{\sqrt{3}}{\pi} \ln 4 k u(t) \right] = \left| x_1 - 2x_3 - \left(2 + \frac{\sqrt{3}}{\pi} \ln 4 \right) k \right|.$$

Using the similar method, we are able to obtain

$$k = \begin{cases} \dfrac{a}{2b_2}, & \text{if } a = \pm b_2 \\ \dfrac{a}{a + b_2}, & \text{if } -b_2 < a < 0 \\ \dfrac{-a}{a - b_2}, & \text{if } 0 < a < b_2, \end{cases}$$

where $a = x_1 - 2x_3$, and $b_2 = 2 + \frac{\sqrt{3}}{\pi}\ln 4 > 0$. The optimal control is

$$u_t^* = sign(a - b_2 k).$$

When $b_1 < a < b_2$, we know

$$b_1 - b_2 + 2a > 3b_1 - b_2 = 4\left(1 - \frac{\sqrt{3}}{\pi}\ln 4\right) > 0,$$

and obviously

$$\frac{-a}{a - b_2} - \frac{a}{a + b_1} = \frac{a(b_1 - b_2 + 2a)}{(b_2 - a)(b_1 + a)} > 0.$$

When $-b_2 < a < -b_1$, similarly we get that

$$\frac{a}{a + b_2} - \frac{-a}{a - b_1} = \frac{a(b_1 - b_2 - 2a)}{(b_2 + a)(b_1 - a)} < 0.$$

Summarily, the optimal control of the problem (3.47) is

$$u_t^* = \begin{cases} sign(a - b_1 k), & \text{if } |a| = b_1,\ 0,\ \text{or } |a| > b_2, \\ sign(a - b_2 k), & \text{if } 0 < |a| < b_1, \text{or } b_1 < |a| \le b_2. \end{cases}$$

References

1. Sheng L, Zhu Y (2013) Optimistic value model of uncertain optimal control. Int J Uncertain Fuzziness Knowl-Based Syst 21(Suppl. 1):75–83
2. Hurwicz L (1951) Some specification problems and application to econometric models. Econometrica 19:343–344
3. Sheng L, Zhu Y, Hamalainen T (2013) An uncertain optimal control model with Hurwicz criterion. Appl Math Comput 224:412–421
4. Sethi S, Thompson G (2000) Optimal control theory: applications to management science and economics, 2nd edn. Springer
5. Li B, Zhu Y (2018) Parametric optimal control of uncertainn systems under optimistic value criterion. Eng Optim 50(1):55–69
6. Shu Y, Zhu Y (2017) Optimistic value based optimal control for uncertain linear singular systems and application to dynamic input-output model. ISA Trans 71(part 2):235–251

Chapter 4
Optimal Control for Multistage Uncertain Systems

In this chapter, we will investigate the following expected value optimal control problem for a multistage uncertain system:

$$\begin{cases} \min\limits_{\substack{u(i)\in U_i \\ 0\le i\le N}} E\left[\sum_{j=0}^{N} f(x(j), u(j), j)\right] \\ \text{subject to:} \\ \quad x(j+1) = \phi(x(j), u(j), j) + \sigma(x(j), u(j), j)\, C_{j+1}, \\ \quad j = 0, 1, 2, \ldots, N-1, \quad x(0) = x_0, \end{cases} \tag{4.1}$$

where $x(j)$ is the state of the system at stage j, $u(j)$ the control variable at stage j, U_j the constraint domain for the control variables $u(j)$ for $j = 0, 1, 2, \ldots, N$, f the objective function, ϕ and σ two functions, and x_0 the initial state of the system. In addition, $C_1, C_2, \ldots, C_N$ are some independent uncertain variables.

4.1 Recurrence Equation

For any $0 < k < N$, let $J(x_k, k)$ be the expected optimal reward obtainable in $[k, N]$ with the condition that at stage k, we are in state $x(k) = x_k$. That is, we have

$$\begin{cases} J(x_k, k) \equiv \min\limits_{\substack{u(i)\in U_i \\ k\le i\le N}} E\left[\sum_{j=k}^{N} f(x(j), u(j), j)\right] \\ \text{subject to:} \\ \quad x(j+1) = \phi(x(j), u(j), j) + \sigma(x(j), u(j), j)\, C_{j+1}, \\ \quad j = k, k+1, \ldots, N-1, \quad x(k) = x_k, \end{cases}$$

Y. Zhu, *Uncertain Optimal Control*, Springer Uncertainty Research,
https://doi.org/10.1007/978-981-13-2134-4_4

Theorem 4.1 *We have the following recurrence equations*

$$J(x_N, N) = \min_{u(N)\in U_N} f(x_N, u(N), N), \tag{4.2}$$
$$J(x_k, k) = \min_{u(k)\in U_k} E[f(x_k, u(k), k) + J(x(k+1), k+1)] \tag{4.3}$$

for $k = N-1, N-2, \ldots, 1, 0$.

Proof It is obvious that $J(x_N, N) = \min_{u(N)\in U_N} f(x_N, u(N), N)$. For any $k = N-1, N-2, \ldots, 1, 0$, we have

$$\begin{aligned}
J(x_k, k) &= \min_{\substack{u(i)\in U_i\\ k\le i\le N}} E\left[\sum_{j=k}^{N} f(x(j), u(j), j)\right]\\
&= \min_{\substack{u(i)\in U_i\\ k\le i\le N}} E\left[f(x(k), u(k), k) + E\left[\sum_{j=k+1}^{N} f(x(j), u(j), j)\right]\right]\\
&\ge \min_{\substack{u(i)\in U_i\\ k\le i\le N}} E\left[f(x_k, u(k), k) + \min_{\substack{u(i)\in U_i\\ k+1\le i\le N}} E\left[\sum_{j=k+1}^{N} f(x(j), u(j), j)\right]\right]\\
&= \min_{u(k)\in U_k} E[f(x_k, u(k), k) + J(x(k+1), k+1)].
\end{aligned}$$

In addition, for any $u(i)$, $k \le i \le N$, we have

$$\begin{aligned}
J(x_k, k) &\le E\left[\sum_{j=k}^{N} f(x(j), u(j), j)\right]\\
&= E\left[f(x_k, u(k), k) + E\left[\sum_{j=k+1}^{N} f(x(j), u(j), j)\right]\right].
\end{aligned}$$

Since $J(x_k, k)$ is independent on $u(i)$ for $k+1 \le i \le N$, we have

$$\begin{aligned}
J(x_k, k) &\le E\left[f(x_k, u(k), k) + \min_{\substack{u(i)\in U_i\\ k+1\le i\le N}} E\left[\sum_{j=k+1}^{N} f(x(j), u(j), j)\right]\right]\\
&= E\left[f(x_k, u(k), k) + J(x(k+1), k+1)\right].
\end{aligned}$$

Taking the minimum for $u(k)$ in the above inequality yields that

$$J(x_k, k) \le \min_{u(k)\in U_k} E[f(x_k, u(k), k) + J(x(k+1), k+1)].$$

The recurrence Eq. (4.3) is proved.

Note that the recurrence Eqs. (4.2) and (4.3) may be reformulated as

$$J(x_N, N) = \min_{u(N)\in U_N} f(x_N, u(N), N), \tag{4.4}$$

$$\begin{aligned} J(x_k, k) = \min_{u(k)\in U_k} & E[f(x_k, u(k), k) + J(\phi(x_k, u(k), k) \\ & +\sigma(x_k, u(k), k)\, C_{k+1}, k+1)] \end{aligned} \tag{4.5}$$

for $k = N-1, N-2, \ldots, 1, 0$.

Theorem 4.1 tells us that the solution of problem (4.1) can be derived from the solution of the simpler problems (4.2) and (4.3) step by step from the last stage to the initial stage or in reverse order.

4.2 Linear Quadratic Model

By using the recurrence Eqs. (4.2) and (4.3), we will obtain the exact solution for the following uncertain optimal control problem with a quadratic objective function subject to an uncertain linear system:

$$\begin{cases} \min\limits_{\substack{u(i)\\ 0\le i\le N}} E\left[\sum\limits_{j=0}^{N} A_j x^2(j) + B_j u^2(j)\right] \\ \text{subject to:} \\ \qquad x(j+1) = a_j\, x(j) + b_j\, u(j) + \sigma_{j+1}\, C_{j+1}, \\ \qquad j = 0, 1, 2, \ldots, N-1, \quad x(0) = x_0, \end{cases} \tag{4.6}$$

where $A_j \ge 0$, $B_j \ge 0$ and $a_j, b_j, \sigma_j \ne 0$ are constants for all j. Generally, $|a_j\, x(j) + b_j\, u(j)| > |\sigma_{j+1}|$ for any j. In addition, $C_1, C_2, \ldots, C_N$ are ordinary linear uncertain variables $\mathcal{L}(-1, 1)$ with the same distribution

$$\Phi(x) = \begin{cases} 0, & \text{if } x \le -1 \\ (x+1)/2, & \text{if } -1 \le x \le 1 \\ 1, & \text{if } x \ge 1. \end{cases}$$

Denote the optimal control for the above problem by $u^*(0), u^*(1), \ldots, u^*(N)$. By the recurrence Eq. (4.4), we have

$$J(x_N, N) = \min_{u(N)}\{A_N x_N^2 + B_N u^2(N)\} = A_N x_N^2,$$

where $u^*(N) = 0$. For $k = N-1$, we have

$$
\begin{aligned}
&J(x_{N-1}, N-1) \\
&= \min_{u(N-1)} E[A_{N-1}x_{N-1}^2 + B_{N-1}u^2(N-1) + J(x(N), N)] \\
&= \min_{u(N-1)} \{A_{N-1}x_{N-1}^2 + B_{N-1}u^2(N-1) + A_N E[x^2(N)]\} \\
&= \min_{u(N-1)} \{A_{N-1}x_{N-1}^2 + B_{N-1}u^2(N-1) \\
&\quad + A_N E[(a_{N-1}x_{N-1} + b_{N-1}u(N-1) + \sigma_N C_N)^2]\} \\
&= \min_{u(N-1)} \{A_{N-1}x_{N-1}^2 + B_{N-1}u^2(N-1) + A_N(a_{N-1}x_{N-1} + b_{N-1}u(N-1))^2 \\
&\quad + A_N E[2\sigma_N(a_{N-1}x_{N-1} + b_{N-1}u(N-1))C_N + \sigma_N^2 C_N^2]\}. \qquad (4.7)
\end{aligned}
$$

Denote $d = 2\sigma_N(a_{N-1}x_{N-1} + b_{N-1}u(N-1))$. It follows from Example 1.6, denoting $b = d/\sigma_N^2$ the absolute value of which is larger than 2 that

$$
\begin{aligned}
E[2\sigma_N(a_{N-1}x_{N-1} + b_{N-1}u(N-1))C_N + \sigma_N^2 C_N^2] &= \sigma_N^2 E[bC_N + C_N^2] \\
&= \frac{1}{3}\sigma_N^2. \qquad (4.8)
\end{aligned}
$$

Substituting (4.8) into (4.7) yields that

$$
\begin{aligned}
&J(x_{N-1}, N-1) \\
&= \min_{u(N-1)} \{A_{N-1}x_{N-1}^2 + B_{N-1}u^2(N-1) \\
&\quad + A_N(a_{N-1}x_{N-1} + b_{N-1}u(N-1))^2 + \frac{1}{3}\sigma_N^2 A_N\bigg\}.
\end{aligned}
$$

Let

$$
H = A_{N-1}x_{N-1}^2 + B_{N-1}u^2(N-1) + A_N(a_{N-1}x_{N-1} + b_{N-1}u(N-1))^2 + \frac{1}{3}\sigma_N^2 A_N.
$$

It follows from

$$
\begin{aligned}
\frac{\partial H}{\partial u(N-1)} &= 2B_{N-1}u(N-1) + 2A_N b_{N-1}[a_{N-1}x_{N-1} + b_{N-1}u(N-1)] \\
&= 0
\end{aligned}
$$

that the optimal control is

$$
u^*(N-1) = -\frac{a_{N-1}b_{N-1}A_N}{B_{N-1} + b_{N-1}^2 A_N} x_{N-1}
$$

which is the minimum point of the function H because

$$\frac{\partial^2 H}{\partial u^2(N-1)} = 2B_{N-1} + 2A_N b_{N-1}^2 \geq 0.$$

Hence,

$$\begin{aligned}
&J(x_{N-1}, N-1)\\
&= A_{N-1}x_{N-1}^2 + \frac{a_{N-1}^2 b_{N-1}^2 A_N^2 B_{N-1}}{(B_{N-1} + b_{N-1}^2 A_N)^2} x_{N-1}^2\\
&\quad + A_N \left(a_{N-1} - \frac{a_{N-1} b_{N-1}^2 A_N}{B_{N-1} + b_{N-1}^2 A_N}\right)^2 x_{N-1}^2 + \frac{1}{3}\sigma_N^2 A_N\\
&= \left(A_{N-1} + \frac{a_{N-1}^2 b_{N-1}^2 B_{N-1} A_N^2}{(B_{N-1} + b_{N-1}^2 A_N)^2}\right.\\
&\quad \left. + \frac{a_{N-1}^2 B_{N-1}^2 A_N}{(B_{N-1} + b_{N-1}^2 A_N)^2}\right) x_{N-1}^2 + \frac{1}{3}\sigma_N^2 A_N.
\end{aligned}$$

Let

$$Q_{N-1} = \left(A_{N-1} + \frac{a_{N-1}^2 b_{N-1}^2 B_{N-1} A_N^2}{(B_{N-1} + b_{N-1}^2 A_N)^2} + \frac{a_{N-1}^2 B_{N-1}^2 A_N}{(B_{N-1} + b_{N-1}^2 A_N)^2}\right).$$

We have

$$J(x_{N-1}, N-1) = Q_{N-1} x_{N-1}^2 + \frac{1}{3}\sigma_N^2 A_N. \tag{4.9}$$

For $k = N-2$, we have

$$\begin{aligned}
&J(x_{N-2}, N-2)\\
&= \min_{u(N-2)} E[A_{N-2}x_{N-2}^2 + B_{N-2}u^2(N-2) + J(x(N-1), N-1)]\\
&= \min_{u(N-2)} \{A_{N-2}x_{N-2}^2 + B_{N-2}u^2(N-2) + E[Q_{N-1}x^2(N-1) + \frac{1}{3}\sigma_N^2 A_N]\\
&= \min_{u(N-2)} \{A_{N-2}x_{N-2}^2 + B_{N-2}u^2(N-2) + Q_{N-1}E[(a_{N-2}x_{N-2} + b_{N-2}u(N-2)\\
&\quad + \sigma_{N-1} C_{N-1})^2] + \frac{1}{3}\sigma_N^2 A_N\Big\}\\
&= \min_{u(N-2)} \{A_{N-2}x_{N-2}^2 + B_{N-2}u^2(N-2) + Q_{N-1}(a_{N-2}x_{N-2} + b_{N-2}u(N-2))^2\\
&\quad + Q_{N-1}E[2\sigma(a_{N-2}x_{N-2} + b_{N-2}u(N-2))C_{N-1} + \sigma_{N-1}^2 C_{N-1}^2] + \frac{1}{3}\sigma_N^2 A_N\Big\}.
\end{aligned}$$

It follows from the similar computation to (4.8) that

$$E[2\sigma(a_{N-2}x_{N-2}+b_{N-2}u(N-2))C_{N-1}+\sigma_{N-1}^2C_{N-1}^2]=\frac{1}{3}\sigma_{N-1}^2.$$

By the similar computation to the case for $k=N-1$, we get

$$\begin{aligned}
&J(x_{N-2},N-2)\\
&=\min_{u(N-2)}\{A_{N-2}x_{N-2}^2+B_{N-2}u^2(N-2)+Q_{N-1}(a_{N-2}x_{N-2}+b_{N-2}u(N-2))^2\\
&\quad+\frac{1}{3}(\sigma_{N-1}^2Q_{N-1}+\sigma_N^2A_N)\Big\}\\
&=\left(A_{N-2}+\frac{a_{N-2}^2b_{N-2}^2B_{N-2}Q_{N-1}^2}{(B_{N-2}+b_{N-2}^2Q_{N-1})^2}+\frac{a_{N-2}^2B_{N-2}^2Q_{N-1}}{(B_{N-2}+b_{N-2}^2Q_{N-1})^2}\right)x_{N-2}^2\\
&\quad+\frac{1}{3}(\sigma_{N-1}^2Q_{N-1}+\sigma_N^2A_N)
\end{aligned}$$

with the optimal control

$$u^*(N-2)=-\frac{a_{N-2}b_{N-2}Q_{N-1}}{B_{N-2}+b_{N-2}^2Q_{N-1}}x_{N-2}.$$

Let

$$Q_{N-2}=\left(A_{N-2}+\frac{a_{N-2}^2b_{N-2}^2B_{N-2}Q_{N-1}^2}{(B_{N-2}+b_{N-2}^2Q_{N-1})^2}+\frac{a_{N-2}^2B_{N-2}^2Q_{N-1}}{(B_{N-2}+b_{N-2}^2Q_{N-1})^2}\right).$$

We have

$$J(x_{N-2},N-2)=Q_{N-2}x_{N-2}^2+\frac{1}{3}(\sigma_{N-1}^2Q_{N-1}+\sigma_N^2A_N). \tag{4.10}$$

By induction, we can obtain the optimal control for problem (4.6) as follows:

$$u^*(N)=0,\quad u^*(k)=-\frac{a_kb_kQ_{k+1}}{B_k+b_k^2Q_{k+1}}x_k$$

where

$$Q_N=A_N,$$

$$Q_k=\left(A_k+\frac{a_k^2b_k^2B_kQ_{k+1}^2}{(B_k+b_k^2Q_{k+1})^2}+\frac{a_k^2B_k^2Q_{k+1}}{(B_k+b_k^2Q_{k+1})^2}\right),$$

and the optimal values are

$$J(x_N, N) = A_N x_N^2,$$

$$J(x_k, k) = Q_k x_k^2 + \frac{1}{3} \sum_{j=k+1}^{N} \sigma_j^2 Q_j$$

for $k = N-1, N-2, \ldots, 1, 0$.

4.3 General Case

In previous section, we studied an optimal control problem for a quadratic objective function subject to an uncertain linear system. For that problem, we can get the exact feedback optimal controls of the state at all stages. If the system is nonlinear, or the objective function is not quadratic, or the uncertain variables C_j's are not linear, the optimal controls may be not displayed exactly by the state of the system at all stages. In such cases, we have to consider the numerical solutions for the problem.

For the uncertain optimal control problem (4.1), assume that the state $x(k)$ of the system is in $[l_k^-, l_k^+]$, and the control variable $u(k)$ is constrained by the set U_k for $k = 0, 1, \ldots, N$. For each k, divide the interval $[l_k^-, l_k^+]$ into n_k subintervals:

$$l_k^- = x(k)_0 < x(k)_1 < \cdots < x(k)_{n_k} = l_k^+.$$

We will numerically compute the optimal controls in outline way for all states $x(k)_i$ $(i = 0, 1, \ldots, n_k, k = 0, 1, \ldots, N)$. Based on these data, we can obtain the optimal controls in online way for any initial state x_0 by an interpolation method.

In practice, for simplicity, it is reasonable to assume that the range of each state variable $x(k)$ is a finite interval, even if it may be a subset of a finite interval. These intervals are set according to the background of the problem. To balance the accuracy of approximations by interpolation and the computational cost, the number of state variables in the range $[l_k^-, l_k^+]$ should be chosen properly.

Next, we will establish two methods to produce the optimal controls for all states $x(k)_i$ $(i = 0, 1, \ldots, n_k,\ k = 0, 1, \ldots, N)$: hybrid intelligent algorithm and finite search method.

4.3.1 Hybrid Intelligent Algorithm

By the recurrence Eqs. (4.4) and (4.5), we first approximate the value $J(x(N), N)$. For each $x(N)_i$ $(i = 0, 1, \ldots, n_N)$, solve the following optimization

$$J(x(N)_i, N) = \min_{u(N) \in U_N} f(x(N)_i, u(N), N)$$

by genetic algorithm to get optimal control $u^*(N)_i$ and optimal objective value $J(x(N)_i, N)$. Then for each $x(N-1)_i$ $(i = 0, 1, \ldots, n_{N-1})$, solve the following optimization

$$J(x(N-1)_i, N-1) = \min_{u(N-1)\in U_{N-1}} E[f(x(N-1)_i, u(N-1), N-1) + J(x(N), N)],$$

where $x(N) = \phi(x(N-1)_i, u(N-1), N-1) + \sigma(x(N-1)_i, u(N-1), N-1)\, C_N$, by hybrid intelligent algorithm (integrating uncertain simulation, neural network, and genetic algorithm) to get optimal control $u^*(N-1)_i$ and optimal objective value $J(x(N-1)_i, N-1)$. Note that the optimal control $u^*(N-1)_i$ is selected in U_{N-1} and the set of $u(N-1)$ such that $x(N) = \phi(x(N-1)_i, u(N-1), N-1) + \sigma(x(N-1)_i, u(N-1), N-1)\, C_N$ is in $[l_N^-, l_N^+]$. The value of $J(x(N), N)$ may be calculated by interpolation based on the values $J(x(N)_i, N)$ $(i = 0, 1, \ldots, n_N)$. In addition, the expected value $E[f(x(N-1)_i, u(N-1), N-1) + J(x(N), N)]$ may be approximated by uncertain simulation established in Sect. 1.4. By induction, we can solve the following optimization

$$J(x(k)_i, k) = \min_{u(k)\in U_k} E[f(x(k)_i, u(k), k) + J(x(k+1), k+1)],$$

by hybrid intelligent algorithm, to get optimal control $u^*(k)_i$ and optimal objective value $J(x(k)_i, k)$ for $k = N-2, N-3, \ldots, 1, 0$.

The method to produce a list of data on the optimal controls and optimal objective values for all states $x(k)_i$ $(i = 0, 1, \ldots, n_k, k = 0, 1, \ldots, N)$ by hybrid intelligent algorithm may be summarized as Algorithm 4.1.

4.3.2 Finite Search Method

At every stage k, the constraint domain U_k of control variable $u(k)$ is assumed to be an interval $[q_k^-, q_k^+]$. Averagely divide the interval $[q_k^-, q_k^+]$ into m_k subintervals:

$$q_k^- = u(k)_0 < u(k)_1 < \cdots < u(k)_{m_k-1} < u(k)_{m_k} = q_k^+.$$

The approximate optimal control $u^*(k)_i$ is searched in the finite set $\{u(k)_j \mid 0 \le j \le m_k\}$. That is,

$$\begin{aligned} &E[f(x(k)_i, u^*(k)_i, k) + J(x(k+1), k+1)] \\ &= \min_{0\le j\le m_k} E[f(x(k)_i, u(k)_j, k) + J(x(k+1), k+1)] \end{aligned} \tag{4.11}$$

where

$$x(k+1) = \phi(x(k)_i, u(k)_j, k) + \sigma(x(k)_i, u(k)_j, k)\, C_{k+1}.$$

Algorithm 4.1 (Data production by hybrid intelligent algorithm)

Step 1. Averagely divide $[l_k^-, l_k^+]$ to generate states $x(k)_i$ as

$$l_k^- = x(k)_0 < x(k)_1 < \cdots < x(j)_{n_k} = l_k^+$$

for $k = 0, 1, \ldots, N$.

Step 2. Solve

$$J(x(N)_i, N) = \min_{u(N)\in U_N} f(x(N)_i, u(N), N)$$

by genetic algorithm to produce $u^*(N)_i$ and $J(x(N)_i, N)$ for $i = 0, 1, \ldots, n_N$.

Step 3. For $k = N - 1$ to 0, perform the next two steps.

Step 4. Approximate the function

$$u(k) \to E[f(x(k)_i, u(k), k) + J(x(k+1), k+1)]$$

by Algorithm 1.3, where

$$x(k+1) = \phi(x(k)_i, u(k), k) + \sigma(x(k)_i, u(k), k)\, C_{k+1}.$$

Step 5. Solve

$$J(x(k)_i, k) = \min_{u(k)\in U_k} E[f(x(k)_i, u(k), k) + J(x(k+1), k+1)],$$

by hybrid intelligent algorithm to produce $u^*(k)_i$ and $J(x(k)_i, k)$ for $i = 0, 1, \ldots, n_k$.

The method to produce a list of data on the optimal controls and optimal objective values for all states $x(k)_i$ ($i = 0, 1, \ldots, n_k, k = 0, 1, \ldots, N$) by finite search method may be summarized as the following Algorithm 4.2.

Remark 4.1 Generally speaking, the optimal controls $u^*(k)_i$ obtained by Algorithm 4.2 is not finer than by Algorithm 4.1. But the perform time by Algorithm 4.2 is much less than by Algorithm 4.1, which will be seen in the next numerical example.

4.3.3 Optimal Controls for Any Initial State

Now, if an initial state $x(0)$ is given, we may online perform the following Algorithm 4.3 to get a state of the system, optimal control and optimal objective value based on the data produced by Algorithms 4.1 or 4.2.

Remark 4.2 The data on the optimal controls and optimal objective values at all given states are produced based on the recurrence equations step by step from the last stage to the initial stage in reverse order, whereas the optimal controls and optimal objective value for any initial state are got, based on the data obtained, step by step from the initial stage to the last stage orderly.

Algorithm 4.2 (Data production by finite search method)

Step 1. Averagely divide $[q_k^-, q_k^+]$ to generate controls $u(k)_j$ as

$$q_k^- = u(k)_0 < u(k)_1 < \cdots < u(k)_{m_k-1} < u(k)_{m_k} = q_k^+.$$

for $k = 0, 1, \ldots, N$.

Step 2. Find $u^*(N)_i \in \{u(N)_j \mid 0 \le j \le m_N\}$ such that

$$J(x(N)_i, N) = f(x(N)_i, u^*(N)_i, N) = \min_{0 \le j \le m_N} f(x(N)_i, u(N)_j, N)$$

for $i = 0, 1, \ldots, n_N$.

Step 3. For $k = N - 1$ to 0, perform the next two steps.

Step 4. Approximate the value

$$E[f(x(k)_i, u(k)_j, k) + J(x(k+1), k+1)]$$

by Algorithm 1.3, where

$$x(k+1) = \phi(x(k)_i, u(k)_j, k) + \sigma(x(k)_i, u(k)_j, k)\, C_{k+1}.$$

Step 5. Find $u^*(k)_i \in \{u(k)_j \mid 0 \le j \le m_k\}$ such that (4.11) holds, and

$$J(x(k)_i, k) = E[f(x(k)_i, u^*(k)_i, k) + J(x(k+1), k+1)].$$

for $i = 0, 1, \ldots, n_k$

Algorithm 4.3 (Online optimal control)

Step 1. For initial state $x(0)$, if $x(0)_i \le x(0) \le x(0)_{i+1}$, compute $u^*(0)$ and $J(x(0), 0)$ by interpolation:

$$u^*(0) = u^*(0)_i + \frac{u^*(0)_{i+1} - u^*(0)_i}{x(0)_{i+1} - x(0)_i}(x(0) - x(0)_i),$$

$$J(x(0), 0) = J(x(0)_i, 0) + \frac{J(x(0)_{i+1}, 0) - J(x(0)_i, 0)}{x(0)_{i+1} - x(0)_i}(x(0) - x(0)_i).$$

Step 2. For $k = 1$ to N, perform the next two steps.

Step 3. Randomly generate a number $r \in [0, 1]$, produce a number $c(k)$ according to the distribution function $\Phi_k(x)$ of uncertain variable C_k such that $\Phi_k(c(k)) = r$. Set

$$x(k) = \phi(x(k-1), u^*(k-1), k-1) + \sigma(x(k-1), u^*(k-1), k-1)\, c(k).$$

Step 4. If $x(k)_i \le x(k) \le x(k)_{i+1}$, compute $u^*(k)$ by interpolation:

$$u^*(k) = u^*(k)_i + \frac{u^*(k)_{i+1} - u^*(k)_i}{x(k)_{i+1} - x(k)_i}(x(k) - x(k)_i).$$

4.4 Example

Consider the following example:

$$\begin{cases} \min\limits_{\substack{u(i)\\0\le i\le 10}} E\left[\sum_{j=0}^{10} A\,x^4(j) + B\,u^2(j)\right] \\ \text{subject to:} \\ \qquad x(j+1) = a\,x(j) + b\,u(j) + \sigma\,C_{j+1}, \\ \qquad j = 0, 1, 2, \ldots, N-1, \quad x(0) = x_0, \end{cases} \tag{4.12}$$

where $A = 2$, $B = 0.01$, $a = 0.8$, $b = 0.09$, $\sigma = 0.0018$, and $-0.5 \le x(j) \le 0.5$, $-1 \le u(j) \le 1$ for $0 \le j \le 10$. In addition, the uncertain variables $C_1, C_2, \ldots, C_{10}$ are independent and normally distributed with expected value 0 and variance 1, whose distribution function is

Table 4.1 Data produced by hybrid intelligent algorithm (Algorithm 4.1)

$x(k)$		−0.50	−0.45	−0.40	−0.35	−0.30	−0.25	−0.20
Stage	$J(\cdot, 10)$	0.1250	0.0820	0.0512	0.0300	0.0162	0.0078	0.0032
10	$u^*(10)$	0.0000	0.0000	0.0000	0.0000	0.0000	0.0000	0.0000
9	$J(\cdot, 9)$	0.1529	0.1019	0.0653	0.0388	0.0215	0.0106	0.0043
	$u^*(9)$	1.0000	1.0000	0.6676	0.4860	0.3459	0.2112	0.1172
8	$J(\cdot, 8)$	0.1587	0.1057	0.0685	0.0410	0.0229	0.0113	0.0047
	$u^*(8)$	1.0000	1.0000	0.8157	0.5903	0.4246	0.2667	0.1492
7	$J(\cdot, 7)$	0.1606	0.1062	0.0689	0.0416	0.0233	0.0116	0.0048
	$u^*(7)$	1.0000	1.0000	0.8831	0.6099	0.4228	0.2778	0.1641
6	$J(\cdot, 6)$	0.1605	0.1067	0.0682	0.0421	0.0235	0.0117	0.0050
	$u^*(6)$	1.0000	1.0000	1.0000	0.6224	0.4337	0.2840	0.1655
5	$J(\cdot, 5)$	0.1616	0.1068	0.0694	0.0419	0.0234	0.0121	0.0051
	$u^*(5)$	1.0000	1.0000	0.8790	0.6247	0.4369	0.2816	0.1754
4	$J(\cdot, 4)$	0.1610	0.1072	0.0697	0.0423	0.0236	0.0118	0.0050
	$u^*(4)$	1.0000	1.0000	0.8490	0.6214	0.4578	0.2982	0.1747
3	$J(\cdot, 3)$	0.1610	0.1070	0.0693	0.0424	0.0236	0.0117	0.0050
	$u^*(3)$	1.0000	1.0000	0.8712	0.6325	0.4564	0.2984	0.1735
2	$J(\cdot, 2)$	0.1617	0.1066	0.0694	0.0419	0.0234	0.0119	0.0049
	$u^*(2)$	1.0000	1.0000	0.8788	0.6238	0.4399	0.2945	0.1750
1	$J(\cdot, 1)$	0.1606	0.1067	0.0696	0.0422	0.0235	0.0119	0.0051
	$u^*(1)$	1.0000	1.0000	0.8816	0.6306	0.4345	0.2831	0.1820
0	$J(\cdot, 0)$	0.1619	0.1068	0.0694	0.0421	0.0236	0.0118	0.0050
	$u^*(0)$	1.0000	1.0000	0.8658	0.6158	0.4587	0.3183	0.1716

$$\Phi(x) = \left(1 + \exp\left(-\frac{\pi x}{\sqrt{3}}\right)\right)^{-1}, \quad x \in R. \tag{4.13}$$

The interval $[-0.5, 0.5]$ of state $x(k)$ is averagely inserted into 21 states $x(k)_i = -0.5 + 0.05 * i$ $(i = 0, 1, \ldots, 20)$ for $k = 0, 1, \ldots, 10$. Algorithm 4.1 (with 4000 cycles in simulation, 2000 training data in neural network and 600 generations in genetic algorithm) is employed to produce a list of data as shown in Tables 4.1, 4.2, and 4.3.

The interval $[-1, 1]$ of control $u(k)$ is averagely inserted into 1001 controls $u(k)_j = -1 + 0.002 * j$ $(j = 0, 1, \ldots, 1000)$ for $k = 0, 1, \ldots, 10$. Algorithm 4.2 is employed to produce a list of data as shown in Tables 4.4, 4.5, and 4.6.

In Tables 4.1, 4.2, 4.3, 4.5, and 4.6, the data in the first rows are the 21 states in range $[-0.5, 0.5]$. In the following each row, reported are the optimal objective values (topmost number) and the optimal controls with respect to corresponding states at the stage indicated with the leftmost number. Note that the optimal controls $u^*(10)$ in the stage 10 are all zero because each of them is the minimal solution of the problem such as

Table 4.2 Data produced by hybrid intelligent algorithm (continuous)

$x(k)$		−0.15	−0.10	−0.05	0	0.05	0.10	0.15
Stage	$J(\cdot, 10)$	0.0010	0.0002	0.0000	0.0000	0.0000	0.0002	0.0010
10	$u^*(10)$	0.0000	0.0000	0.0000	0.0000	0.0000	0.0000	0.0000
9	$J(\cdot, 9)$	0.0014	0.0002	0.0000	0.0000	0.0000	0.0003	0.0014
	$u^*(9)$	0.0499	0.0139	−0.0033	−0.0088	−0.0066	−0.0182	−0.0519
8	$J(\cdot, 8)$	0.0016	0.0003	0.0001	0.0001	0.0000	0.0004	0.0016
	$u^*(8)$	0.0707	0.0197	−0.0013	−0.0084	−0.0110	−0.0247	−0.0725
7	$J(\cdot, 7)$	0.0016	0.0004	0.0000	0.0001	0.0000	0.0004	0.0018
	$u^*(7)$	0.0794	0.0234	0.0043	−0.0008	−0.0106	−0.0352	−0.0829
6	$J(\cdot, 6)$	0.0019	0.0004	0.0001	0.0001	0.0000	0.0004	0.0018
	$u^*(6)$	0.0826	0.0323	0.0064	−0.0109	−0.0109	−0.0358	−0.0884
5	$J(\cdot, 5)$	0.0017	0.0002	0.0002	0.0001	0.0000	0.0004	0.0018
	$u^*(5)$	0.1002	0.0401	0.0023	−0.0080	−0.0106	−0.0374	−0.0906
4	$J(\cdot, 4)$	0.0016	0.0002	0.0002	0.0001	0.0000	0.0004	0.0017
	$u^*(4)$	0.0944	0.0336	−0.0072	−0.0137	−0.0056	−0.0347	−0.0915
3	$J(\cdot, 3)$	0.0016	0.0002	0.0001	0.0002	0.0000	0.0004	0.0018
	$u^*(3)$	0.0888	0.0356	−0.0020	−0.0150	−0.0108	−0.0381	−0.0867
2	$J(\cdot, 2)$	0.0018	0.0002	0.0001	0.0002	0.0000	0.0004	0.0018
	$u^*(2)$	0.0901	0.0297	0.0023	−0.0128	−0.0152	−0.0410	−0.0882
1	$J(\cdot, 1)$	0.0015	0.0003	0.0002	0.0002	0.0001	0.0004	0.0018
	$u^*(1)$	0.0959	0.0352	0.0005	−0.0099	−0.0190	−0.0385	−0.0922
0	$J(\cdot, 0)$	0.0015	0.0002	0.0002	0.0002	0.0001	0.0004	0.0018
	$u^*(0)$	0.0832	0.0334	0.0011	−0.0090	−0.0097	−0.0422	−0.0917

Table 4.3 Data produced by hybrid intelligent algorithm (continuous)

$x(k)$		0.20	0.25	0.30	0.35	0.40	0.45	0.50
Stage	$J(\cdot, 10)$	0.0032	0.0078	0.0162	0.0300	0.0512	0.0820	0.1250
10	$u^*(10)$	0.0000	0.0000	0.0000	0.0000	0.0000	0.0000	0.0000
9	$J(\cdot, 9)$	0.0044	0.0107	0.0215	0.0388	0.0653	0.1027	0.1527
	$u^*(9)$	−0.1138	−0.2163	−0.3462	−0.4806	−0.6624	−0.9041	−1.0000
8	$J(\cdot, 8)$	0.0048	0.0114	0.0229	0.0411	0.0682	0.1058	0.1584
	$u^*(8)$	−0.1521	−0.2752	−0.4249	−0.5916	−1.0000	−1.0000	−1.0000
7	$J(\cdot, 7)$	0.0050	0.0117	0.0234	0.0416	0.0688	0.1067	0.1608
	$u^*(7)$	−0.1598	−0.2903	−0.4412	−0.6079	−0.8489	−1.0000	−1.0000
6	$J(\cdot, 6)$	0.0052	0.0119	0.0235	0.0420	0.0691	0.1068	0.1614
	$u^*(6)$	−0.1676	−0.2883	−0.4430	−0.6065	−0.8755	−1.0000	−1.0000
5	$J(\cdot, 5)$	0.0052	0.0121	0.0237	0.0422	0.0698	0.1066	0.1613
	$u^*(5)$	−0.1770	−0.2989	−0.4509	−0.6109	−0.8571	−1.0000	−1.0000
4	$J(\cdot, 4)$	0.0053	0.0119	0.0238	0.0423	0.0695	0.1071	0.1609
	$u^*(4)$	−0.1797	−0.3046	−0.4459	−0.6215	−0.8693	−1.0000	−1.0000
3	$J(\cdot, 3)$	0.0051	0.0120	0.0237	0.0422	0.0696	0.1068	0.1614
	$u^*(3)$	−0.1751	−0.3101	−0.4505	−0.6077	−0.8794	−1.0000	−1.0000
2	$J(\cdot, 2)$	0.0052	0.0118	0.0236	0.0423	0.0692	0.1066	0.1613
	$u^*(2)$	−0.1716	−0.2921	−0.4576	−0.6185	−1.0000	−1.0000	−1.0000
1	$J(\cdot, 1)$	0.0052	0.0120	0.0238	0.0419	0.0691	0.1071	0.1619
	$u^*(1)$	−0.1779	−0.3055	−0.4400	−0.6018	−0.8686	−1.0000	−1.0000
0	$J(\cdot, 0)$	0.0053	0.0120	0.0238	0.0422	0.0687	0.1075	0.1616
	$u^*(0)$	−0.1751	−0.3042	−0.4463	−0.6147	−1.0000	−1.0000	−1.0000

$$\min_{u(10)\in U_{10}} f(x(10)_i, u(10), 10) = \min_{-1\le u(10)\le 1} \{2\,x^4(10)_i + 0.01\,u^2(10)\}.$$

If we have six initial states $x_0 = -0.435, -0.365, -0.126, 0.09, 0.275,$ and 0.488, performing Algorithm 4.3 for every initial state yields optimal objective value and optimal controls which are listed in Tables 4.7 and 4.8. The data in the second rows in these two tables are the optimal objective values of the problem for initial states given in the first rows. In the third rows are the optimal controls at initial stage. In the following each row, reported are the optimal controls (topmost number) and the realized states at the corresponding stage.

All computations are processed with C programming in a PC (Intel(R) Core(TM) 2 Duo CPU P8600@2.40GHz). Note that performing Algorithm 4.3 very quick (less than one second), but performing Algorithms 4.1 or 4.2 is time-consuming. Performing Step 5 in Algorithm 4.1 each time needs about 175 seconds, and then completing the data in Tables 4.1, 4.2, and 4.3 by Algorithm 4.2 needs about 175 × 210 seconds. However performing Step 5 in Algorithm 4.2 each time needs about 75 seconds, and then completing the data in Tables 4.4, 4.5, and 4.6 by Algorithm 4.2

Table 4.4 Data produced by finite search method (Algorithm 4.2)

$x(k)$		−0.50	−0.45	−0.40	−0.35	−0.30	−0.25	−0.20
Stage	$J(\cdot, 10)$	0.12500	0.08201	0.05120	0.03001	0.01620	0.00781	0.00320
10	$u^*(10)$	0.000	0.000	0.000	0.000	0.000	0.000	0.000
9	$J(\cdot, 9)$	0.15388	0.10252	0.06501	0.03883	0.02134	0.01061	0.00435
	$u^*(9)$	0.994	0.746	0.738	0.396	0.392	0.190	0.106
8	$J(\cdot, 8)$	0.15979	0.10682	0.06789	0.04116	0.02256	0.01148	0.00482
	$u^*(8)$	0.998	0.984	0.774	0.542	0.444	0.256	0.132
7	$J(\cdot, 7)$	0.16130	0.10781	0.06879	0.04185	0.02304	0.01182	0.00501
	$u^*(7)$	0.994	1.000	0.780	0.550	0.444	0.268	0.138
6	$J(\cdot, 6)$	0.16176	0.10823	0.06910	0.04211	0.02324	0.01196	0.00510
	$u^*(6)$	1.000	0.986	0.770	0.662	0.442	0.306	0.120
5	$J(\cdot, 5)$	0.16197	0.10844	0.06923	0.04222	0.02332	0.01202	0.00515
	$u^*(5)$	0.998	0.990	0.792	0.614	0.442	0.294	0.126
4	$J(\cdot, 4)$	0.16211	0.10850	0.06932	0.04230	0.02335	0.01206	0.00517
	$u^*(4)$	0.996	0.962	0.778	0.608	0.446	0.304	0.142
3	$J(\cdot, 3)$	0.16208	0.10848	0.06934	0.04231	0.02340	0.01208	0.00520
	$u^*(3)$	1.000	1.000	0.808	0.648	0.444	0.282	0.148
2	$J(\cdot, 2)$	0.16215	0.10856	0.06937	0.04233	0.02341	0.01211	0.00521
	$u^*(2)$	0.994	1.000	0.780	0.602	0.444	0.260	0.136
1	$J(\cdot, 1)$	0.16216	0.10853	0.06938	0.04237	0.02343	0.01211	0.00523
	$u^*(1)$	0.998	0.994	0.780	0.654	0.442	0.278	0.154
0	$J(\cdot, 0)$	0.16216	0.10855	0.06941	0.04237	0.02343	0.01213	0.00523
	$u^*(0)$	1.000	0.966	0.776	0.594	0.450	0.332	0.152

needs about 75 × 210 s. Therefore, perform time by Algorithm 4.2 is much less than by Algorithm 4.1.

Generally, if the length of the state variable range $[l_k^-, l_k^+]$ is thought to be larger or the precision of approximations by interpolation (Algorithm 4.3) is required to improve, the number of state variables in range $[l_k^-, l_k^+]$ will be increased, and then this will increase the perform time. In the example, one more of number of state variables results in about 175 × 10 s more of perform time by Algorithm 4.1, and about 75 × 10 s more of perform time by Algorithm 4.2.

It follows from Tables 4.7 and 4.8 that for problem (4.12), optimal solutions obtained based on the data produced by hybrid intelligent algorithm are near to optimal solutions obtained based on the data produced by finite search method. The difference of the optimal objective values obtained by two methods and listed at the second rows in Tables 4.7 and 4.8 may be seen in Table 4.9. Each absolute difference is small (the first not larger than 0.0013, and the others not larger than 0.0007). So, the efficiency of two proposed methods to solve the problem presented in the paper is comparative.

Table 4.5 Data produced by finite search method (continuous)

$x(k)$		−0.15	−0.10	−0.05	0	0.05	0.10	0.15
Stage	$J(\cdot, 10)$	0.00101	0.00020	0.00001	0.000000	0.00001	0.00020	0.00101
10	$u^*(10)$	0.000	0.000	0.000	0.000	0.000	0.000	0.000
9	$J(\cdot, 9)$	0.00148	0.00032	0.00002	0.000000	0.00002	0.00032	0.00148
	$u^*(9)$	0.068	0.022	0.000	0.000	0.000	−0.014	−0.074
8	$J(\cdot, 8)$	0.00168	0.00039	0.00003	0.000001	0.00003	0.00039	0.00168
	$u^*(8)$	0.086	0.022	0.004	0.000	−0.002	−0.026	−0.100
7	$J(\cdot, 7)$	0.00178	0.00043	0.00004	0.000001	0.00004	0.00044	0.00178
	$u^*(7)$	0.100	0.024	0.004	0.000	−0.004	−0.034	−0.110
6	$J(\cdot, 6)$	0.00183	0.00046	0.00004	0.000002	0.00004	0.00046	0.00183
	$u^*(6)$	0.116	0.034	0.004	0.000	−0.004	−0.036	−0.122
5	$J(\cdot, 5)$	0.00186	0.00048	0.00005	0.000003	0.00005	0.00048	0.00186
	$u^*(5)$	0.128	0.040	0.002	0.000	−0.002	−0.040	−0.120
4	$J(\cdot, 4)$	0.00188	0.00049	0.00005	0.000004	0.00005	0.00049	0.00188
	$u^*(4)$	0.110	0.034	0.006	0.000	0.000	−0.040	−0.146
3	$J(\cdot, 3)$	0.00189	0.00050	0.00005	0.000004	0.00005	0.00050	0.00189
	$u^*(3)$	0.128	0.042	0.004	0.000	−0.006	−0.034	−0.150
2	$J(\cdot, 2)$	0.00191	0.00050	0.00006	0.000005	0.00006	0.00050	0.00190
	$u^*(2)$	0.122	0.030	0.006	0.000	−0.004	−0.042	−0.130
1	$J(\cdot, 1)$	0.00191	0.00051	0.00006	0.000006	0.00006	0.00051	0.00191
	$u^*(1)$	0.136	0.034	0.002	0.000	−0.002	−0.034	−0.140
0	$J(\cdot, 0)$	0.00191	0.00051	0.00006	0.000007	0.00006	0.00051	0.00191
	$u^*(0)$	0.130	0.038	0.002	0.000	−0.004	−0.046	−0.118

4.5 Indefinite LQ Optimal Control with Equality Constraint

4.5.1 Problem Setting

Consider the indefinite LQ optimal control with equality constraint for discrete-time uncertain systems as follows.

$$\begin{cases} \inf\limits_{\substack{u_k \\ 0\le k\le N-1}} J(\boldsymbol{x}_0, \boldsymbol{u}) = \sum\limits_{k=0}^{N-1} E\left[\boldsymbol{x}_k^\tau Q_k \boldsymbol{x}_k + \boldsymbol{u}_k^\tau R_k \boldsymbol{u}_k\right] + E\left[\boldsymbol{x}_N^\tau Q_N \boldsymbol{x}_N\right] \\ \text{subject to} \\ \qquad \boldsymbol{x}_{k+1} = A_k \boldsymbol{x}_k + B_k \boldsymbol{u}_k + \lambda_k (A_k \boldsymbol{x}_k + B_k \boldsymbol{u}_k)\xi_k, \;\; k = 0, 1, \ldots, N-1, \\ \qquad F\boldsymbol{x}_N = \boldsymbol{\eta}, \end{cases} \tag{4.14}$$

Table 4.6 Data produced by finite search method (continuous)

$x(k)$		0.20	0.25	0.30	0.35	0.40	0.45	0.50
Stage	$J(\cdot, 10)$	0.00320	0.00781	0.01620	0.03001	0.05120	0.08201	0.12500
10	$u^*(10)$	0.000	0.000	0.000	0.000	0.000	0.000	0.000
9	$J(\cdot, 9)$	0.00435	0.01061	0.02134	0.03884	0.06504	0.10251	0.15390
	$u^*(9)$	−0.124	−0.200	−0.404	−0.410	−0.748	−0.706	−0.994
8	$J(\cdot, 8)$	0.00481	0.01147	0.02256	0.04117	0.06790	0.10681	0.15980
	$u^*(8)$	−0.132	−0.272	−0.442	−0.576	−0.766	−0.956	−1.000
7	$J(\cdot, 7)$	0.00500	0.01180	0.02304	0.04185	0.06876	0.10787	0.16129
	$u^*(7)$	−0.136	−0.264	−0.430	−0.582	−0.778	−0.986	−1.000
6	$J(\cdot, 6)$	0.00510	0.01195	0.02321	0.04211	0.06907	0.10825	0.16177
	$u^*(6)$	−0.138	−0.268	−0.440	−0.626	−0.780	−0.984	−0.996
5	$J(\cdot, 5)$	0.00516	0.01203	0.02331	0.04223	0.06923	0.10841	0.16192
	$u^*(5)$	−0.140	−0.312	−0.450	−0.604	−0.782	−0.990	−0.992
4	$J(\cdot, 4)$	0.00517	0.01207	0.02338	0.04231	0.06930	0.10849	0.16210
	$u^*(4)$	−0.142	−0.278	−0.450	−0.648	−0.792	−0.994	−1.000
3	$J(\cdot, 3)$	0.00520	0.01209	0.02340	0.04233	0.06935	0.10851	0.16209
	$u^*(3)$	−0.142	−0.298	−0.454	−0.630	−0.788	−0.996	−1.000
2	$J(\cdot, 2)$	0.00521	0.01210	0.02340	0.04235	0.06936	0.10859	0.16220
	$u^*(2)$	−0.136	−0.320	−0.444	−0.638	−0.798	−0.954	−0.998
1	$J(\cdot, 1)$	0.00522	0.01211	0.02343	0.04235	0.06941	0.10859	0.16214
	$u^*(1)$	−0.144	−0.298	−0.432,	−0.624	−0.786	−0.968	−1.000
0	$J(\cdot, 0)$	0.00523	0.01212	0.02344	0.04236	0.06941	0.10856	0.16212
	$u^*(0)$	−0.150	−0.290	−0.458	−0.608	−0.774	−0.978	−1.000

where $\lambda_k \in R$ and $0 \leq |\lambda_k| \leq 1$. The vector $\boldsymbol{x}_k$ is an uncertain state with the initial state $\boldsymbol{x}_0 \in R^n$ and $\boldsymbol{u}_k$ is a control vector subject to a constraint set $\boldsymbol{U}_k \subset R^m$. Denote $\boldsymbol{u} = (\boldsymbol{u}_0, \boldsymbol{u}_1, \ldots, \boldsymbol{u}_{N-1})$. Moreover, $Q_0, Q_1, \ldots, Q_N$ and $R_0, R_1, \ldots, R_{N-1}$ are real symmetric matrices with appropriate dimensions. In addition, the coefficients $A_0, A_1, \ldots, A_{N-1}$ and $B_0, B_1, \ldots, B_{N-1}$ are assumed to be crisp matrices with appropriate dimensions. Let $F \in R^{r \times n}$, $\boldsymbol{\eta} = (\eta_1, \eta_2, \cdots, \eta_r)^\tau$, where η_i $(i = 1, 2, \ldots, r)$ are uncertain variables. Besides, the noises $\xi_0, \xi_1, \cdots, \xi_{N-1}$ are independent ordinary linear uncertain variables $\mathcal{L}(-1, 1)$ with the distribution

$$\Phi(x) = \begin{cases} 0, & \text{if } x \leq -1 \\ (x+1)/2, & \text{if } -1 \leq x \leq 1 \\ 1, & \text{if } x \geq 1. \end{cases}$$

Note that we allow the cost matrices to be singular or indefinite. We need to give the following definitions.

Table 4.7 Optimal controls for some initial states based on the data of Tables 4.1, 4.2, and 4.3

x_0	−0.435	−0.365	−0.126	0.009	0.275	0.448
$J(x_0, 0)$	0.095564	0.050303	0.000883	0.000292	0.017896	0.148606
$u^*(0)$	0.959749	0.690801	0.059299	−0.035692	−0.375246	−1.000000
$u^*(1)$	0.296624	0.242283	0.030932	−0.028243	−0.157106	−0.458041
$x(1)$	−0.254466	−0.229805	−0.093857	0.073750	0.187872	0.305566
$u^*(2)$	0.126329	0.108482	0.012816	−0.019077	−0.080182	−0.184313
$x(2)$	−0.171344	−0.160842	−0.069204	0.057567	0.141459	0.205290
$u^*(3)$	0.059373	0.054106	−0.001613	−0.011138	−0.046155	−0.091623
$x(3)$	−0.122335	−0.117385	−0.050455	0.045503	0.108312	0.152794
$u^*(4)$	0.026281	0.022893	−0.008531	−0.007961	−0.025339	−0.050701
$x(4)$	−0.091021	−0.086875	−0.039909	0.035416	0.083896	0.114076
$u^*(5)$	0.017204	0.014000	−0.001317	−0.009621	−0.019786	−0.032067
$x(5)$	−0.069729	−0.065485	−0.032442	0.031272	0.067137	0.090007
$u^*(6)$	0.007410	0.005394	−0.002676	−0.010923	−0.012836	−0.020756
$x(6)$	−0.052029	−0.047215	−0.023890	0.026305	0.053875	0.069773
$u^*(7)$	0.003291	0.002394	0.001081	−0.005021	−0.009098	−0.013228
$x(7)$	−0.040252	−0.031524	−0.018764	0.021334	0.042184	0.055299
$u^*(8)$	−0.004590	−0.004989	−0.006749	−0.009412	−0.010224	−0.010783
$x(8)$	−0.026844	−0.024069	−0.011800	0.018692	0.034371	0.045162
$u^*(9)$	−0.006487	−0.006937	−0.007993	−0.008140	−0.007547	−0.007113
$x(9)$	−0.021046	−0.016910	−0.007208	0.014432	0.027833	0.037639
$u^*(10)$	0.000000	0.000000	0.000000	0.000000	0.000000	0.000000
$x(10)$	−0.014109	−0.012527	−0.005027	0.013130	0.024977	0.031152

Definition 4.1 The uncertain LQ problem (4.14) is called well posed if

$$V(\boldsymbol{x}_0) = \inf_{\substack{u_k \\ 0 \le k \le N-1}} J(\boldsymbol{x}_0, \boldsymbol{u}) > -\infty, \forall\, \boldsymbol{x}_0 \in R^n.$$

Definition 4.2 A well-posed problem is called solvable, if for $\boldsymbol{x}_0 \in R^n$, there is a control sequence $(\boldsymbol{u}_0^*, \boldsymbol{u}_1^*, \cdots, \boldsymbol{u}_{N-1}^*)$ that achieves $V(\boldsymbol{x}_0)$. In this case, the control sequence $(\boldsymbol{u}_0^*, \boldsymbol{u}_1^*, \cdots, \boldsymbol{u}_{N-1}^*)$ is called an optimal control sequence.

4.5.2 An Equivalent Deterministic Optimal Control

We transform the uncertain LQ problem (4.14) into an equivalent deterministic optimal control problem.

Let $X_k = E[\boldsymbol{x}_k \boldsymbol{x}_k^\tau]$. Since state $\boldsymbol{x}_k \in R^n$, we know that $\boldsymbol{x}_k \boldsymbol{x}_k^\tau$ is a $n \times n$ matrix which elements are uncertain variables, and X_k is a symmetric crisp matrix

Table 4.8 Optimal controls for some initial states based on the data of Tables 4.4, 4.5, and 4.6

x_0	−0.435	−0.365	−0.126	0.009	0.275	0.488
$J(x_0, 0)$	0.096807	0.050480	0.001238	0.000419	0.017781	0.149269
$u^*(0)$	0.909000	0.648600	0.085840	−0.037600	−0.374000	−0.994720
$u^*(1)$	0.307629	0.237337	0.028540	−0.017090	−0.143039	−0.455197
$x(1)$	−0.259033	−0.233604	−0.091468	0.073578	0.187984	0.306041
$u^*(2)$	0.128722	0.126011	0.014404	−0.010409	−0.117354	−0.157808
$x(2)$	−0.174007	−0.164326	−0.067509	0.058433	0.142815	0.205926
$u^*(3)$	0.083711	0.073983	0.003916	−0.005637	−0.048038	−0.149090
$x(3)$	−0.124250	−0.118594	−0.048956	0.046977	0.106051	0.155688
$u^*(4)$	0.028603	0.026190	0.004585	0.000000	−0.025534	−0.063784
$x(4)$	−0.090363	−0.086053	−0.038212	0.037090	0.081918	0.111219
$u^*(5)$	0.016435	0.013044	0.001196	−0.001333	−0.013808	−0.029773
$x(5)$	−0.068993	−0.064532	−0.029904	0.033328	0.065537	0.086544
$u^*(6)$	0.004906	0.003723	0.001731	−0.002296	−0.006005	−0.015014
$x(6)$	−0.051510	−0.046538	−0.021633	0.028695	0.053133	0.067209
$u^*(7)$	0.003205	0.002491	0.001325	−0.001922	−0.003376	−0.006259
$x(7)$	−0.040062	−0.031133	−0.016562	0.024023	0.042205	0.053765
$u^*(8)$	0.002136	0.001900	0.000801	−0.000845	−0.001396	−0.001782
$x(8)$	−0.026700	−0.023748	−0.010016	0.021122	0.034903	0.044562
$u^*(9)$	0.000000	0.000000	0.000000	0.000000	0.000000	0.000000
$x(9)$	−0.020326	−0.016033	−0.005101	0.017147	0.029053	0.037969
$u^*(10)$	0.000000	0.000000	0.000000	0.000000	0.000000	0.000000
$x(10)$	−0.012949	−0.011201	−0.002623	0.016035	0.026632	0.032056

Table 4.9 Absolute difference of the optimal values obtained by two methods

Initial state x_0	−0.435	−0.365	−0.126	0.009	0.275	0.488
Optimal value $J(x_0, 0)$ in Table 4.7	0.095564	0.050303	0.000883	0.000292	0.017896	0.148606
Optimal value $J(x_0, 0)$ in Table 4.8	0.096807	0.050480	0.001238	0.000419	0.017781	0.149269
Absolute difference	0.001243	0.000177	0.000355	0.000127	0.000115	0.000663

($k = 0, 1, \ldots, N$). Denote $\boldsymbol{K} = (K_0, K_1, \ldots, K_{N-1})$, where K_i are matrices for $i = 0, 1, \ldots, N-1$.

Theorem 4.2 ([1]) *If the uncertain LQ problem* (4.14) *is solvable by a feedback control sequence*

$$\boldsymbol{u}_k = K_k \boldsymbol{x}_k \textit{ for } k = 0, 1, \ldots, N-1,$$

where $K_0, K_1, \ldots, K_{N-1}$ are constant crisp matrices, then the uncertain LQ problem (4.14) *is equivalent to the following deterministic optimal control problem*

$$\begin{cases} \min\limits_{\substack{K_k \\ 0\le k\le N-1}} J(X_0, \boldsymbol{K}) = \sum\limits_{k=0}^{N-1} tr\left[(Q_k + K_k^\tau R_k K_k)X_k\right] + tr\left[Q_N X_N\right] \\ \text{subject to} \\ \quad X_{k+1} = (1+\frac{1}{3}\lambda_k^2)(A_k X_k A_k^\tau + A_k X_k K_k^\tau B_k^\tau + B_k K_k X_k A_k^\tau \\ \qquad + B_k K_k X_k K_k^\tau B_k^\tau), \ \ k = 0, 1, \ldots, N-1, \\ \quad X_0 = \boldsymbol{x}_0 \boldsymbol{x}_0^\tau, \\ \quad F X_N F^\tau = G, \ G = E[\eta\eta^\tau]. \end{cases} \tag{4.15}$$

Proof Assume that the uncertain LQ problem (4.14) is solvable by a feedback control sequence

$$\boldsymbol{u}_k = K_k \boldsymbol{x}_k \text{ for } k = 0, 1, \ldots, N-1.$$

Considering the dynamical equation of the uncertain LQ problem (4.14), we have

$$\begin{aligned} X_{k+1} &= E[\boldsymbol{x}_{k+1}\boldsymbol{x}_{k+1}^\tau] \\ &= E[(A_k + B_k K_k + \lambda_k(A_k + B_k K_k)\xi_k)\boldsymbol{x}_k \boldsymbol{x}_k^\tau (A_k^\tau + K_k^\tau B_k^\tau \\ &\quad + \lambda_k(A_k^\tau + K_k^\tau B_k^\tau)\xi_k)] \\ &= A_k X_k A_k^\tau + A_k X_k K_k^\tau B_k^\tau + B_k K_k X_k A_k^\tau + B_k K_k X_k K_k^\tau B_k^\tau \\ &\quad + E[S_k \xi_k + V_k \xi_k^2], \end{aligned} \tag{4.16}$$

where

$$S_k = 2\lambda_k(A_k X_k A_k^\tau + A_k X_k K_k^\tau B_k^\tau + B_k K_k X_k A_k^\tau + B_k K_k X_k K_k^\tau B_k^\tau),$$

$$V_k = \lambda_k^2(A_k X_k A_k^\tau + A_k X_k K_k^\tau B_k^\tau + B_k K_k X_k A_k^\tau + B_k K_k X_k K_k^\tau B_k^\tau).$$

It is easily found that $\lambda_k S_k = 2V_k$. Now, we compute $E[S_k\xi_k + V_k\xi_k^2]$ as follows.

(i) If $V_k = \boldsymbol{0}$, we obtain

$$E[S_k\xi_k + V_k\xi_k^2] = E[S_k\xi_k] = S_k E[\xi_k] = \boldsymbol{0}.$$

(ii) If $V_k \neq \boldsymbol{0}$, we know that $\lambda_k \neq 0$ and $|\frac{2}{\lambda_k}| \geq 2$. According to Example 1.6, we have

$$E\left[S_k\xi_k + V_k\xi_k^2\right] = E\left[\frac{2}{\lambda_k}V_k\xi_k + V_k\xi_k^2\right] = V_k E\left[\frac{2}{\lambda_k}\xi_k + \xi_k^2\right] = \frac{1}{3}V_k.$$

Based on the above analysis, we conclude that

$$E\left[S_k\xi_k + V_k\xi_k^2\right] = \frac{1}{3}V_k. \tag{4.17}$$

Substituting (4.17) into (4.16), we know that (4.16) can be written as

$$X_{k+1} = (1+\frac{1}{3}\lambda_k^2)(A_kX_kA_k^\tau + A_kX_kK_k^\tau B_k^\tau + B_kK_kX_kA_k^\tau + B_kK_kX_kK_k^\tau B_k^\tau). \tag{4.18}$$

Moreover, the associated cost function is expressed equivalently as

$$\min_{\substack{K_k \\ 0\le k\le N-1}} J(X_0, \boldsymbol{K}) = \min_{\substack{K_k \\ 0\le k\le N-1}} \sum_{k=0}^{N-1} tr\left[(Q_k + K_k^\tau R_k K_k)X_k\right] + tr\left[Q_N X_N\right].$$

Note that

$$F\boldsymbol{x}_N\boldsymbol{x}_N^\tau F^\tau = \boldsymbol{\eta}\boldsymbol{\eta}^\tau. \tag{4.19}$$

Taking expectations in (4.19), we have

$$FX_NF^\tau = G,\ G = E[\boldsymbol{\eta}\boldsymbol{\eta}^\tau].$$

Therefore, the uncertain LQ problem (4.14) is equivalent to the deterministic optimal control problem (4.15).

Remark 4.3 Obviously, if the uncertain LQ problem (4.14) has a linear feedback optimal control solution $\boldsymbol{u}_k^* = K_k^*\boldsymbol{x}_k$ $(k = 0, 1, \ldots, N-1)$, then K_k^* $(k = 0, 1, \ldots, N-1)$ is the optimal solution of the deterministic LQ problem (4.15).

4.5.3 A Necessary Condition for State Feedback Control

We apply the deterministic matrix minimum principle [2] to get a necessary condition for the optimal linear state feedback control with deterministic gains to the uncertain LQ optimal control problem (4.14).

Theorem 4.3 ([1]) *If the uncertain LQ problem* (4.14) *is solvable by a feedback control*

$$\boldsymbol{u}_k = K_k\boldsymbol{x}_k \tag{4.20}$$

for $k = 0, 1, \ldots, N-1$, *where* $K_0, K_1, \ldots, K_{N-1}$ *are constant crisp matrices, then there exist symmetric matrices* H_k, *and a matrix* $\boldsymbol{\rho} \in R^{r\times r}$ *solving the following constrained difference equation*

$$\begin{cases} H_k = Q_k + (1 + \frac{1}{3}\lambda_k^2)A_k^\tau H_{k+1}A_k - M_k^\tau L_k^+ M_k \\ L_k L_k^+ M_k - M_k = 0, \ \textit{and} \ L_k \geq 0 \\ L_k = R_k + (1 + \frac{1}{3}\lambda_k^2)B_k^\tau H_{k+1}B_k \\ M_k = (1 + \frac{1}{3}\lambda_k^2)B_k^\tau H_{k+1}A_k \\ H_N = Q_N + F^\tau \rho F \end{cases} \tag{4.21}$$

for $k = 0, 1, \ldots, N-1$. *Moreover*

$$K_k = -L_k^+ M_k + Y_k - L_k^+ L_k Y_k \tag{4.22}$$

with $Y_k \in R^{m\times n}$, $k = 0, 1, \ldots, N-1$, *being any given crisp matrices.*

Proof Assume the uncertain LQ problem (4.14) is solvable by

$$\boldsymbol{u}_k = K_k \boldsymbol{x}_k \text{ for } k = 0, 1, \ldots, N-1,$$

where the matrices $K_0, \ldots, K_{N-1}$ are viewed as the control to be determined. It is obvious that problem (4.15) is a matrix dynamical optimization problem. Next, we will deal with this class of problems by minimum principle. Introducc thc Lagrangian function associated with problem (4.15) as follows

$$\mathscr{L} = J(X_0, \boldsymbol{K}) + \sum_{k=0}^{N-1} tr[H_{k+1}g_{k+1}(X_k, K_k)] + tr[\rho g(X_N)],$$

where

$$\begin{cases} J(X_0, \boldsymbol{K}) = \sum_{k=0}^{N-1} tr\left[(Q_k + K_k^\tau R_k K_k)X_k\right] + tr\left[Q_N X_N\right], \\ g_{k+1}(X_k, K_k) = (1 + \frac{1}{3}\lambda_k^2)(A_k X_k A_k^\tau + A_k X_k K_k^\tau B_k^\tau + B_k K_k X_k A_k^\tau \\ \qquad + B_k K_k X_k K_k^\tau B_k^\tau) - X_{k+1} \\ g(X_N) = F X_N F^\tau - G, \end{cases}$$

and the matrices $H_0, \ldots, H_{k+1}$ as well as $\rho \in R^{r\times r}$ are the Lagrangian multipliers.

By the matrix minimum principle [2], the optimal feedback gains and Lagrangian multipliers satisfy the following first-order necessary conditions

$$\frac{\partial \mathscr{L}}{\partial K_k} = 0 \; (k = 0, 1, \ldots, N-1), \tag{4.23}$$

$$H_k = \frac{\partial \mathscr{L}}{\partial X_k} \; (k = 0, 1, \ldots, N). \tag{4.24}$$

Based on the partial rule of gradient matrices, (4.23) can be transformed into

$$[R_k + (1 + \frac{1}{3}\lambda_k^2) B_k^\tau H_{k+1} B_k] K_k + (1 + \frac{1}{3}\lambda_k^2) B_k^\tau H_{k+1} A_k = 0. \tag{4.25}$$

Let

$$\begin{cases} L_k = R_k + (1 + \frac{1}{3}\lambda_k^2) B_k^\tau H_{k+1} B_k \\ M_k = (1 + \frac{1}{3}\lambda_k^2) B_k^\tau H_{k+1} A_k. \end{cases} \tag{4.26}$$

Then, (4.25) can be rewritten as $L_k K_k + M_k = 0$. The solution of (4.25) is given by

$$K_k = -L_k^+ M_k + Y_k - L_k^+ L_k Y_k, \; Y_k \in R^{m \times n}. \tag{4.27}$$

if and only if $L_k L_k^+ M_k = M_k$, where L_k^+ is the Moor–Penrose inverse of the matrix L_k. By (4.24), first we have

$$H_N = \frac{\partial \mathscr{L}}{\partial X_N}, \tag{4.28}$$

that is

$$H_N = Q_N + F^\tau \rho F.$$

Second, we have

$$H_k = \frac{\partial \mathscr{L}}{\partial X_k} \; (k = 0, 1, \ldots, N-1),$$

which is

$$\begin{aligned} H_k = Q_k &+ (1 + \frac{1}{3}\lambda_k^2) A_k^\tau H_{k+1} A_k + K_k^\tau [R_k + (1 + \frac{1}{3}\lambda_k^2) B_k^\tau H_{k+1} B_k] K_k \\ &+ (1 + \frac{1}{3}\lambda_k^2) A_k^\tau H_{k+1} B_k K_k + (1 + \frac{1}{3}\lambda_k^2) K_k^\tau B_k^\tau H_{k+1} A_k. \end{aligned} \tag{4.29}$$

Substituting (4.27) into (4.29) gets

$$H_k = Q_k + (1 + \frac{1}{3}\lambda_k^2) A_k^\tau H_{k+1} A_k - M_k^\tau L_k^+ M_k. \tag{4.30}$$

The objective function is

$$\begin{aligned}
& J(\boldsymbol{x}_0, \boldsymbol{u}) \\
&= \sum_{k=0}^{N-1} E\left[\boldsymbol{x}_k^\tau Q_k \boldsymbol{x}_k + \boldsymbol{u}_k^\tau R_k \boldsymbol{u}_k\right] + E\left[\boldsymbol{x}_N^\tau Q_N \boldsymbol{x}_N\right] \\
&= \sum_{k=0}^{N-1} \left\{E\left[\boldsymbol{x}_k^\tau Q_k \boldsymbol{x}_k + \boldsymbol{u}_k^\tau R_k \boldsymbol{u}_k\right] + E\left[\boldsymbol{x}_{k+1}^\tau H_{k+1} \boldsymbol{x}_{k+1}\right] - E\left[\boldsymbol{x}_k^\tau H_k \boldsymbol{x}_k\right]\right\} \\
& +E\left[\boldsymbol{x}_N^\tau Q_N \boldsymbol{x}_N\right] - E\left[\boldsymbol{x}_N^\tau H_N \boldsymbol{x}_N\right] + \boldsymbol{x}_0^\tau H_0 \boldsymbol{x}_0 \\
&= \sum_{k=0}^{N-1} \left\{tr\left[(Q_k + K_k^\tau R_k K_k) X_k\right] + tr\left[H_{k+1} X_{k+1}\right] - tr\left[H_k X_k\right]\right\} \\
& +tr\left[(Q_N - H_N) X_N\right] + \boldsymbol{x}_0^\tau H_0 \boldsymbol{x}_0.
\end{aligned} \tag{4.31}$$

Substituting (4.18) into (4.31), we can rewrite the cost function as follows

$$\begin{aligned}
& J(X_0, \boldsymbol{K}) \\
&= \sum_{k=0}^{N-1} \Big\{ tr\,[\,(Q_k + K_k^\tau R_k K_k) + (1 + \frac{1}{3}\lambda_k^2)(A_k^\tau H_{k+1} A_k + B_k^\tau H_{k+1} A_k K_k \\
& + A_k^\tau H_{k+1} B_k K_k + K_k^\tau B_k^\tau H_{k+1} B_k K_k) - H_k\,]\,X_k\,\} + tr\,[(Q_N - H_N) X_N] \\
& + \boldsymbol{x}_0^\tau H_0 \boldsymbol{x}_0 \\
&= \sum_{k=0}^{N-1} tr\,\{ \left[Q_k + (1 + \frac{1}{3}\lambda_k^2) A_k^\tau H_{k+1} A_k - H_k \right] + 2(1 + \frac{1}{3}\lambda_k^2) B_k^\tau H_{k+1} A_k K_k \\
& + K_k^\tau \left[R_k + (1 + \frac{1}{3}\lambda_k^2) B_k^\tau H_{k+1} B_k \right] K_k\,\}\,X_k + tr\,[(Q_N - H_N) X_N] \\
& + \boldsymbol{x}_0^\tau H_0 \boldsymbol{x}_0.
\end{aligned} \tag{4.32}$$

Substituting (4.26) and (4.30) into (4.32), a completion of square implies

$$\begin{aligned}
J(X_0, \boldsymbol{K}) = & \sum_{k=0}^{N-1} tr\left[(K_k + L_k^+ M_k)^\tau L_k (K_k + L_k^+ M_k) X_k\right] \\
& + tr\,[(Q_N - H_N) X_N] + \boldsymbol{x}_0^\tau H_0 \boldsymbol{x}_0.
\end{aligned} \tag{4.33}$$

Next, we will prove that L_k $(k = 0, 1, \ldots, N-1)$ satisfies

$$L_k = R_k + (1 + \frac{1}{3}\lambda_k^2) B_k^\tau H_{k+1} B_k \geq 0. \tag{4.34}$$

If it is not so, there is a L_p for $p \in \{0, 1, \ldots, N-1\}$ with a negative eigenvalue λ. Denote the unitary eigenvector with respect to λ as $\boldsymbol{v}_\lambda$ (i.e., $\boldsymbol{v}_\lambda^\tau \boldsymbol{v}_\lambda = 1$ and $L_p \boldsymbol{v}_\lambda = \lambda \boldsymbol{v}_\lambda$). Let $\delta \neq 0$ be an arbitrary scalar. We construct a control sequence $\tilde{\boldsymbol{u}} = (\tilde{\boldsymbol{u}}_1, \tilde{\boldsymbol{u}}_2, \cdots, \tilde{\boldsymbol{u}}_{N-1})$ as follows

$$\tilde{\boldsymbol{u}}_k = \begin{cases} -L_k^+ M_k \boldsymbol{x}_k, & k \neq p \\ \delta |\lambda|^{-\frac{1}{2}} \boldsymbol{v}_\lambda - L_k^+ M_k \boldsymbol{x}_k, & k = p. \end{cases} \tag{4.35}$$

By (4.33), the associated cost function becomes

$$\begin{aligned}
& J(\boldsymbol{x}_0, \tilde{\boldsymbol{u}}) \\
&= \sum_{k=0}^{N-1} tr\left[(\tilde{K}_k + L_k^+ M_k)^\tau L_k (\tilde{K}_k + L_k^+ M_k) X_k \right] + tr\left[(Q_N - H_N) X_N \right] + \boldsymbol{x}_0^\tau H_0 \boldsymbol{x}_0 \\
&= \sum_{k=0}^{N-1} E\left[(\tilde{\boldsymbol{u}}_k + L_k^+ M_k \boldsymbol{x}_k)^\tau L_k (\tilde{\boldsymbol{u}}_k + L_k^+ M_k \boldsymbol{x}_k) \right] + tr\left[(Q_N - H_N) X_N \right] + \boldsymbol{x}_0^\tau H_0 \boldsymbol{x}_0 \\
&= \left[\frac{\delta}{|\lambda|^{\frac{1}{2}}} \boldsymbol{v}_\lambda \right]^\tau L_p \left[\frac{\delta}{|\lambda|^{\frac{1}{2}}} \boldsymbol{v}_\lambda \right] + tr\left[(Q_N - H_N) X_N \right] + \boldsymbol{x}_0^\tau H_0 \boldsymbol{x}_0 \\
&= -\delta^2 + tr\left[(Q_N - H_N) X_N \right] + \boldsymbol{x}_0^\tau H_0 \boldsymbol{x}_0.
\end{aligned}$$

Letting $\delta \to \infty$, it yields $J(\boldsymbol{x}_0, \tilde{\boldsymbol{u}}) \to -\infty$, which contradicts the solvability of the uncertain LQ problem (4.14).

4.5.4 Well Posedness of the Uncertain LQ Problem

Next, we will show that the solvability of Eq. (4.21) is sufficient for the well posedness of the uncertain LQ problem (4.14). Moreover, any optimal control can be obtained via the solution to Eq. (4.21).

Theorem 4.4 ([1]) *The uncertain LQ problem* (4.14) *is well posed if there exist symmetric matrices H_k solving the constrained difference Eq.* (4.21). *Moreover, the uncertain LQ problem* (4.14) *is solvable by*

$$\boldsymbol{u}_k = -[R_k + (1 + \frac{1}{3}\lambda_k^2) B_k^\tau H_{k+1} B_k]^+ [(1 + \frac{1}{3}\lambda_k^2) B_k^\tau H_{k+1} A_k] \boldsymbol{x}_k, \tag{4.36}$$

for $k = 0, 1, \ldots, N-1$. Furthermore, the optimal cost of the uncertain LQ problem (4.14) *is*

$$V(\boldsymbol{x}_0) = \boldsymbol{x}_0^\tau H_0 \boldsymbol{x}_0 - tr(\rho G).$$

Proof Let H_k solve Eq. (4.21). Then, we have

$$
\begin{aligned}
&J(\boldsymbol{x}_0, \boldsymbol{u})\\
&= \sum_{k=0}^{N-1} E\left[\boldsymbol{x}_k^\tau Q_k \boldsymbol{x}_k + \boldsymbol{u}_k^\tau R_k \boldsymbol{u}_k\right] + E\left[\boldsymbol{x}_N^\tau Q_N \boldsymbol{x}_N\right]\\
&= \sum_{k=0}^{N-1} \left\{E\left[\boldsymbol{x}_k^\tau Q_k \boldsymbol{x}_k + \boldsymbol{u}_k^\tau R_k \boldsymbol{u}_k\right] + E\left[\boldsymbol{x}_{k+1}^\tau H_{k+1} \boldsymbol{x}_{k+1}\right] - E\left[\boldsymbol{x}_k^\tau H_k \boldsymbol{x}_k\right]\right\}\\
&\quad + E\left[\boldsymbol{x}_N^\tau Q_N \boldsymbol{x}_N\right] - E\left[\boldsymbol{x}_N^\tau H_N \boldsymbol{x}_N\right] + \boldsymbol{x}_0^\tau H_0 \boldsymbol{x}_0\\
&= \sum_{k=0}^{N-1} \left\{tr\left[(Q_k + K_k^\tau R_k K_k) X_k\right] + tr\left[H_{k+1} X_{k+1}\right] - tr\left[H_k X_k\right]\right\}\\
&\quad + tr\left[(Q_N - H_N) X_N\right] + \boldsymbol{x}_0^\tau H_0 \boldsymbol{x}_0\\
&= \sum_{k=0}^{N-1} tr\left\{\left[Q_k + (1 + \frac{1}{3}\lambda_k^2) A_k^\tau H_{k+1} A_k - H_k\right] + 2(1 + \frac{1}{3}\lambda_k^2) B_k^\tau H_{k+1} A_k K_k\right.\\
&\quad \left. + K_k^\tau \left[R_k + (1 + \frac{1}{3}\lambda_k^2) B_k^\tau H_{k+1} B_k\right] K_k\right\} X_k + tr\left[(Q_N - H_N) X_N\right] + \boldsymbol{x}_0^\tau H_0 \boldsymbol{x}_0\\
&= \sum_{k=0}^{N-1} tr\left[M_k^\tau L_k^+ M_k + 2 M_k K_k + K_k^\tau L_k K_k\right] X_k + tr\left[(Q_N - H_N) X_N\right] + \boldsymbol{x}_0^\tau H_0 \boldsymbol{x}_0
\end{aligned}
$$

A completion of square implies

$$
\begin{aligned}
J(X_0, \boldsymbol{K}) = &\sum_{k=0}^{N-1} tr\left[(K_k + L_k^+ M_k)^\tau L_k (K_k + L_k^+ M_k) X_k\right]\\
&+ tr\left[(Q_N - H_N) X_N\right] + \boldsymbol{x}_0^\tau H_0 \boldsymbol{x}_0.
\end{aligned} \tag{4.37}
$$

Because of $L_k \geq 0$, we obtain that the cost function of problem (4.14) is bounded from below by

$$
V(\boldsymbol{x}_0) \geq tr\left[(Q_N - H_N) X_N\right] + \boldsymbol{x}_0^\tau H_0 \boldsymbol{x}_0 > -\infty, \ \forall\, \boldsymbol{x}_0 \in R^n.
$$

Hence, the uncertain LQ problem (4.14) is well posed. It is clear that it is solvable by the feedback control

$$
\boldsymbol{u}_k = -K_k \boldsymbol{x}_k = -L_k^+ M_k \boldsymbol{x}_k, \ k = 0, 1, \ldots, N-1.
$$

Furthermore, (4.37) indicates that the optimal value equals

$$
V(\boldsymbol{x}_0) = tr\left[(Q_N - H_N) X_N\right] + \boldsymbol{x}_0^\tau H_0 \boldsymbol{x}_0.
$$

Since

$$\begin{cases} H_N = Q_N + F^\tau \rho F \\ F X_N F^\tau = G, \end{cases}$$

and

$$X_N = E[\boldsymbol{x}_N \boldsymbol{x}_N^\tau],$$

we obtain

$$V(\boldsymbol{x}_0) = \boldsymbol{x}_0^\tau H_0 \boldsymbol{x}_0 - tr(\rho G).$$

Remark 4.4 We have shown that the solvability of the constrained difference Eq. (4.21) is sufficient for the existence of an optimal linear state feedback control.

As a special case, we consider the following indefinite LQ optimal control without constraint for the discrete-time uncertain systems.

$$\begin{cases} \inf_{\substack{\boldsymbol{u}_k \\ 0\le k\le N-1}} J(\boldsymbol{x}_0, \boldsymbol{u}) = \sum_{k=0}^{N-1} E\left[\boldsymbol{x}_k^\tau Q_k \boldsymbol{x}_k + \boldsymbol{u}_k^\tau R_k \boldsymbol{u}_k\right] + E\left[\boldsymbol{x}_N^\tau Q_N \boldsymbol{x}_N\right] \\ \text{subject to} \\ \boldsymbol{x}_{k+1} = A_k \boldsymbol{x}_k + B_k \boldsymbol{u}_k + \lambda_k (A_k \boldsymbol{x}_k + B_k \boldsymbol{u}_k)\xi_k, \ k = 0, 1, \ldots, N-1. \end{cases} \tag{4.38}$$

Corollary 4.1 *If the uncertain LQ problem* (4.38) *is solvable by a feedback control*

$$\boldsymbol{u}_k = K_k \boldsymbol{x}_k, \ \ for\ k = 0, 1, \ldots, N-1, \tag{4.39}$$

where $K_0, K_1, \ldots, K_{N-1}$ are constant crisp matrices, then there exist symmetric matrices H_k that solve the following constrained difference equation

$$\begin{cases} H_k = Q_k + (1 + \frac{1}{3}\lambda_k^2) A_k^\tau H_{k+1} A_k - M_k^\tau L_k^+ M_k \\ L_k L_k^+ M_k - M_k = 0, \ and\ L_k \ge 0 \\ L_k = R_k + (1 + \frac{1}{3}\lambda_k^2) B_k^\tau H_{k+1} B_k \\ M_k = (1 + \frac{1}{3}\lambda_k^2) B_k^\tau H_{k+1} A_k \\ H_N = Q_N \end{cases} \tag{4.40}$$

for $k = 0, 1, \ldots, N-1$. Moreover

$$K_k = -L_k^+ M_k + Y_k - L_k^+ L_k Y_k \tag{4.41}$$

with $Y_k \in R^{m\times n}$, $k = 0, 1, \ldots, N-1$, being any given crisp matrices. Furthermore, the uncertain LQ problem (4.14) *is solvable by*

$$u_k = -[R_k + (1+\frac{1}{3}\lambda_k^2)B_k^\tau H_{k+1}B_k]^+[(1+\frac{1}{3}\lambda_k^2)B_k^\tau H_{k+1}A_k]x_k,\ k = 0, 1, \ldots, N-1,$$

the optimal cost of the uncertain LQ problem (4.38) *is given by*

$$V(x_0) = x_0^\tau H_0 x_0.$$

Proof Let $F = \mathbf{0}$ and $\eta = \mathbf{0}$ in the unconstrained uncertain LQ problem (4.14). Then, the constrained uncertain LQ problem (4.14) becomes the unconstrained uncertain LQ problem (4.38). The conclusions in the corollary directly follow by similar approach as in Theorems 4.3 and 4.4.

4.5.5 Example

Present a two-dimensional indefinite LQ optimal control with equality constraint for discrete-time uncertain systems to illustrate the effectiveness of our result. In the constrained discrete-time uncertain LQ control problem (4.14), we give out a set of specific parameters of the coefficients:

$$x_0 = \begin{pmatrix} 0 \\ 1 \end{pmatrix},\quad F = \begin{pmatrix} \frac{1}{2}, \frac{1}{2} \end{pmatrix},\quad N = 2,\quad \eta \sim \mathcal{L}(0, 5\sqrt{3}/4),$$

and

$$A_0 = \begin{pmatrix} 1\ 0 \\ 0\ 0 \end{pmatrix},\ A_1 = \begin{pmatrix} 1\ 0 \\ 1\ 0 \end{pmatrix},\ B_0 = \begin{pmatrix} 1 \\ 0 \end{pmatrix},\ B_1 = \begin{pmatrix} 1 \\ 1 \end{pmatrix},\ \lambda_0 = 0.2,\quad \lambda_1 = -0.1.$$

The state weights and the control weights are as follows

$$Q_0 = \begin{pmatrix} -1 & 0 \\ 0 & -1 \end{pmatrix},\ Q_1 = \begin{pmatrix} -1 & 0 \\ 0 & 0 \end{pmatrix},\ Q_2 = \begin{pmatrix} 0\ 0 \\ 0\ 0 \end{pmatrix},\ R_0 = -1,\quad R_1 = 4.$$

Note that in this example, the state weight Q_0 is negative definite, Q_1 is negative semidefinite, Q_2 is positive semidefinite, and the control weight R_0 is negative definite.

The constraint is given as follows

$$FX_2F^\tau = FE[x_2x_2^\tau]F^\tau = G = E[\eta^2] = \frac{25}{4}.$$

First, it follows from

$$\begin{cases} H_k = Q_k + (1 + \frac{1}{3}\lambda_k^2)A_k^\tau H_{k+1}A_k - M_k^\tau L_k^+ M_k \\ L_k L_k^+ M_k - M_k = 0 \\ L_k = R_k + (1 + \frac{1}{3}\lambda_k^2)B_k^\tau H_{k+1}B_k \geq 0 \\ M_k = (1 + \frac{1}{3}\lambda_k^2)B_k^\tau H_{k+1}A_k, \ k = 0, 1. \\ H_2 = Q_2 + F^\tau \boldsymbol{\rho} F \\ X_{k+1} = (1 + \frac{1}{3}\lambda_k^2)(A_k X_k A_k^\tau + A_k X_k K_k^\tau B_k^\tau + B_k K_k X_k A_k^\tau + B_k K_k X_k K_k^\tau B_k^\tau), \\ \quad k = 0, 1, \ X_0 = \boldsymbol{x}_0 \boldsymbol{x}_0^\tau, \\ F X_2 F^\tau = F E[\boldsymbol{x}_2 \boldsymbol{x}_2^\tau] F^\tau = G = \frac{25}{4}, \end{cases}$$

that $\rho = 8$. Then, we have

$$H_2 = Q_2 + F^\tau \rho F = \begin{pmatrix} 2 & 2 \\ 2 & 2 \end{pmatrix}.$$

Second, applying Theorem 4.3, we obtain the optimal controls and optimal cost value as follows.

For $k = 1$, we obtain

$$L_1 = R_1 + (1 + \frac{1}{3}\lambda_1^2)B_1^\tau H_2 B_1 = 12.1067, \ M_1 = (1 + \frac{1}{3}\lambda_1^2)B_1^\tau H_2 A_1 = (8.1067, 0),$$

$$H_1 = Q_1 + (1 + \frac{1}{3}\lambda_1^2)A_1^\tau H_2 A_1 - M_1^\tau L_1^+ M_1 = \begin{pmatrix} 1.6784 & -5.4283 \\ -5.4283 & -5.4283 \end{pmatrix}.$$

The optimal feedback control is $\boldsymbol{u}_1 = K_1 \boldsymbol{x}_1$ where

$$K_1 = -L_1^+ M_1 = (-0.6696, 0).$$

For $k = 0$, we obtain

$$L_0 = R_0 + (1 + \frac{1}{3}\lambda_0^2)B_0^\tau H_1 B_0 = 0.6840, \ M_0 = (1 + \frac{1}{3}\lambda_0^2)B_0^\tau H_1 A_0 = (1.6840, 0),$$

$$H_0 = Q_0 + (1 + \frac{1}{3}\lambda_0^2)A_0^\tau H_1 A_0 - M_0^\tau L_0^+ M_0 = \begin{pmatrix} -2.9620 & -4.1460 \\ -4.1460 & -3.9460 \end{pmatrix}.$$

The optimal feedback control is $\boldsymbol{u}_0 = K_0\boldsymbol{x}_0$ where

$$K_0 = -L_0^+ M_0 = (-2.4620, 0).$$

Finally, the optimal cost value is

$$V(\boldsymbol{x}_0) = \boldsymbol{x}_0^\tau H_0 \boldsymbol{x}_0 - tr(\rho G) = -53.9460.$$

References

1. Chen Y, Zhu Y (2016) Indefinite LQ optimal control with equality constraint for discrete-time uncertain systems. Jpn J Ind Appl Math 33(2):361–378
2. Athans M (1968) The matrix minimum principle. Inf Control, 11:592–606

Chapter 5
Bang–Bang Control for Uncertain Systems

If the optimal control of a problem takes the maximum value or minimum value in its admissible field, the problem is called a bang–bang control problem.

5.1 Bang–Bang Control for Continuous Uncertain Systems

Now, we consider the following problem:

$$\begin{cases} J(0, \boldsymbol{x}_0) \equiv \max\limits_{\boldsymbol{u}_s} E\left[\int_0^T f(\boldsymbol{X}_s, s)\mathrm{d}s + h(\boldsymbol{X}_T, T)\right] \\ \text{subject to} \\ \qquad \mathrm{d}\boldsymbol{X}_s = (\alpha(\boldsymbol{X}_s, s) + \beta(\boldsymbol{X}_s, s)\boldsymbol{u}_s)\mathrm{d}s + \sigma(\boldsymbol{X}_s, \boldsymbol{u}_s, s)\mathrm{d}\boldsymbol{C}_s \\ \qquad \boldsymbol{X}_0 = \boldsymbol{x}_0 \\ \qquad \boldsymbol{u}_s \in [-1, 1]^r, \end{cases} \tag{5.1}$$

where $\boldsymbol{X}_s$ is the state vector of dimension n with the initial condition that at time 0 we are in state $\boldsymbol{X}_0 = \boldsymbol{x}_0$, $\boldsymbol{u}_s$ the decision vector of dimension r in a domain $[-1, 1]^r$, $f : R^n \times [0, +\infty) \to R$ the objective function, and $h : R^n \times [0, +\infty) \to R$ the function of terminal reward. In addition, $\alpha : R^n \times [0, +\infty) \to R^n$ is a column-vector function, $\beta : R^n \times [0, +\infty) \to R^n \times R^r$ and $\sigma : R^n \times R^r \times [0, +\infty) \to R^n \times R^k$ are matrix functions, and $\boldsymbol{C}_s = (C_{s1}, C_{s2}, \cdots, C_{sk})^\tau$, where $C_{s1}, C_{s2}, \ldots, C_{sk}$ are independent canonical Liu processes. The final time $T > 0$ is fixed or free.

The model (5.1) may be suitable to the fuel and time problems when the system $\mathrm{d}\boldsymbol{X}_s = (\alpha(\boldsymbol{X}_s, s) + \beta(\boldsymbol{X}_s, s)\boldsymbol{u}_s)\mathrm{d}s$ is disturbed by an uncertain factor and then is the form of uncertain differential equation $\mathrm{d}\boldsymbol{X}_s = (\alpha(\boldsymbol{X}_s, s) + \beta(\boldsymbol{X}_s, s)\boldsymbol{u}_s)\mathrm{d}s + \sigma(\boldsymbol{X}_s, \boldsymbol{u}_s, s)\mathrm{d}\boldsymbol{C}_s$. For any $0 < t < T$, $J(t, \boldsymbol{x})$ is the expected optimal reward obtainable in $[t, T]$ with the condition that at time t we are in state $X_t = \boldsymbol{x}$.

Y. Zhu, *Uncertain Optimal Control*, Springer Uncertainty Research,
https://doi.org/10.1007/978-981-13-2134-4_5

Theorem 5.1 ([1]) *Assume that $J(t, \boldsymbol{x})$ is a twice differentiable function on $[0, T] \times R^n$. Then the optimal control of (5.1) is a bang–bang control.*

Proof It follows from the equation of optimality (2.15) that

$$-J_t(t, \boldsymbol{x}) = \max_{\boldsymbol{u}_t \in [-1,1]^r} \{f(\boldsymbol{x}, t) + (\alpha(\boldsymbol{x}, t) + \beta(\boldsymbol{x}, t)\boldsymbol{u}_t)^\tau \nabla_{\boldsymbol{x}} J(t, \boldsymbol{x})\}. \tag{5.2}$$

On the right side of (5.2), let $\boldsymbol{u}_t^*$ make it the maximum,

$$\begin{aligned}&\max_{\boldsymbol{u}_t \in [-1,1]^r} \{f(\boldsymbol{x}, t) + (\alpha(\boldsymbol{x}, t) + \beta(\boldsymbol{x}, t)\boldsymbol{u}_t)^\tau \nabla_{\boldsymbol{x}} J(t, \boldsymbol{x})\} \\ &= f(\boldsymbol{x}, t) + (\alpha(\boldsymbol{x}, t) + \beta(\boldsymbol{x}, t)\boldsymbol{u}_t^*)^\tau \nabla_{\boldsymbol{x}} J(t, \boldsymbol{x}),\end{aligned}$$

that is,

$$\max_{\boldsymbol{u}_t \in [-1,1]^r} \{\nabla_{\boldsymbol{x}} J(t, \boldsymbol{x})^\tau \beta(\boldsymbol{x}, t)\boldsymbol{u}_t\} = \nabla_{\boldsymbol{x}} J(t, \boldsymbol{x})^\tau \beta(\boldsymbol{x}, t)\boldsymbol{u}_t^*. \tag{5.3}$$

Denote

$$\boldsymbol{u}_t^* = (u_1^*(t), u_2^*(t), \cdots, u_r^*(t))^\tau,$$

and

$$\nabla_{\boldsymbol{x}} J(t, \boldsymbol{x})^\tau \beta(\boldsymbol{x}, t) = (g_1(t, \boldsymbol{x}), g_2(t, \boldsymbol{x}), \cdots, g_r(t, \boldsymbol{x})), \tag{5.4}$$

which is called a switching vector. Then,

$$u_i^*(t) = \begin{cases} 1, & \text{if } g_i(t, \boldsymbol{x}) > 0 \\ -1, & \text{if } g_i(t, \boldsymbol{x}) < 0 \\ \text{undetermined}, & \text{if } g_i(t, \boldsymbol{x}) = 0 \end{cases} \tag{5.5}$$

for $i = 1, 2, \ldots, r$, which is a bang–bang control as shown in Fig. 5.1.

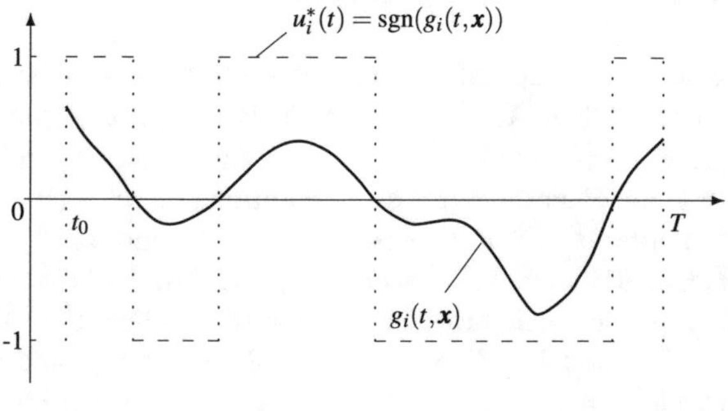

Fig. 5.1 Bang–bang control

5.1.1 An Uncertain Bang–Bang Model

Consider a special case of the model (5.1) as follows.

$$\begin{cases} J(0, \boldsymbol{x}_0) \equiv \max\limits_{\boldsymbol{u}_s} E\left[\int_0^T f(s)^\tau \boldsymbol{X}_s \mathrm{d}s + S_T^\tau \boldsymbol{X}_T\right] \\ \text{subject to} \\ \qquad \mathrm{d}\boldsymbol{X}_s = (A(s)\boldsymbol{X}_s + B(s)\boldsymbol{u}_s)\mathrm{d}s + \sigma(\boldsymbol{X}_s, \boldsymbol{u}_s, s)\mathrm{d}\boldsymbol{C}_s \\ \qquad \boldsymbol{X}_0 = \boldsymbol{x}_0 \\ \qquad \boldsymbol{u}_s \in [-1, 1]^n, \end{cases} \tag{5.6}$$

where $f : [0, +\infty) \to R^n$ and $A, B : [0, +\infty) \to R^{n\times n}$ are some twice continuously differentiable functions, and $S_T \in R^n$. Denote $B(s) = (b_{ij}(s))_{n\times n}$. We have the following conclusion.

Theorem 5.2 ([1]) *The optimal control* $\boldsymbol{u}_t^* = (u_1^*(t), u_2^*(t), \cdots, u_n^*(t))^\tau$ *of* (5.6) *is a bang–bang control*

$$u_j^*(t) = sgn\{(b_{1j}(t), b_{2j}(t), \cdots, b_{nj}(t))p(t)\}$$

for $j = 1, 2, \ldots, n$, *where* $p(t) \in R^n$ *satisfies*

$$\frac{\mathrm{d}p(t)}{\mathrm{d}t} = -f(t) - A(t)^\tau p(t), \quad p(T) = S_T. \tag{5.7}$$

The optimal value of (5.6) *is*

$$J(0, \boldsymbol{x}_0) = p(0)^\tau \boldsymbol{x}_0 + \int_0^T p(s)^\tau B(s)\boldsymbol{u}_s^* \mathrm{d}s. \tag{5.8}$$

Proof It follows from the equation of optimality (2.7) that

$$-J_t(t, \boldsymbol{x}) = \max_{\boldsymbol{u}_t\in[-1,1]^r} \{f(t)^\tau \boldsymbol{x} + (A(t)\boldsymbol{x} + B(t)\boldsymbol{u}_t)^\tau \nabla_{\boldsymbol{x}} J(t, \boldsymbol{x})\}. \tag{5.9}$$

Since $J(T, \boldsymbol{x}_T) = S_T^\tau \boldsymbol{x}_T$, we guess

$$J(t, \boldsymbol{x}) = p(t)^\tau \boldsymbol{x} + c(t) \quad \text{and} \quad p(T) = S_T, \quad c(T) = 0.$$

So

$$\nabla_{\boldsymbol{x}} J(t, \boldsymbol{x}) = p(t), \ \ J_t(t, \boldsymbol{x}) = \frac{\mathrm{d}p(t)^\tau}{\mathrm{d}t}\boldsymbol{x} + \frac{\mathrm{d}c(t)}{\mathrm{d}t}. \tag{5.10}$$

Substituting (5.10) into (5.9) that

$$-\frac{\mathrm{d}p(t)^\tau}{\mathrm{d}t}\boldsymbol{x} + \frac{\mathrm{d}c(t)}{\mathrm{d}t} = f(t)^\tau \boldsymbol{x} + (A(t)\boldsymbol{x} + B(t)\boldsymbol{u}_t^*)^\tau p(t).$$

Therefore,

$$-\frac{\mathrm{d}p(t)^\tau}{\mathrm{d}t} = f(t)^\tau + p(t)^\tau A(t), \quad p(T) = S_T,$$

and

$$\frac{\mathrm{d}c(t)}{\mathrm{d}t} = p(t)^\tau B(t)\boldsymbol{u}_t^*.$$

Thus, it follows from (5.5) that

$$u_j^*(t) = \mathrm{sgn}\{g_j(t, \boldsymbol{x})\} = \mathrm{sgn}\{p(t)^\tau (b_{1j}(t), b_{2j}(t), \cdots, b_{nj}(t))^\tau\}.$$

Furthermore,

$$J(t, \boldsymbol{x}) = p(t)^\tau \boldsymbol{x} + c(t) = p(t)^\tau \boldsymbol{x} + \int_t^T p(s)^\tau B(s)\boldsymbol{u}_s^* \mathrm{d}s.$$

The theorem is proved.

5.1.2 *Example*

Consider the following example of uncertain optimal control model

$$\begin{cases} J(0, \boldsymbol{x}_0) \equiv \max\limits_u E[2X_1(1) - X_2(1)] \\ \text{subject to} \\ \quad \mathrm{d}X_1(s) = X_2(s)\mathrm{d}s \\ \quad \mathrm{d}X_2(s) = u(s)\mathrm{d}s + \sigma \mathrm{d}C_s, \quad \sigma \in R \\ \quad \boldsymbol{X}(0) = (X_1(0), X_2(0)) = \boldsymbol{x}_0 \\ \quad |u(s)| \le 1, 0 \le s \le 1. \end{cases} \tag{5.11}$$

We have

$$A(s) = \begin{pmatrix} 0 & 1 \\ 0 & 0 \end{pmatrix}, \quad B(s) = \begin{pmatrix} 0 & 0 \\ 0 & 1 \end{pmatrix}, \quad S_1 = \begin{pmatrix} 2 \\ -1 \end{pmatrix}.$$

It follows from (5.7) that

$$\frac{\mathrm{d}p(t)}{\mathrm{d}t} = -\begin{pmatrix} 0 & 0 \\ 1 & 0 \end{pmatrix} p(t), \quad p(1) = \begin{pmatrix} 2 \\ -1 \end{pmatrix},$$

which has the solution

$$p(t) = \begin{pmatrix} 2 \\ -2t + 1 \end{pmatrix}.$$

The switching vector is

$$\nabla_{\boldsymbol{x}} J(t, \boldsymbol{x})^\tau \beta(\boldsymbol{x}, t) = p(t)^\tau B(t) = (0, -2t + 1).$$

So we have the switching function

$$g(t, \boldsymbol{x}) = -2t + 1.$$

Hence,

$$u^*(t) = \mathrm{sgn}\{-2t + 1\} = \begin{cases} 1, & \text{if } 0 \le t < \frac{1}{2} \\ -1, & \text{if } \frac{1}{2} < t \le 1. \end{cases}$$

We can find out the switching time at 0.5 as shown in Fig. 5.2.

Next, we will find out the optimal trajectory $(X_1(t), X_2(t))^\tau$. Denote $\boldsymbol{x}_0 = (x_{01}, x_{02})^\tau$. It follows from $\mathrm{d}X_2(s) = u^*(s)\mathrm{d}s + \sigma \mathrm{d}C_s$ that

$$X_2(t) = x_{02} + t \wedge (1 - t) + \sigma C_t.$$

It follows from $\mathrm{d}X_1(s) = X_2(s)\mathrm{d}s$ that

$$\begin{aligned} X_1(t) &= x_{01} + \int_0^t X_2(s)\mathrm{d}s \\ &= x_{01} + x_{02}t + \int_0^t s \wedge (1 - s)\mathrm{d}s + \sigma \int_0^t C_s \mathrm{d}s \\ &= \begin{cases} x_{01} + x_{02}t + \frac{1}{2}t^2 + \sigma t C_t + \sigma \int_0^t s\mathrm{d}C_s, & \text{if } 0 \le t < \frac{1}{2} \\ x_{01} + x_{02}t - \frac{1}{2}t^2 + t - \frac{1}{4} + \sigma t C_t + \sigma \int_0^t s\mathrm{d}C_s, & \text{if } \frac{1}{2} < t \le 1 \end{cases} \\ &= \begin{cases} x_{01} + x_{02}t + \frac{1}{2}t^2 + \sigma \xi(t), & \text{if } 0 \le t < \frac{1}{2} \\ x_{01} + x_{02}t - \frac{1}{2}t^2 + t - \frac{1}{4} + \sigma \xi(t), & \text{if } \frac{1}{2} < t \le 1, \end{cases} \end{aligned}$$

Fig. 5.2 Optimal control

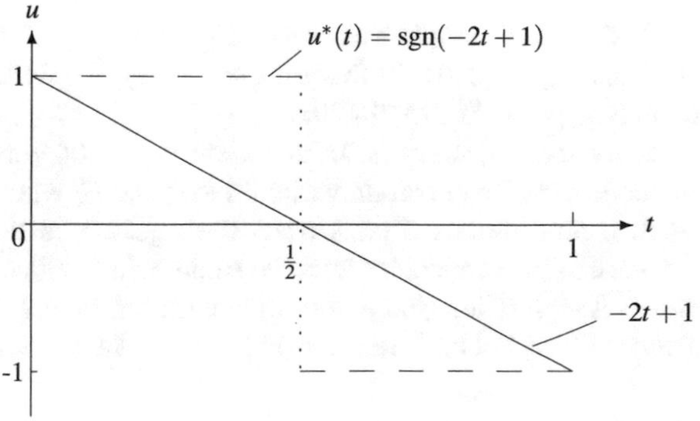

where ξ is an uncertain variable such that

$$\xi(t) = tC_t + \int_0^t s\mathrm{d}C_s \sim t\mathcal{N}(0, t) + \mathcal{N}\left(0, \int_0^t s\mathrm{d}s\right) = \mathcal{N}\left(0, \frac{3}{2}t^2\right).$$

We can see that $X_1(1) = x_{01} + x_{02} + 0.25 + \sigma\xi(1)$ and $X_2(1) = x_{02} + \sigma C_1$. Thus, $E[2X_1(1) - X_2(1)] = 2x_{01} + x_{02} + 0.5 + \sigma E[2\xi(1) - C_1] = 2x_{01} + x_{02} + 0.5$, which is coincident with the optimal value provided by (5.8). The switching point is $(X_1(0.5), X_2(0.5)) = (x_{01} + x_{02} + 0.125 + \sigma\xi(0.5), x_{02} + 0.5 + \sigma C_{0.5})$.

The trajectory of the system is an uncertain vector $(X_1(t), X_2(t))^\tau$. In practice, a realization of the system is a sample trajectory which formed by sample points. We try to provide sample point and sample trajectory as follows.

Since the distribution function of $\xi(t)$ is

$$\Phi(x) = \left(1 + \exp\left(\frac{-2\pi x}{3\sqrt{3}t^2}\right)\right)^{-1}, \quad x \in R,$$

we may get a sample point $\tilde{\xi}(t)$ of $\xi(t)$ from $\tilde{\xi}(t) = \Phi^{-1}(\text{rand}(0, 1))$ that

$$\tilde{\xi}(t) = \frac{3\sqrt{3}t^2}{-2\pi} \ln\left(\frac{1}{\text{rand}(0, 1)} - 1\right).$$

Similarly, we may get a sample point $\tilde{c}_t$ of C_t by

$$\tilde{c}_t = \frac{\sqrt{3}t}{-\pi} \ln\left(\frac{1}{\text{rand}(0, 1)} - 1\right).$$

A sample trajectory of $(X_1(t), X_2(t))^\tau$ may be given by

$$X_1(t) = \begin{cases} x_{01} + x_{02}t + \frac{1}{2}t^2 + \sigma\tilde{\xi}(t), & \text{if } 0 \le t < \frac{1}{2} \\ x_{01} + x_{02}t - \frac{1}{2}t^2 + t - \frac{1}{4} + \sigma\tilde{\xi}(t), & \text{if } \frac{1}{2} < t \le 1, \end{cases}$$
$$X_2(t) = x_{02} + t \wedge (1 - t) + \sigma\tilde{c}_t.$$

A simulated sample trajectory of $(X_1(t), X_2(t))^\tau$ is shown in Fig. 5.3 with $\sigma = 0.01$ and $\boldsymbol{x}_0 = (0, 0)$. In this sample, $X_1(1) = 0.2482$, $X_2(1) = -0.0012$, and the switching point $(0.125, 0.499)$.

Since the trajectory is an uncertain vector of dimension two, its realization is dependent on the uncertain variables $\xi(t)$ and C_t whose sample points are produced by their distributions. That is to say, the trajectory is disturbed by an uncertain vector. So, each sample trajectory is not so smooth. In practice, the system may realize many times. A realization (including sample trajectory and switching point) may different from another, but each realization has only one switching point.

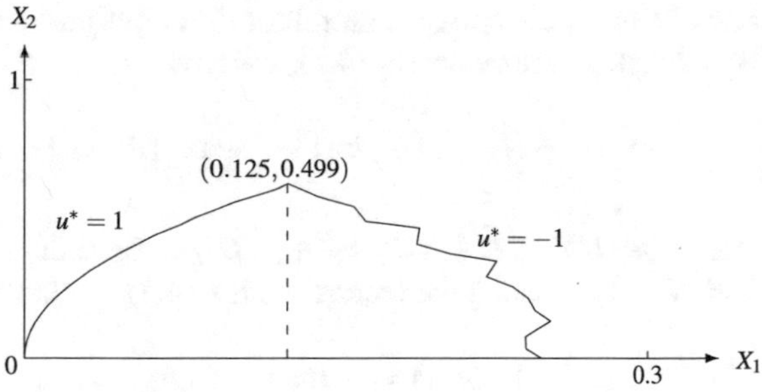

Fig. 5.3 A sample trajectory

5.2 Bang–Bang Control for Multistage Uncertain Systems

Consider the following uncertain optimal control problem with a linear objective function subject to an uncertain linear system:

$$\begin{cases} J(0, x_0) = \max\limits_{\substack{|u(i)|\le 1 \\ 0\le i\le N}} E\left[\sum_{j=0}^{N} A_j x(j)\right] \\ \text{subject to} \\ \quad x(j+1) = a_j x(j) + b_j u(j) + \sigma_{j+1} C_{j+1}, \\ \qquad \text{for } \ j = 0, 1, 2, \ldots, N-1, \\ \quad x(0) = x_0, \end{cases} \tag{5.12}$$

where $A_j > 0$, $a_j > 0$ and b_j, σ_j are constants for all j. In addition, $C_1, C_2, \ldots, C_N$ are independent uncertain variables with expected values $e_1, e_2, \ldots, e_N$, respectively.

Theorem 5.3 ([2]) *The optimal controls $u^*(k)$ of (5.12) are provided by*

$$|u^*(N)| \le 1,$$
$$u^*(k) = \begin{cases} sgn\{b_k\}, & \text{if } b_k \ne 0 \\ undetermined, & otherwise \end{cases}$$

and the optimal values are

$$J(N, x_N) = P_N x_N + Q_N,$$
$$J(k, x_k) = P_k x_k + \sum_{i=k}^{N} Q_i + \sum_{i=k+1}^{N} P_i \sigma_i e_i,$$

where

$$P_N = A_N, \quad P_k = A_k + P_{k+1} a_k, \quad Q_N = 0, \quad Q_k = P_{k+1}|b_k|,$$

for $k = N-1, N-2, \ldots, 1, 0$.

Proof Denote the optimal control for the above problem by $u^*(0), u^*(1), \ldots, u^*(N)$. By using the recurrence Eq. (4.2), we have

$$J(N, x_N) = \max_{|u(N)|\le 1} \{A_N x_N\} = A_N x_N,$$

where $|u^*(N)| \le 1$. Let $P_N = A_N,\ Q_N = 0$. Then $J(N, x_N) = P_N x_N + Q_N$. For $k = N - 1$, by using the recurrence Eq. (4.3), we have

$$\begin{aligned}
J(N-1, x_{N-1}) &= \max_{|u(N-1)|\le 1} E[A_{N-1}x_{N-1} + J(x(N), N)] \\
&= \max_{|u(N-1)|\le 1} \{A_{N-1}x_{N-1} + P_N E[x(N)] + Q_N\} \\
&= \max_{|u(N-1)|\le 1} \{A_{N-1}x_{N-1} + P_N E[a_{N-1}x_{N-1} + \\
&\qquad b_{N-1}u(N-1) + \sigma_N C_N] + Q_N\} \\
&= \max_{|u(N-1)|\le 1} \{(A_{N-1} + P_N a_{N-1})x_{N-1} + \\
&\qquad P_N b_{N-1}u(N-1) + P_N \sigma_N e_N + Q_N\}.
\end{aligned}$$

Hence,

$$P_N b_{N-1} u^*(N-1) = \max_{|u(N-1)|\le 1} P_N b_{N-1} u(N-1).$$

Therefore, we have

$$\begin{aligned}
u^*(N-1) &= \begin{cases} 1, & \text{if } b_{N-1} > 0; \\ -1, & \text{if } b_{N-1} < 0; \\ \text{undetermined}, & \text{if } b_{N-1} = 0, \end{cases} \\
&= \begin{cases} \text{sgn}\{b_{N-1}\}, & \text{if } b_{N-1} \ne 0; \\ \text{undetermined}, & \text{otherwise}. \end{cases}
\end{aligned}$$

Hence, we have

$$J(N-1, x_{N-1}) = (A_{N-1} + P_N a_{N-1})x_{N-1} + P_N b_{N-1} u^*(N-1) + P_N \sigma_N e_N + Q_N.$$

When $b_{N-1} = 0$, we have

$$J(N-1, x_{N-1}) = (A_{N-1} + P_N a_{N-1})x_{N-1} + P_N \sigma_N e_N + Q_N.$$

Denote

$$P_{N-1} = A_{N-1} + P_N a_{N-1}, \quad Q_{N-1} = 0.$$

Then

$$J(N-1, x_{N-1}) = P_{N-1}x_{N-1} + Q_{N-1} + Q_N + P_N \sigma_N e_N.$$

When $b_{N-1} > 0$, we have $u^*(N-1) = 1$, and then

$$J(N-1, x_{N-1}) = (A_{N-1} + P_N a_{N-1})x_{N-1} + P_N b_{N-1} + Q_N + P_N \sigma_N e_N.$$

Denote

$$P_{N-1} = A_{N-1} + P_N a_{N-1}, \quad Q_{N-1} = P_N b_{N-1}.$$

Then

$$J(N-1, x_{N-1}) = P_{N-1}x_{N-1} + Q_{N-1} + Q_N + P_N \sigma_N e_N.$$

When $b_{N-1} < 0$, we have $u^*(N-1) = -1$, and then

$$J(N-1, x_{N-1}) = (A_{N-1} + P_N a_{N-1})x_{N-1} - P_N b_{N-1} + Q_N + P_N \sigma_N e_N.$$

Denote

$$P_{N-1} = A_{N-1} + P_N a_{N-1}, \quad Q_{N-1} = -P_N b_{N-1}.$$

Then

$$J(N-1, x_{N-1}) = P_{N-1}x_{N-1} + Q_{N-1} + Q_N + P_N \sigma_N e_N.$$

By induction, we can obtain the conclusion of the theorem. The theorem is proved.

By Theorem 5.3, we can get the exact bang–bang optimal controls and the optimal objective values with the state of the system at all stages for a linear objective function subject to an uncertain linear system. If the system is nonlinear in control variable, we consider the following problem.

$$\begin{cases} J(0, x_0) = \max\limits_{\substack{|u(i)| \le 1 \\ 0 \le i \le N}} E\left[\sum\limits_{j=0}^{N} A_j x(j)\right] \\ \text{subject to} \\ \quad x(j+1) = a_j x(j) + b_j u(j) + d_j u^2(j) + \sigma_{j+1} C_{j+1}, \\ \qquad \text{for} \quad j = 0, 1, 2, \ldots, N-1, \\ \quad x(0) = x_0, \end{cases} \tag{5.13}$$

where $d_j < 0$ for $0 \le j \le N$, and other parameters have the same meaning as in (5.12).

Theorem 5.4 ([2]) *The optimal controls $u^*(k)$ of* (5.13) *are provided by*

$$|u^*(N)| \le 1,$$
$$u^*(k) = \begin{cases} -\frac{b_k}{2d_k}, & \text{if } 2d_k \le b_k \le -2d_k \\ -sgn\left\{\frac{b_k}{2d_k}\right\}, & \text{otherwise,} \end{cases}$$

and the optimal values are

$$J(N, x_N) = P_N x_N + Q_N, \quad J(k, x_k) = P_k x_k + \sum_{i=k}^{N} Q_i + \sum_{i=k+1}^{N} P_i \sigma_i e_i,$$

where

$$P_N = A_N, \quad P_k = A_k + P_{k+1} a_k;$$

and

$$Q_N = 0, \quad Q_k = \begin{cases} P_{k+1}(d_k + b_k), & \text{if } u^*(k) = 1 \\ P_{k+1}(d_k - b_k), & \text{if } u^*(k) = -1 \\ -\frac{P_{k+1} b_k^2}{4d_k}, & \text{if } u^*(k) = -\frac{b_k}{2d_k}, \end{cases}$$

for $k = N-1, N-2, \ldots, 1, 0.$

Proof Denote the optimal control for the above problem by $u^*(0), u^*(1), \ldots, u^*(N)$. By using the recurrence Eq. (4.2), we have

$$J(N, x_N) = \max_{|u(N)| \le 1} \{A_N x_N\} = A_N x_N,$$

where $|u^*(N)| \le 1$. Let $P_N = A_N$, $Q_N = 0$. Then $J(N, x_N) = P_N x_N + Q_N$.

For $k = N-1$, by using the recurrence Eq. (4.3), we have

$$\begin{aligned} & J(N-1, x_{N-1}) \\ = & \max_{|u(N-1)| \le 1} E[A_{N-1} x_{N-1} + J(x(N), N)] \\ = & \max_{|u(N-1)| \le 1} \{A_{N-1} x_{N-1} + P_N E[x(N)] + Q_N\} \\ = & \max_{|u(N-1)| \le 1} \{A_{N-1} x_{N-1} + P_N E[a_{N-1} x_{N-1} \\ & + b_{N-1} u(N-1) + d_{N-1} u^2(N-1) + \sigma_N C_N] + Q_N\} \\ = & (A_{N-1} + P_N a_{N-1}) x_{N-1} + P_N \sigma_N e_N + Q_N \\ & + \max_{|u(N-1)| \le 1} \{P_N b_{N-1} u(N-1) + P_N d_{N-1} u^2(N-1)\}. \end{aligned} \quad (5.14)$$

Let $H(u(N-1)) = P_N b_{N-1} u(N-1) + P_N d_{N-1} u^2(N-1)$. It follows from

$$\frac{\mathrm{d}H(u(N-1))}{\mathrm{d}u(N-1)} = P_N b_{N-1} + 2 P_N d_{N-1} u(N-1) = 0$$

that $u(N-1) = -\frac{b_{N-1}}{2d_{N-1}}$. If $|-b_{N-1}/(2d_{N-1})| \le 1$, then $u^*(N-1) = -b_{N-1}/(2d_{N-1})$ is the maximum point of $H(u(N-1))$ (its trace is as $H1$ in Fig. 5.4) because

$$\frac{\mathrm{d}^2 H(u(N-1))}{\mathrm{d}u(N-1)^2} = 2 P_N d_{N-1} < 0.$$

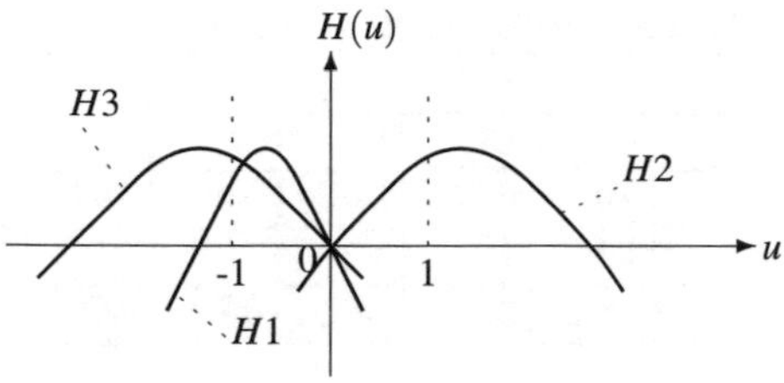

Fig. 5.4 Three types of functions $H(u)$

That is, if $2d_{N-1} \le b_{N-1} \le -2d_{N-1}$, then the optimal control at $(N-1)$th stage is $u^*(N-1) = -b_{N-1}/(2d_{N-1})$. Otherwise, since $H(u(N-1))$ (its trace is as $H2$ in Fig. 5.4) is increasing in $u(N-1) \in [-1, 1]$ if $-b_{N-1}/(2d_{N-1}) > 1$, and $H(u(N-1))$ (its trace is as $H3$ in Fig. 5.4) is decreasing in $u(N-1) \in [-1, 1]$ if $-b_{N-1}/(2d_{N-1}) < -1$, we know that the optimal control at $(N-1)$th stage is 1 if $-b_{N-1}/(2d_{N-1}) > 1$ and -1 if $-b_{N-1}/(2d_{N-1}) < -1$. Hence

$$\max_{|u(N-1)|\le 1} H(u(N-1)) = \begin{cases} -\frac{P_N b_{N-1}^2}{4d_{N-1}}, & \text{if } 2d_k \le b_k \le -2d_k \\ P_N(d_{N-1} + b_{N-1}), & \text{if } b_{N-1} > -2d_{N-1} \\ P_N(d_{N-1} - b_{N-1}), & \text{if } b_{N-1} < 2d_{N-1}. \end{cases}$$

Substituting it into (5.14) gets the result of $J(N-1, x_{N-1})$. By induction, we can obtain the conclusion of the theorem. The theorem is proved.

5.2.1 Example

Consider the following example:

$$\begin{cases} J(0, x_0) = \max\limits_{\substack{|u(i)|\le 1 \\ 0\le i\le 10}} E\left[\sum_{j=0}^{10} A_j x(j)\right] \\ \text{subject to} \\ \qquad x(j+1) = a_j x(j) + b_j u(j) + \sigma_{j+1} C_{j+1}, \\ \qquad\quad \text{for} \quad j = 0, 1, 2, \ldots, 9, \\ \qquad x(0) = x_0, \end{cases} \tag{5.15}$$

where coefficients are listed in Table 5.1. In addition, $C_1, C_2, \ldots, C_{10}$ are independent zigzag uncertain variables $(-1, 0, 1)$, and then $E[C_j] = 0$ for $j = 1, 2, \ldots, 10$.

The optimal controls and optimal values are obtained by Theorem 5.3 and listed in Table 5.2. The data in the fourth column of Table 5.2 is the corresponding states which are derived from $x(k+1) = a_k x(k) + b_k u(k) + \sigma_{k+1} c_{k+1}$ for initial stage $x(0) = 1$,

Table 5.1 Coefficients of the example

j	0	1	2	3	4	5	6	7	8	9	10
A_j	8	13	10	11	11	16	8	11	12	10	15
σ_j	0.01	0.02	0.01	0.01	0.01	0.02	0.01	0.01	0.01	0.02	0.01
a_j	1.2	1.5	1.2	1	1.2	1.3	1.4	1	1.2	1.1	1.3
b_j	−0.2	0	−0.3	0.1	−0.3	−0.2	0.2	0.2	−0.2	0.3	−0.4

Table 5.2 The optimal results

Stage	r_k	c_k	$x(k)$	$u^*(k)$	$J(k, x_k)$
0			1	−1	636.625
1	0.761345	0.522691	1.41045	$\lvert u^*(N)\rvert \le 1$	632.18
2	0.053407	−0.893185	2.10675	−1	611.897
3	0.831812	0.663625	2.83473	1	591.98
4	0.315439	−0.369121	2.93104	−1	560.198
5	0.560045	0.12009	3.81965	−1	528.26
6	0.784173	0.568346	5.17123	1	467.627
7	0.604602	0.209204	7.44182	1	426.372
8	0.064882	−0.870235	7.63312	−1	344.131
9	0.21952	−0.560961	9.34852	1	252.236
10	0.748283	0.496567	10.5883	$\lvert u^*(N)\rvert \le 1$	158.825

where c_{k+1} is the realization of uncertain variable C_{k+1}, and may be generated by $c_{k+1} = 2r_{k+1} - 1$ for a random number $r_{k+1} \in [0, 1]$ $(k = 0, 1, 2, \ldots, 9)$.

5.3 Equation of Optimality for Saddle Point Problem

A saddle point problem concerns on a situation that one control vector aims at minimizing some given objective function while the other control vector tries to maximize it. This problem often arises in the military and security fields. When launching a missile to pursue the target, we hope to minimize the distance between the missile and the target. Meanwhile, the target tries to increase the distance so that it can evade. The policemen do their best to catch the terrorists to reduce the loss while the terrorists do the opposite. This is why we need to make a study of the saddle point problem.

The research is based on an uncertain dynamic system as follows:

$$\mathrm{d}\boldsymbol{X}_s = \boldsymbol{f}(s, \boldsymbol{u}_1, \boldsymbol{u}_2, \boldsymbol{X}_s)\mathrm{d}s + \boldsymbol{g}(s, \boldsymbol{u}_1, \boldsymbol{u}_2, \boldsymbol{X}_s)\mathrm{d}\boldsymbol{C}_s \quad \text{and} \quad \boldsymbol{X}_0 = \boldsymbol{x}_0.$$

In the above equation, $\boldsymbol{X}_s$ is the state variable of dimension n with the initial state $\boldsymbol{X}_0 = \boldsymbol{x}_0$, $\boldsymbol{u}_1 \in \mathcal{D}_1 \subset R^p$ is a control vector which maximizes some given objective function, and $\boldsymbol{u}_2 \in \mathcal{D}_2 \subset R^q$ is to minimize the objective function. $\boldsymbol{C}_t = (C_{t1}, C_{t2}, \cdots, C_{tk})^\tau$ where $C_{t1}, C_{t2}, \cdots, C_{tk}$ are independent canonical Liu processes. In addition, $\boldsymbol{f} : [0, T] \times R^p \times R^q \times R^n \to R^n$ is a vector value function, $\boldsymbol{g} : [0, T] \times R^p \times R^q \times R^n \to R^{n\times k}$ is a matrix value function.

For any $0 < t < T$ and some given confidence level $\alpha \in (0, 1)$, we choose the objective function as follows:

$$V(\boldsymbol{u}_1, \boldsymbol{u}_2) = H_{\sup}(\alpha)$$

where $H_{\sup}(\alpha) = \sup\{\bar{H} \mid \mathcal{M}\{H \geq \bar{H}\} \geq \alpha\}$ and

$$H = \int_t^T h(s, \boldsymbol{u}_1, \boldsymbol{u}_2, \boldsymbol{X}_s)\mathrm{d}s + G(\boldsymbol{X}_T, T).$$

Besides, $h : [0, T] \times R^p \times R^q \times R^n \to R$ is an integrand function of state and control, and $G : [0, T] \times R^p \times R^q \times R^n \to R$ is a function of terminal reward. In addition, all the functions mentioned above are continuous. Then we consider the following saddle point problem.

$$\begin{cases} \text{Find } (\boldsymbol{u}_1^*, \boldsymbol{u}_2^*) \text{ such that} \\ V(\boldsymbol{u}_1, \boldsymbol{u}_2^*) \leq V(\boldsymbol{u}_1^*, \boldsymbol{u}_2^*) \leq V(\boldsymbol{u}_1^*, \boldsymbol{u}_2) \\ \text{subject to:} \\ \quad \mathrm{d}\boldsymbol{X}_s = \boldsymbol{f}(s, \boldsymbol{u}_1, \boldsymbol{u}_2, \boldsymbol{X}_s)\mathrm{d}s + \boldsymbol{g}(s, \boldsymbol{u}_1, \boldsymbol{u}_2, \boldsymbol{X}_s)\mathrm{d}\boldsymbol{C}_s \ , \ t \leq s \leq T \\ \quad \boldsymbol{X}_t = \boldsymbol{x}. \end{cases} \tag{5.16}$$

In fact, the optimal value will change as long as the initial time t and the initial state $\boldsymbol{x}$ change. Thus we can denote the optimal value $V(\boldsymbol{u}_1^*, \boldsymbol{u}_2^*)$ as $J(t, \boldsymbol{x})$. Now we present the equation of optimality for saddle point problem under uncertain environment.

Theorem 5.5 ([3]) *Let $J(t, \boldsymbol{x})$ be twice differentiable on $[0, T] \times R^n$. Then we have*

$$-J_t(t, \boldsymbol{x}) = \max_{\boldsymbol{u}_1} \min_{\boldsymbol{u}_2} \Big\{\nabla_{\boldsymbol{x}} J(t, \boldsymbol{x})^\tau \boldsymbol{f}(t, \boldsymbol{u}_1, \boldsymbol{u}_2, \boldsymbol{x}) + h(t, \boldsymbol{u}_1, \boldsymbol{u}_2, \boldsymbol{x})$$
$$+\frac{\sqrt{3}}{\pi} \ln \frac{1-\alpha}{\alpha} \|\nabla_{\boldsymbol{x}} J(t, \boldsymbol{x})^\tau \boldsymbol{g}(t, \boldsymbol{u}_1, \boldsymbol{u}_2, \boldsymbol{x})\|_1 \Big\} \tag{5.17}$$

$$= \min_{\boldsymbol{u}_2} \max_{\boldsymbol{u}_1} \Big\{\nabla_{\boldsymbol{x}} J(t, \boldsymbol{x})^\tau \boldsymbol{f}(t, \boldsymbol{u}_1, \boldsymbol{u}_2, \boldsymbol{x}) + h(t, \boldsymbol{u}_1, \boldsymbol{u}_2, \boldsymbol{x})$$
$$+\frac{\sqrt{3}}{\pi} \ln \frac{1-\alpha}{\alpha} \|\nabla_{\boldsymbol{x}} J(t, \boldsymbol{x})^\tau \boldsymbol{g}(t, \boldsymbol{u}_1, \boldsymbol{u}_2, \boldsymbol{x})\|_1 \Big\} \tag{5.18}$$

Proof Assume that $(\boldsymbol{u}_1^*, \boldsymbol{u}_2^*)$ is the optimal control function pair for saddle point problem (5.16). We know that $\boldsymbol{u}_1^*$ and $\boldsymbol{u}_2^*$ are the solutions to the following problems:

$$
(\text{P1}) \begin{cases} J(t, \boldsymbol{x}) \equiv \max\limits_{\boldsymbol{u}_1 \in \mathcal{D}_1} V(\boldsymbol{u}_1, \boldsymbol{u}_2^*) \\ \text{subject to:} \\ \quad \mathrm{d}\boldsymbol{X}_s = \boldsymbol{f}(s, \boldsymbol{u}_1, \boldsymbol{u}_2^*, \boldsymbol{X}_s)\mathrm{d}s + \boldsymbol{g}(s, \boldsymbol{u}_1, \boldsymbol{u}_2^*, \boldsymbol{X}_s)\mathrm{d}\boldsymbol{C}_s \quad \text{and} \quad \boldsymbol{X}_t = \boldsymbol{x}, \end{cases}
$$

and

$$
(\text{P2}) \begin{cases} J(t, \boldsymbol{x}) \equiv \min\limits_{\boldsymbol{u}_2 \in \mathcal{D}_2} V(\boldsymbol{u}_1^*, \boldsymbol{u}_2) \\ \text{subject to:} \\ \quad \mathrm{d}\boldsymbol{X}_s = \boldsymbol{f}(s, \boldsymbol{u}_1^*, \boldsymbol{u}_2, \boldsymbol{X}_s)\mathrm{d}s + \boldsymbol{g}(s, \boldsymbol{u}_1^*, \boldsymbol{u}_2, \boldsymbol{X}_s)\mathrm{d}\boldsymbol{C}_s \quad \text{and} \quad \boldsymbol{X}_t = \boldsymbol{x}. \end{cases}
$$

Applying Theorem 3.2 to (P1) and (P2), we have

$$
\begin{aligned} -J_t(t, \boldsymbol{x}) = \max_{\boldsymbol{u}_1} \Big\{ & \nabla_{\boldsymbol{x}} J(t, \boldsymbol{x})^\tau \boldsymbol{f}(t, \boldsymbol{u}_1, \boldsymbol{u}_2^*, \boldsymbol{x}) + h(t, \boldsymbol{u}_1, \boldsymbol{u}_2^*, \boldsymbol{x}) \\ & + \frac{\sqrt{3}}{\pi} \ln \frac{1-\alpha}{\alpha} \|\nabla_{\boldsymbol{x}} J(t, \boldsymbol{x})^\tau \boldsymbol{g}(t, \boldsymbol{u}_1, \boldsymbol{u}_2^*, \boldsymbol{x})\|_1 \Big\} \end{aligned} \tag{5.19}
$$

$$
\begin{aligned} = \min_{\boldsymbol{u}_2} \Big\{ & \nabla_{\boldsymbol{x}} J(t, \boldsymbol{x})^\tau \boldsymbol{f}(t, \boldsymbol{u}_1^*, \boldsymbol{u}_2, \boldsymbol{x}) + h(t, \boldsymbol{u}_1^*, \boldsymbol{u}_2, \boldsymbol{x}) \\ & + \frac{\sqrt{3}}{\pi} \ln \frac{1-\alpha}{\alpha} \|\nabla_{\boldsymbol{x}} J(t, \boldsymbol{x})^\tau \boldsymbol{g}(t, \boldsymbol{u}_1^*, \boldsymbol{u}_2, \boldsymbol{x})\|_1 \Big\}. \end{aligned} \tag{5.20}
$$

From (5.19), we know that

$$
\begin{aligned} -J_t(t, \boldsymbol{x}) \ge \max_{\boldsymbol{u}_1} \min_{\boldsymbol{u}_2} \Big\{ & \nabla_{\boldsymbol{x}} J(t, \boldsymbol{x})^\tau \boldsymbol{f}(t, \boldsymbol{u}_1, \boldsymbol{u}_2, \boldsymbol{x}) + h(t, \boldsymbol{u}_1, \boldsymbol{u}_2, \boldsymbol{x}) \\ & + \frac{\sqrt{3}}{\pi} \ln \frac{1-\alpha}{\alpha} \|\nabla_{\boldsymbol{x}} J(t, \boldsymbol{x})^\tau \boldsymbol{g}(t, \boldsymbol{u}_1, \boldsymbol{u}_2, \boldsymbol{x})\|_1 \Big\}. \end{aligned} \tag{5.21}
$$

Similarly, from (5.20), we can also get

$$
\begin{aligned} -J_t(t, \boldsymbol{x}) \le \min_{\boldsymbol{u}_2} \max_{\boldsymbol{u}_1} \Big\{ & \nabla_{\boldsymbol{x}} J(t, \boldsymbol{x})^\tau \boldsymbol{f}(t, \boldsymbol{u}_1, \boldsymbol{u}_2, \boldsymbol{x}) + h(t, \boldsymbol{u}_1, \boldsymbol{u}_2, \boldsymbol{x}) \\ & + \frac{\sqrt{3}}{\pi} \ln \frac{1-\alpha}{\alpha} \|\nabla_{\boldsymbol{x}} J(t, \boldsymbol{x})^\tau \boldsymbol{g}(t, \boldsymbol{u}_1, \boldsymbol{u}_2, \boldsymbol{x})\|_1 \Big\}. \end{aligned} \tag{5.22}
$$

Let

$$\begin{aligned}\sigma(\boldsymbol{u}_1, \boldsymbol{u}_2) = & \nabla_{\boldsymbol{x}} J(t, \boldsymbol{x})^\tau \boldsymbol{f}(t, \boldsymbol{u}_1, \boldsymbol{u}_2, \boldsymbol{x}) + h(t, \boldsymbol{u}_1, \boldsymbol{u}_2, \boldsymbol{x}) \\ & + \frac{\sqrt{3}}{\pi} \ln \frac{1-\alpha}{\alpha} \|\nabla_{\boldsymbol{x}} J(t, \boldsymbol{x})^\tau \boldsymbol{g}(t, \boldsymbol{u}_1, \boldsymbol{u}_2, \boldsymbol{x})\|_1.\end{aligned}$$

We note that

$$\max_{\boldsymbol{u}_1} \min_{\boldsymbol{u}_2} \sigma(\boldsymbol{u}_1, \boldsymbol{u}_2) \geq \min_{\boldsymbol{u}_2} \sigma(\boldsymbol{u}_1, \boldsymbol{u}_2), \quad \forall \boldsymbol{u}_1.$$

Thus

$$\max_{\boldsymbol{u}_1} \min_{\boldsymbol{u}_2} \sigma(\boldsymbol{u}_1, \boldsymbol{u}_2) \geq \min_{\boldsymbol{u}_2} \max_{\boldsymbol{u}_1} \sigma(\boldsymbol{u}_1, \boldsymbol{u}_2). \tag{5.23}$$

Together with (5.21) and (5.22), we prove the theorem.

Remark 5.1 The equation of optimality (5.17) for saddle point problem gives a sufficient condition. If it has solutions, the saddle point is determined. Specially, if the max and min operators are interchangeable and $\sigma(\boldsymbol{u}_1, \boldsymbol{u}_2)$ is concave (convex respectively) in $\boldsymbol{u}_1$ ($\boldsymbol{u}_2$ respectively), then the system can reach a saddle point equilibrium.

Remark 5.2 The conclusion we obtained is different from that in the case of stochastic saddle point problem which has an extra term

$$\frac{1}{2} tr\{\boldsymbol{g}^\tau(t, \boldsymbol{u}_1, \boldsymbol{u}_2, \boldsymbol{x}) \nabla_{\boldsymbol{xx}} J(t, \boldsymbol{x}) \boldsymbol{g}(t, \boldsymbol{u}_1, \boldsymbol{u}_2, \boldsymbol{x})\}$$

on the right side of the equation comparing to the deterministic case. Here, we have one additional term $\frac{\sqrt{3}}{\pi} \ln \frac{1-\alpha}{\alpha} \|\nabla_{\boldsymbol{x}} J(t, \boldsymbol{x})^\tau \boldsymbol{g}(t, \boldsymbol{u}_1, \boldsymbol{u}_2, \boldsymbol{x})\|_1$ on the right side of the equation comparing to the deterministic case. Note that the first-order derivative here may make it easier to calculate than the stochastic case.

5.4 Bang–Bang Control for Saddle Point Problem

For a given confidence level $\alpha \in (0, 1)$, consider the following model.

$$\begin{cases} \text{Find } (\boldsymbol{u}_1^*, \boldsymbol{u}_2^*) \text{ such that} \\ V(\boldsymbol{u}_1, \boldsymbol{u}_2^*) \leq V(\boldsymbol{u}_1^*, \boldsymbol{u}_2^*) \leq V(\boldsymbol{u}_1^*, \boldsymbol{u}_2) \\ \text{subject to:} \\ \quad \mathrm{d}\boldsymbol{X}_s = [\boldsymbol{a}(\boldsymbol{X}_s, s) + \boldsymbol{b}(\boldsymbol{X}_s, s)\boldsymbol{u}_1 + \boldsymbol{c}(\boldsymbol{X}_s, s)\boldsymbol{u}_2]\mathrm{d}s \\ \qquad\qquad + \boldsymbol{\sigma}(\boldsymbol{X}_s, s)\mathrm{d}\boldsymbol{C}_s \quad \text{and} \quad X_0 = x_0 \\ \quad \boldsymbol{u}_1 \in [-1, 1]^p, \ \boldsymbol{u}_2 \in [-1, 1]^q \end{cases} \tag{5.24}$$

where

$$V(u_1, u_2) = \left[\int_0^T f(\boldsymbol{X}_s, s)\mathrm{d}s + G(\boldsymbol{X}_T, T)\right]_{\sup}(\alpha).$$

In the above model, $\boldsymbol{a} : R^n \times [0, T]) \to R^n$ is a column-vector function, $\boldsymbol{b} : R^n \times [0, T] \to R^{n\times p}$, $\boldsymbol{c} : R^n \times [0, T] \to R^{n\times q}$ and $\boldsymbol{\sigma} : R^n \times [0, T] \to R^{n\times k}$ are matrix functions. We still use $J(t, \boldsymbol{x})$ to denote the optimal reward obtainable in $[t, T]$ with the condition that we have state $\boldsymbol{X}_t = \boldsymbol{x}$ at time t.

Theorem 5.6 ([3]) *Let $J(t, \boldsymbol{x})$ be a twice differentiable function on* $[0, T] \times R^n$. *Then the optimal control pair of problem* (5.24) *are bang–bang controls.*

Proof It follows from the equation of optimality (5.17) that

$$
\begin{aligned}
-J_t(t,\boldsymbol{x}) &= \max_{\boldsymbol{u}_1}\min_{\boldsymbol{u}_2}\left\{\nabla_{\boldsymbol{x}}J(t,\boldsymbol{x})^\tau(\boldsymbol{a}(\boldsymbol{x},t)+\boldsymbol{b}(\boldsymbol{x},t)\boldsymbol{u}_1+\boldsymbol{c}(\boldsymbol{x},t)\boldsymbol{u}_2)+f(\boldsymbol{x},t)\right.\\
&\qquad\left.+\frac{\sqrt{3}}{\pi}\ln\frac{1-\alpha}{\alpha}\|\nabla_{\boldsymbol{x}}J(t,\boldsymbol{x})^\tau\boldsymbol{\sigma}(\boldsymbol{x},t)\|_1\right\}\\
&= \max_{\boldsymbol{u}_1}\nabla_{\boldsymbol{x}}J(t,\boldsymbol{x})^\tau\boldsymbol{b}(\boldsymbol{x},t)\boldsymbol{u}_1+\min_{\boldsymbol{u}_2}\nabla_{\boldsymbol{x}}J(t,\boldsymbol{x})^\tau\boldsymbol{c}(\boldsymbol{x},t)\boldsymbol{u}_2\\
&\quad+\nabla_{\boldsymbol{x}}J(t,\boldsymbol{x})^\tau\boldsymbol{a}(\boldsymbol{x},t)+f(\boldsymbol{x},t)+\frac{\sqrt{3}}{\pi}\ln\frac{1-\alpha}{\alpha}\|\nabla_{\boldsymbol{x}}J(t,\boldsymbol{x})^\tau\boldsymbol{\sigma}(\boldsymbol{x},t)\|_1.
\end{aligned}
$$

Assume that $\boldsymbol{u}_1^*$ and $\boldsymbol{u}_2^*$ are the optimal controls. We have

$$\max_{\boldsymbol{u}_1}\nabla_{\boldsymbol{x}}J(t,\boldsymbol{x})^\tau\boldsymbol{b}(\boldsymbol{x},t)\boldsymbol{u}_1=\nabla_{\boldsymbol{x}}J(t,\boldsymbol{x})^\tau\boldsymbol{b}(\boldsymbol{x},t)\boldsymbol{u}_1^*,$$

$$\min_{\boldsymbol{u}_2}\nabla_{\boldsymbol{x}}J(t,\boldsymbol{x})^\tau\boldsymbol{c}(\boldsymbol{x},t)\boldsymbol{u}_2=\nabla_{\boldsymbol{x}}J(t,\boldsymbol{x})^\tau\boldsymbol{c}(\boldsymbol{x},t)\boldsymbol{u}_2^*.$$

Let

$$\nabla_{\boldsymbol{x}}J(t,\boldsymbol{x})^\tau\boldsymbol{b}(\boldsymbol{x},t)=(g_1(t,\boldsymbol{x}),g_2(t,\boldsymbol{x}),\cdots,g_p(t,\boldsymbol{x})),\tag{5.25}$$

$$\nabla_{\boldsymbol{x}}J(t,\boldsymbol{x})^\tau\boldsymbol{c}(\boldsymbol{x},t)=(h_1(t,\boldsymbol{x}),h_2(t,\boldsymbol{x}),\cdots,h_q(t,\boldsymbol{x})),\tag{5.26}$$

and

$$\boldsymbol{u}_1^*=(u_{11}^*(t),u_{12}^*(t),\cdots,u_{1p}^*(t))^\tau,$$

$$\boldsymbol{u}_2^*=(u_{21}^*(t),u_{22}^*(t),\cdots,u_{2q}^*(t))^\tau.$$

Then, we can easily obtain that

$$u_{1i}^*(t)=\begin{cases}1, & \text{if } g_i(t,\boldsymbol{x})>0\\ -1, & \text{if } g_i(t,\boldsymbol{x})<0\\ \text{undetermined}, & \text{if } g_i(t,\boldsymbol{x})=0\end{cases}\tag{5.27}$$

for $i = 1, 2, \ldots, p$, and

$$u_{2j}^*(t)=\begin{cases}-1, & \text{if } h_j(t,\boldsymbol{x})>0\\ 1, & \text{if } h_j(t,\boldsymbol{x})<0\\ \text{undetermined}, & \text{if } h_j(t,\boldsymbol{x})=0\end{cases}\tag{5.28}$$

for $j = 1, 2, \ldots, q$. They are bang–bang controls and Eqs. (5.25) and (5.26) are called switching vectors.

5.4.1 A Special Bang-Bang Control Model

Consider the following special bang–bang control model.

$$\begin{cases} \text{Find } (\boldsymbol{u}_1^*, \boldsymbol{u}_2^*) \text{ such that} \\ V(\boldsymbol{u}_1, \boldsymbol{u}_2^*) \le V(\boldsymbol{u}_1^*, \boldsymbol{u}_2^*) \le V(\boldsymbol{u}_1^*, \boldsymbol{u}_2) \\ \text{subject to:} \\ \quad \mathrm{d}\boldsymbol{X}_s = [\boldsymbol{a}(s)\boldsymbol{X}_s + \boldsymbol{b}(s)\boldsymbol{u}_1 + \boldsymbol{c}(s)\boldsymbol{u}_2]\mathrm{d}s \\ \qquad\qquad +\boldsymbol{\sigma}(s)\mathrm{d}\boldsymbol{C}_s \quad \text{and} \quad X_0 = x_0 \\ \quad \boldsymbol{u}_1 \in [-1, 1]^p, \ \boldsymbol{u}_2 \in [-1, 1]^q \end{cases} \tag{5.29}$$

where

$$V(\boldsymbol{u}_1, \boldsymbol{u}_2) = \left[\int_0^T \boldsymbol{f}^\tau(s)\boldsymbol{X}_s \mathrm{d}s + \boldsymbol{g}_T^\tau \boldsymbol{X}_T \right]_{\sup} (\alpha).$$

In addition, $\boldsymbol{a} : [0, T] \to R^{n\times n}$, $\boldsymbol{b} : [0, T] \to R^{n\times p}$, $\boldsymbol{c} : [0, T] \to R^{n\times q}$ and $\boldsymbol{\sigma} : [0, T] \to R^{n\times k}$ are all matrix functions. Besides, $\boldsymbol{f} : [0, T] \to R^n$ is a continuously differential function and $\boldsymbol{g}_T \in R^n$. We denote $\boldsymbol{b}(s) = (b_{li}(s))_{n\times p}$ and $\boldsymbol{c}(s) = (c_{lj}(s))_{n\times p}$. Then, we have the conclusion below.

Theorem 5.7 ([3]) *Let $J(t, \boldsymbol{x})$ be a twice differentiable function on $[0, T] \times R^n$. Then the optimal control pair of problem (5.29) are:*

$$u_{1i}^*(t) = \mathrm{sgn}\{(b_{1i}(t), b_{2i}(t), \cdots, b_{ni}(t))\boldsymbol{p}(t)\} \quad \text{for} \quad i = 1, 2, \ldots, p, \tag{5.30}$$
$$u_{2j}^*(t) = -\mathrm{sgn}\{(c_{1j}(t), c_{2j}(t), \cdots, c_{nj}(t))\boldsymbol{p}(t)\} \quad \text{for} \quad j = 1, 2, \ldots, q, \tag{5.31}$$

where $\boldsymbol{p}(t) \in R^n$ satisfies the following equation:

$$\dot{\boldsymbol{p}}(t) = -\boldsymbol{f}(t) - \boldsymbol{a}^\tau(t)\boldsymbol{p}(t), \qquad \boldsymbol{p}(T) = \boldsymbol{g}_T. \tag{5.32}$$

And the optimal value is

$$\begin{aligned} J(0, x_0) = \ & \boldsymbol{p}(0)^\tau x_0 + \int_0^T \boldsymbol{p}^\tau(s)(\boldsymbol{b}(s)\boldsymbol{u}_1^* + \boldsymbol{c}(s)\boldsymbol{u}_2^*)\mathrm{d}s \\ & + \frac{\sqrt{3}}{\pi} \ln \frac{1-\alpha}{\alpha} \int_0^T \|\boldsymbol{p}^\tau(s)\boldsymbol{\sigma}(s)\|_1 \mathrm{d}s. \end{aligned}$$

Proof Applying the equation of optimality (5.17), we have

$$\begin{aligned}
- J_t(t, \boldsymbol{x}) = & \max_{\boldsymbol{u}_1} \min_{\boldsymbol{u}_2} \{\nabla_{\boldsymbol{x}} J(t, \boldsymbol{x})^\tau (\boldsymbol{a}(\boldsymbol{x}, t) + \boldsymbol{b}(\boldsymbol{x}, t)\boldsymbol{u}_1 + \boldsymbol{c}(\boldsymbol{x}, t)\boldsymbol{u}_2) \\
& \left. + \boldsymbol{f}(\boldsymbol{x}, t) + \frac{\sqrt{3}}{\pi} \ln \frac{1-\alpha}{\alpha} \|\nabla_{\boldsymbol{x}} J(t, \boldsymbol{x})^\tau \boldsymbol{\sigma}(t)\|_1 \right\} \\
= & \max_{\boldsymbol{u}_1} \{\nabla_{\boldsymbol{x}} J(t, \boldsymbol{x})^\tau \boldsymbol{b}(\boldsymbol{x}, t)\boldsymbol{u}_1\} + \min_{\boldsymbol{u}_2} \{\nabla_{\boldsymbol{x}} J(t, \boldsymbol{x})^\tau \boldsymbol{c}(\boldsymbol{x}, t)\boldsymbol{u}_2\} + \boldsymbol{f}(\boldsymbol{x}, t) \\
& + \nabla_{\boldsymbol{x}} J(t, \boldsymbol{x})^\tau \boldsymbol{a}(\boldsymbol{x}, t) + \frac{\sqrt{3}}{\pi} \ln \frac{1-\alpha}{\alpha} \|\nabla_{\boldsymbol{x}} J(t, \boldsymbol{x}) \tau \boldsymbol{\sigma}(\boldsymbol{x}, t)\|_1. \quad (5.33)
\end{aligned}$$

Since $J(T, \boldsymbol{X}_T) = \boldsymbol{g}_T^\tau \boldsymbol{X}_T$, we conjuncture that $J(t, \boldsymbol{x}) = \boldsymbol{p}^\tau(t)\boldsymbol{x} + q(t)$ and $\boldsymbol{p}(T) = \boldsymbol{g}_T, q(T) = 0$. Then $J_t(t, \boldsymbol{x}) = \dot{\boldsymbol{p}}^\tau(t)\boldsymbol{x} + \dot{q}(t)$ and $\nabla_{\boldsymbol{x}} J(t, \boldsymbol{x}) = \boldsymbol{p}(t)$. Substituting them into (5.33) yields

$$\begin{aligned}
& -\dot{\boldsymbol{p}}^\tau(t)\boldsymbol{x} - \dot{q}(t) \\
& = \boldsymbol{p}^\tau(t)\boldsymbol{b}(t)\boldsymbol{u}_1^* + \boldsymbol{p}^\tau(t)\boldsymbol{c}(t)\boldsymbol{u}_2^* + \boldsymbol{f}^\tau(t)\boldsymbol{x} + \boldsymbol{p}^\tau(t)\boldsymbol{a}(t)\boldsymbol{x} + \frac{\sqrt{3}}{\pi} \ln \frac{1-\alpha}{\alpha} \|\boldsymbol{p}^\tau(t)\boldsymbol{\sigma}(t)\|_1.
\end{aligned}$$

Thus, we have

$$\dot{\boldsymbol{p}}(t) = -\boldsymbol{f}(t) - \boldsymbol{a}^\tau(t)\boldsymbol{p} \quad (5.34)$$

$$\dot{q}(t) = -\boldsymbol{p}^\tau(t)(\boldsymbol{b}(t)\boldsymbol{u}_1^* + \boldsymbol{c}(t)\boldsymbol{u}_2^*) - \frac{\sqrt{3}}{\pi} \ln \frac{1-\alpha}{\alpha} \|\boldsymbol{p}^\tau(t)\boldsymbol{\sigma}(t)\|_1. \quad (5.35)$$

According to Theorem 5.6, we could obtain the bang–bang controls:

$$u_{1i}^*(t) = \operatorname{sgn}\{\boldsymbol{p}(t)^\tau (b_{1i}(t), b_{2i}(t), \cdots, b_{ni}(t))^\tau\}$$

for $i = 1, 2, \ldots, p$, and

$$u_{2j}^*(t) = -\operatorname{sgn}\{\boldsymbol{p}(t)^\tau (c_{1j}(t), c_{2j}(t), \cdots, c_{nj}(t))^\tau\}$$

for $j = 1, 2, \ldots, q$. Integrating (5.35) from t to T, we have

$$q(t) = \int_t^T \boldsymbol{p}^\tau(s)(\boldsymbol{b}(s)\boldsymbol{u}_1^* + \boldsymbol{c}(s)\boldsymbol{u}_2^*)\mathrm{d}s + \frac{\sqrt{3}}{\pi} \ln \frac{1-\alpha}{\alpha} \int_t^T \|\boldsymbol{p}^\tau(s)\boldsymbol{\sigma}(s)\|_1 \mathrm{d}s.$$

The conclusions are proved.

5.4.2 Example

Consider the following example of the bang–bang control model for saddle point problem. We have the system equations as follows:

$$\begin{cases} \mathrm{d}X_1(s) = (X_1(s) + X_2(s) + u_1(s))\mathrm{d}s + \sigma \mathrm{d}C_s \\ \mathrm{d}X_2(s) = 2u_2(s)\mathrm{d}s \end{cases}$$

where $\sigma \in R$, $\boldsymbol{X}(0) = (X_1(0), X_2(0)) = \boldsymbol{x}_0$ and $u_1(s), u_2(s) \in [-1, 1], 0 \le s \le 1$. The performance index is

$$V(u_1, u_2) = \left[\int_0^1 (X_1(s) + X_2(s))\mathrm{d}s + X_1(1) - X_2(1)\right]_{\sup}(\alpha),$$

in which u_1 aims at maximizing the performance index while u_2 does the minimizing job.

Suppose

$$\boldsymbol{a}(s) = \begin{pmatrix} 1 & 1 \\ 0 & 0 \end{pmatrix}, \quad \boldsymbol{b}(s) = \begin{pmatrix} 1 \\ 0 \end{pmatrix}, \quad \boldsymbol{c}(s) = \begin{pmatrix} 0 \\ 2 \end{pmatrix}, \quad \boldsymbol{f}(s) = \begin{pmatrix} 1 \\ 1 \end{pmatrix}, \quad \boldsymbol{g}_1 = \begin{pmatrix} 1 \\ -1 \end{pmatrix}.$$

It follows from (5.32) that

$$\dot{\boldsymbol{p}} = -\begin{pmatrix} 1 \\ 1 \end{pmatrix} - \begin{pmatrix} 1 & 0 \\ 1 & 0 \end{pmatrix} \boldsymbol{p}, \quad \boldsymbol{p}(1) = \begin{pmatrix} 1 \\ -1 \end{pmatrix}.$$

We can obtain the solution

$$\boldsymbol{p}(t) = \begin{pmatrix} 2e^{1-t} - 1 \\ 2e^{1-t} - 3 \end{pmatrix}.$$

Thus according to Theorem 5.7, we find the bang–bang controls:

$$u_1 = 1, \quad u_2 = -\mathrm{sgn}\{2e^{1-t} - 3\}$$

which are shown in Fig. 5.5.

Denote $\boldsymbol{x}_0 = (x_{01}, x_{02})^\tau$. We can obtain the system states as follows:

$$X_1(t) = \begin{cases} e^t(x_{01} + x_{02} - 1) + e^t\sigma\xi(t) + 1 + 2t - x_{02}, & \text{for } 0 \le t < 1 + \ln 2 - \ln 3, \\ e^t(x_{01} + x_{02} - 1) + e^t\sigma\xi(t) + 6e^{t-1} - 2t & \\ \quad + 4\ln 2 - 4\ln 3 + 1 - x_{02}, & \text{for } 1 + \ln 2 - \ln 3 < t \le 1, \end{cases}$$

$$X_2(t) = x_{02} - 2t \wedge (4 + 4\ln 2 - 4\ln 3 - 2t),$$

where $\xi(t)$ is an uncertain process which is subject to the normal distribution $\mathcal{N}(0, 1 - e^{-t})$ when t is fixed. The optimal value $J(0, \boldsymbol{x}_0)$ is

$$(2e - 1)x_{01} + (2e - 3)x_{02} + 11 - 2e - 12\ln 3 + 12\ln 2 + \frac{\sqrt{3}}{\pi}\ln\frac{1-\alpha}{\alpha}(2e - 3)\sigma.$$

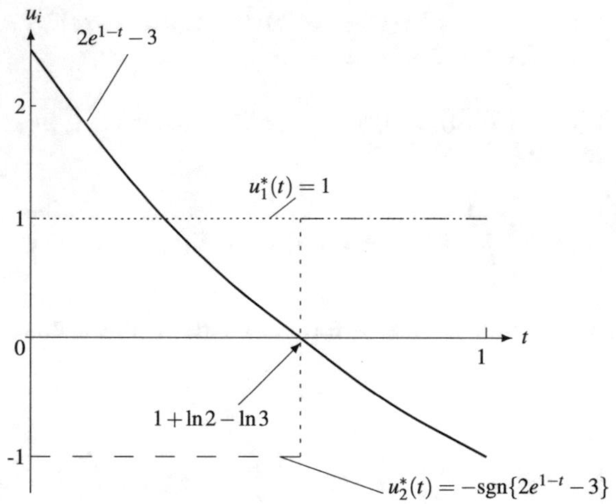

Fig. 5.5 Bang–bang control for two variables

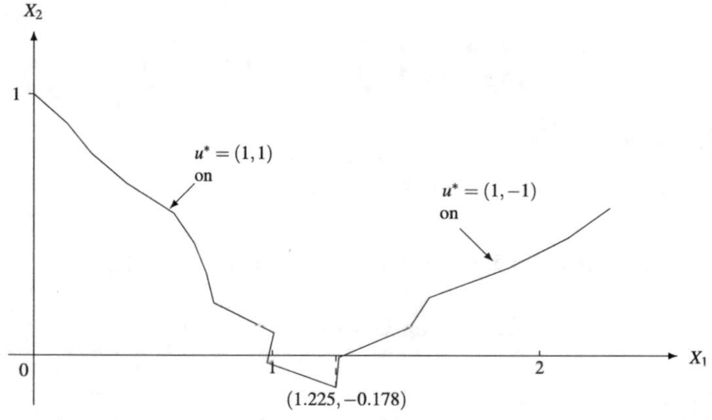

Fig. 5.6 A sample trajectory

It can be seen that the trajectory of the system is an uncertain vector. If we set $\sigma = 0.1$, $\boldsymbol{x}_0 = (0, 1)$ and $\alpha = 0.85$, we could simulate a sample trajectory of the uncertain vector. From Fig. 5.6, we can observe the roughness of the curve which is caused by the uncertain vector in the system. In this sample, the switching point is $(1.225, -0.178)$. But this does not mean the control always switches at this point. The switching point always changes at every simulation for the uncertain vector. And the optimal value is 2.9014 when $\alpha = 0.85$.

References

1. Xu X, Zhu Y (2012) Uncertain bang-bang control for continuous time model. Cybern Syst Int J 43(6):515–527
2. Kang Y, Zhu Y (2012) Bang-bang optimal control for multi-stage uncertain systems. Inf Int Interdiscip J 15(8): 3229–3237
3. Sun Y, Zhu Y (2017) Bang-bang property for an uncertain saddle point problem. J Intell Manufact 28(3):605–613

Chapter 6
Optimal Control for Switched Uncertain Systems

Many practical systems operate by switching between different subsystems or modes. They are called switched systems. The optimal control problems of switched systems arise naturally when the control systems under consideration have multiple operating modes. A powertrain system [1] can also be viewed as a switched system which needs switching between different gears to achieve an objective such as fast and smooth acceleration response to the driver's commands, low fuel consumption, and low levels of pollutant emissions. For switched systems, the aim of optimal control is to seek both the optimal switching law and the optimal continuous input to optimize a certain performance criterion. Many successful algorithms have already been developed to seek the optimal control of switched systems. It is worth mentioning that Xu and Antsaklis [2] considered the optimal control of continuous-time switched systems. A two-stage optimization strategy was proposed in [2]. Stage (a) is a conventional optimal control problem under a given switching law, and Stage (b) is a constrained nonlinear optimization problem that finds the local optimal switching instants. A general continuous-time switching problem was investigated in [3] based on the maximum principle and an embedding method. Furthermore, Teo et al. proposed a control parameterization technique [4] and the time scaling transform method [5] to find the approximate optimal control inputs and switching instants, which have been used extensively.

For continuous-time switched systems with subsystems perturbed by uncertainty, our aim is to seek both the switching instants and the optimal continuous input to optimize a certain performance criterion. In this chapter, we will study such problem based on different criterions and provide suitable solution methods.

Y. Zhu, *Uncertain Optimal Control*, Springer Uncertainty Research,
https://doi.org/10.1007/978-981-13-2134-4_6

6.1 Switched Uncertain Model

Considering a switched uncertain system consisting of the following subsystems:

$$\begin{cases} \mathrm{d}X_s = (A_i(s)X_s + B_i(s)u_s)\mathrm{d}s + \sigma(s, u_s, X_s)\mathrm{d}C_s, \quad s \in [0, T] \\ i \in I = \{1, 2, \ldots, M\} \\ X_0 = x_0 \end{cases} \tag{6.1}$$

where $X_s \in R^n$ is the state vector and $u_s \in R^r$ is the decision vector in a domain U, $A_i : [0, T] \to R^{n\times n}$, $B_i : [0, T] \to R^{n\times r}$ are some twice continuously differentiable functions for $i \in I$, $C_s = (C_{s1}, C_{s2}, \cdots, C_{sk})^\tau$, where $C_{s1}, C_{s2}, \cdots, C_{sk}$ are independent canonical Liu processes.

An optimal control problem of such system involves finding an optimal control u_t^* and an optimal switching law such that a given cost function is minimized. A switching law in $[0, T]$ for system (6.1) is defined as

$$\Lambda = ((t_0, i_0), (t_1, i_1), \ldots, (t_K, i_K))$$

where t_k $(k = 0, 1, \ldots, K)$ satisfying $0 = t_0 \le t_1 \le \cdots \le t_K \le t_{K+1} = T$ are the switching instants and $i_k \in I$ for $k = 0, 1, \ldots, K$. Here (t_k, i_k) indicates that at instants t_k, the system switches from subsystem i_{k-1} to i_k. During the time interval $[t_k, t_{k+1})$ ($[t_K, T]$ if $k = K$), subsystem i_k is active. Since many practical problems only involve optimizations in which a prespecified order of active subsystems is given, for convenience, we assume subsystem i is active in $[t_{i-1}, t_i)$.

6.2 Expected Value Model

Consider the following uncertain expected value optimal control model of a switched uncertain system.

$$\begin{cases} \min\limits_{t_i} \min\limits_{\boldsymbol{u}_s \in [-1,1]^r} E\left[\int_0^T f(s)^\tau X_s \mathrm{d}s + S_T^\tau X_T\right] \\ \text{subject to} \\ \quad \mathrm{d}X_s = (A_i(s)X_s + B_i(s)\boldsymbol{u}_s)\mathrm{d}s + \sigma(s, \boldsymbol{u}_s, X_s)\mathrm{d}C_s \\ \qquad\qquad s \in [t_{i-1}, t_i), \; i = 1, 2, \ldots, K+1 \\ \quad X_0 = \boldsymbol{x}_0. \end{cases} \tag{6.2}$$

In the above model, f is the objective function of dimension n and $S_T \in R^n$. For given $t_1, t_2, \ldots, t_K$, use $J(t, x)$ to denote the optimal value obtained in $[t, T]$ with the condition that at time t we are in state $X_t = \boldsymbol{x}$. That is

$$\begin{cases} J(t, \boldsymbol{x}) = \min\limits_{\boldsymbol{u}_s \in [-1,1]^r} E[\int_t^T f(s)^\tau X_s \mathrm{d}s + S_T^\tau X_T] \\ \text{subject to} \\ \quad \mathrm{d}X_s = (A_i(s)X_s + B_i(s)\boldsymbol{u}_s)\mathrm{d}s + \sigma(s, \boldsymbol{u}_s, X_s)\mathrm{d}C_s \\ \qquad\qquad s \in [t_{i-1}, t_i),\ i = 1, 2, \ldots, K+1 \\ \quad X_t = \boldsymbol{x}. \end{cases} \tag{6.3}$$

By the equation of optimality (2.15) to deal with the model (6.2), the following conclusion can be obtained.

Theorem 6.1 *Let $J(t, \boldsymbol{x})$ be twice differentiable on $[t_{i-1}, t_i) \times R^n$. Then we have*

$$-J_t(t, \boldsymbol{x}) = \min_{\boldsymbol{u}_t \in [-1,1]^r} \{f(t)^\tau \boldsymbol{x} + (A_i(t)\boldsymbol{x} + B_i(t)\boldsymbol{u}_t)^\tau \nabla_{\boldsymbol{x}} J(t, \boldsymbol{x})\}, \tag{6.4}$$

where $J_t(t, \boldsymbol{x})$ is the partial derivatives of the function $J(t, \boldsymbol{x})$ in t, and $\nabla_{\boldsymbol{x}} J(t, \boldsymbol{x})$ is the gradient of $J(t, \boldsymbol{x})$ in $\boldsymbol{x}$.

An optimal control problem of switched uncertain systems given by (6.2) is to choose the best switching instants and the optimal inputs such that an expected value is optimized subject to a switched uncertain system.

6.2.1 Two-Stage Algorithm

In order to solve the problem (6.2), we decompose it into two stages. Stage (a) is an uncertain optimal control problem which seeks the optimal value under a given switching sequence. Stage (b) is an optimization problem in switching instants.

6.2.2 Stage (a)

In this stage, we need to solve the following model and find the optimal value.

$$\begin{cases} J(0, \boldsymbol{x}_0, t_1, \cdots, t_K) = \min\limits_{\boldsymbol{u}_s \in [-1,1]^r} E\left[\int_0^T f(s)^\tau X_s \mathrm{d}s + S_T^\tau X_T\right] \\ \text{subject to} \\ \quad \mathrm{d}X_s = (A_i(s)X_s + B_i(s)\boldsymbol{u}_s)\mathrm{d}s + \sigma(s, \boldsymbol{u}_s, X_s)\mathrm{d}C_s \\ \qquad\qquad s \in [t_{i-1}, t_i),\ i = 1, 2, \ldots, K+1 \\ \quad X_0 = \boldsymbol{x}_0 \end{cases} \tag{6.5}$$

where $t_1, t_2, \ldots, t_K$ are fixed and $t_0 = 0$, $t_{K+1} = T$. Denote $B_i(s) = (b_{lj}^{(i)}(s))_{n \times r}$. We have the following conclusion.

Theorem 6.2 ([6]) *Let $J(t, \boldsymbol{x})$ be twice differentiable on $[t_{i-1}, t_i) \times R^n$ ($i = 1, 2, \ldots, K+1$). The optimal control $\boldsymbol{u}_t^{(i)*} = (u_1^{(i)*}(t), u_2^{(i)*}(t), \cdots, u_r^{(i)*}(t))^\tau$ of* (6.5) *is a bang–bang control*

$$u_j^{(i)*}(t) = sgn\{-(b_{1j}^{(i)}(t), b_{2j}^{(i)}(t), \cdots, b_{nj}^{(i)}(t))p_i(t)\} \tag{6.6}$$

for $i = 1, 2, \ldots, K+1$; $j = 1, 2, \ldots, r$, where $p_i(t) \in R^n$, $t \in [t_{i-1}, t_i)$, satisfies

$$\begin{cases} \dfrac{\mathrm{d}p_i(t)}{\mathrm{d}t} = -f(t) - A_i(t)^\tau p_i(t) \\ p_{K+1}(T) = S_T \text{ and } p_i(t_i) = p_{i+1}(t_i) \text{ for } i \leq K. \end{cases} \tag{6.7}$$

The optimal value of model (6.5) *is*

$$J(0, \boldsymbol{x}_0, t_1, \ldots, t_K) = p_1(0)^\tau \boldsymbol{x}_0 + \sum_{i=1}^{K+1} \int_{t_{i-1}}^{t_i} p_i(t)^\tau B_i(t) \boldsymbol{u}_t^{(i)*} \mathrm{d}t. \tag{6.8}$$

Proof First we prove the optimal control of model (6.5) is a bang–bang control. It follows from the equation of optimality (6.4) that

$$-J_t(t, \boldsymbol{x}) = \min_{\boldsymbol{u}_t \in [-1,1]^r} \{f(t)^\tau \boldsymbol{x} + (A_i(t)\boldsymbol{x} + B_i(t)\boldsymbol{u}_t)^\tau \nabla_{\boldsymbol{x}} J(t, \boldsymbol{x})\}. \tag{6.9}$$

On the right side of (6.9), let $\boldsymbol{u}_t^{(i)*}$ make it the minimum. We have

$$\begin{aligned} &\min_{\boldsymbol{u}_t \in [-1,1]^r} \left\{f(t)^T \boldsymbol{x} + (A_i(t)\boldsymbol{x} + B_i(t)\boldsymbol{u}_t)^\tau \nabla_{\boldsymbol{x}} J(t, \boldsymbol{x})\right\} \\ &= f(t)^\tau \boldsymbol{x} + (A_i(t)\boldsymbol{x} + B_i(t)\boldsymbol{u}_t^{(i)*})^\tau \nabla_{\boldsymbol{x}} J(t, \boldsymbol{x}). \end{aligned}$$

That is,

$$\min_{\boldsymbol{u}_t \in [-1,1]^r} \{\nabla_{\boldsymbol{x}} J(t, \boldsymbol{x})^\tau B_i(t)\boldsymbol{u}_t\} = \nabla_{\boldsymbol{x}} J(t, \boldsymbol{x})^\tau B_i(t)\boldsymbol{u}_t^{(i)*}.$$

Denote

$$\boldsymbol{u}_t^{(i)*} = (u_1^{(i)*}(t), u_2^{(i)*}(t), \ldots, u_r^{(i)*}(t))^\tau \tag{6.10}$$

and

$$\nabla_{\boldsymbol{x}} J(t, \boldsymbol{x})^\tau B_i(t) = (g_1^{(i)}(t, \boldsymbol{x}), g_2^{(i)}(t, \boldsymbol{x}), \ldots, g_r^{(i)}(t, \boldsymbol{x})). \tag{6.11}$$

Then,

$$u_j^{(i)*}(t) = \begin{cases} 1, & \text{if } g_j^{(i)}(t, \boldsymbol{x}) < 0 \\ -1, & \text{if } g_j^{(i)}(t, \boldsymbol{x}) > 0 \\ \text{undetermined}, & \text{if } g_j^{(i)}(t, \boldsymbol{x}) = 0 \end{cases} \tag{6.12}$$

for $i = 1, 2, \ldots, K+1$; $j = 1, 2, \ldots, r$, which is a bang–bang control. The functions $g_j^{(i)}(t, \boldsymbol{x})$ are called switching functions. If at least one switching function equal to zero in some interval, we call it singular control. But here we only consider switching functions equal to zero at most in some discrete points. According to (6.9), when $t \in [t_K, T]$, we have

$$-J_t(t,\boldsymbol{x}) = \min_{\boldsymbol{u}_t \in [-1,1]^r} \{f(t)^\tau \boldsymbol{x} + (A_{K+1}(t)\boldsymbol{x} + B_{K+1}(t)\boldsymbol{u}_t)^\tau \nabla_x J(t,\boldsymbol{x})\}. \tag{6.13}$$

Since $J(T, \boldsymbol{x}_T) = S_T^\tau \boldsymbol{x}_T$, we guess

$$J(t,\boldsymbol{x}) = p_{K+1}(t)^\tau \boldsymbol{x} + q_{K+1}(t) \quad \text{and} \quad p_{K+1}(T) = S_T, \quad q_{K+1}(T) = 0.$$

So

$$\nabla_x J(t,\boldsymbol{x}) = p_{K+1}(t), \ J_t(t,\boldsymbol{x}) = \frac{\mathrm{d}p_{K+1}(t)^\tau}{\mathrm{d}t}\boldsymbol{x} + \frac{\mathrm{d}q_{K+1}(t)}{\mathrm{d}t}. \tag{6.14}$$

Thus, it follows from (6.12) that

$$u_j^{(K+1)*}(t) = \mathrm{sgn}\{-(b_{1j}^{(K+1)}(t), b_{2j}^{(K+1)}(t), \cdots, b_{nj}^{(K+1)}(t))p_{K+1}(t)\}. \tag{6.15}$$

Substituting (6.14) into (6.13) gets

$$-\frac{\mathrm{d}p_{K+1}(t)^\tau}{\mathrm{d}t}\boldsymbol{x} - \frac{\mathrm{d}q_{K+1}(t)}{\mathrm{d}t} = f(t)^\tau \boldsymbol{x} + (A_{K+1}(t)\boldsymbol{x} + B_{K+1}(t)\boldsymbol{u}_t^{(K+1)*})^\tau p_{K+1}(t).$$

Therefore,

$$-\frac{\mathrm{d}p_{K+1}(t)^\tau}{\mathrm{d}t} = f(t)^\tau + p_{K+1}(t)^\tau A_{K+1}(t), \quad p_{K+1}(T) = S_T, \tag{6.16}$$

and

$$-\frac{\mathrm{d}q_{K+1}(t)}{\mathrm{d}t} = p_{K+1}(t)^\tau B_{K+1}(t)\boldsymbol{u}_t^{(K+1)*}, \quad q_{K+1}(T) = 0. \tag{6.17}$$

From (6.17), we have

$$q_{K+1}(t) = \int_t^T p_{K+1}(t)^\tau B_{K+1}(t)\boldsymbol{u}_t^{(K+1)*}\mathrm{d}t. \tag{6.18}$$

Furthermore,

$$\begin{aligned} J(t,\boldsymbol{x}) &= p_{K+1}(t)^\tau \boldsymbol{x} + q_{K+1}(t) \\ &= p_{K+1}(t)^\tau \boldsymbol{x} + \int_t^T p_{K+1}(t)^\tau B_{K+1}(t)\boldsymbol{u}_t^{(K+1)*}\mathrm{d}t, \quad t \in [t_K, T]. \end{aligned} \tag{6.19}$$

where $p_{K+1}(t)$ satisfies the Riccati differential equation and boundary condition (6.16).

When $t \in [t_{i-1}, t_i)$ for $i \le K$, assume

$$J(t, \boldsymbol{x}) = p_i(t)^\tau \boldsymbol{x} + q_i(t), \quad \text{and} \quad p_i(t_i) = p_{i+1}(t_i), \quad q_i(t_i) = q_{i+1}(t_i).$$

By the same method as the above procedure, we can get

$$\begin{cases} -\dfrac{\mathrm{d}p_i(t)^\tau}{\mathrm{d}t} = f(t)^\tau + p_i(t)^\tau A_i(t), \quad p_i(t_i) = p_{i+1}(t_i) \\ -\dfrac{\mathrm{d}q_i(t)}{\mathrm{d}t} = p_i(t)^\tau B_i(t)\boldsymbol{u}_t^{(i)*}, \quad q_i(t_i) = q_{i+1}(t_i). \end{cases} \tag{6.20}$$

Hence,

$$\begin{aligned} J(t, \boldsymbol{x}) &= p_i(t)^\tau \boldsymbol{x} + q_i(t) \\ &= p_i(t)^\tau \boldsymbol{x} + \int_t^{t_i} p_i(t)^\tau B_i(t)\boldsymbol{u}_t^{(i)*}\mathrm{d}t + q_{i+1}(t_i), \quad t \in [t_{i-1}, t_i). \end{aligned}$$

Hence, the optimal value of the model (6.5) is

$$J(0, \boldsymbol{x}_0, t_1, \cdots, t_K) = p_1(0)^\tau \boldsymbol{x}_0 + \sum_{i=1}^{K+1} \int_{t_{i-1}}^{t_i} p_i(t)^\tau B_i(t)\boldsymbol{u}_t^{(i)*}\mathrm{d}t.$$

The theorem is proved.

If there are two subsystems only, the model is as follows:

$$\begin{cases} J(0, \boldsymbol{x}_0, t_1) = \min\limits_{\boldsymbol{u}_s \in [-1,1]^2} E\left[\int_0^T f(s)^\tau X_s \mathrm{d}s + S_T^\tau X_T\right] \\ \text{subject to} \\ \quad \mathrm{d}X_s = (A_1(s)X_s + B_1(s)u_s)\mathrm{d}s + \sigma(s, u_s, X_s)\mathrm{d}C_s, \; s \in [t_0, t_1) \\ \quad \mathrm{d}X_s = (A_2(s)X_s + B_2(s)u_s)\mathrm{d}s + \sigma(s, u_s, X_s)\mathrm{d}C_s, \; s \in [t_1, T] \\ \quad X_0 = \boldsymbol{x}_0. \end{cases} \tag{6.21}$$

According to Theorem 6.2, two Riccati differential equations have to be solved in order to solve the model (6.21). Then the optimal cost $J(0, \boldsymbol{x}_0, t_1)$ can be obtained as follows:

$$J(0, \boldsymbol{x}_0, t_1) = p_1(0)^\tau \boldsymbol{x}_0 + \int_0^{t_1} p_1(t)^\tau B_1(t)\boldsymbol{u}_t^{(1)*}\mathrm{d}t + \int_{t_1}^{T} p_2(t)^\tau B_2(t)\boldsymbol{u}_t^{(2)*}\mathrm{d}t.$$

Denote $\tilde{J}(t_1) = J(0, \boldsymbol{x}_0, t_1)$. The next stage is to solve an optimization problem

$$\min_{0 \le t_1 \le T} \tilde{J}(t_1). \tag{6.22}$$

6.2.3 *Stage (b)*

For the model (6.21), we cannot obtain the analytical expressions of solutions according to Theorem 6.2 which leads to the unavailability of an explicit form of the first-order derivative and the second-order derivative of the cost function in t_1. Because the cost functions of optimal control problems are not multimodal in practice, the modified golden section method [7], which does not require derivatives of cost functions, can be carried to solve the optimization problem (6.22). This method is usually used to solve one dimension optimization problems. Its basic idea for minimizing a function over an interval is iteratively reducing the length of the interval by comparing the function values of the observations. When the length of the interval is reduced to some acceptable degree, the points on the interval can be regarded as approximations of minimizer. We can use the following algorithm to solve the optimization problem.

Algorithm 6.1 (Modified golden section method for solving (6.22))

Step 1. Give the iteration precision $\varepsilon > 0$. Set

$$a_1 = 0, \quad b_1 = T,$$

$$\lambda_1 = a_1 + 0.382(b_1 - a_1) = 0.382T, \quad \mu_1 = a_1 + 0.618(b_1 - a_1) = 0.618T.$$

Calculate $\tilde{J}(a_1), \tilde{J}(b_1), \tilde{J}(\lambda_1), \tilde{J}(\mu_1)$. Put $k = 1$.

Step 2. If $|b_k - a_k| < \varepsilon$, end. The optimal solution $t_1 \in [a_k, b_k]$. Let $t_1 = \frac{1}{2}(a_k + b_k)$.

Step 3. Let $\tilde{J} = \min\{\tilde{J}(a_k), \tilde{J}(b_k), \tilde{J}(\lambda_k), \tilde{J}(\mu_k)\}$. If $\tilde{J} = \tilde{J}(a_k)$ or $\tilde{J} = \tilde{J}(\lambda_k)$, go to step 4; otherwise, go to step 5.

Step 4. Let

$$a_{k+1} := a_k, \quad \mu_{k+1} := \lambda_k, \quad b_{k+1} := \mu_k,$$

$$\tilde{J}(a_{k+1}) := \tilde{J}(a_k), \quad \tilde{J}(\mu_{k+1}) := \tilde{J}(\lambda_k), \quad \tilde{J}(b_{k+1}) := \tilde{J}(\mu_k),$$

$$\lambda_{k+1} = a_{k+1} + 0.382(b_{k+1} - a_{k+1}).$$

Calculate $\tilde{J}(\lambda_{k+1})$. Turn to step 6.

Step 5. Let

$$a_{k+1} := \lambda_k, \quad \lambda_{k+1} := \mu_k, \quad b_{k+1} := b_k,$$

$$\tilde{J}(a_{k+1}) := \tilde{J}(\lambda_k), \quad \tilde{J}(\lambda_{k+1}) := \tilde{J}(\mu_k), \quad \tilde{J}(b_{k+1}) := \tilde{J}(b_k),$$

$$\mu_{k+1} = a_{k+1} + 0.618(b_{k+1} - a_{k+1}).$$

Calculate $\tilde{J}(\mu_{k+1})$.

Step 6. Let $k := k + 1$. Turn to step 2.

From Algorithm 6.1, we can see after nth iteration that the length of the interval is $(0.618)^n T$. Therefore, the convergence rate of this method is linear.

6.2.4 An Example

Consider the following example of optimal control model for switched uncertain systems

$$\begin{cases} J(0,\boldsymbol{x}_0)=\min\limits_{t_1}\min\limits_{\boldsymbol{u}_s} E\left[\dfrac{1}{2}X_1(1)-\dfrac{2}{3}X_2(1)\right] \\ \text{subject to} \\ \quad \textit{subsystem } 1: \begin{cases} \mathrm{d}X_1(s)=u_1(s)\mathrm{d}s \\ \mathrm{d}X_2(s)=(X_1(s)+u_2(s))\mathrm{d}s+\sigma\mathrm{d}C_s, \quad s\in[0,t_1) \end{cases} \\ \quad \textit{subsystem } 2: \begin{cases} \mathrm{d}X_1(s)=(2X_2(s)-u_1(s))\mathrm{d}s \\ \mathrm{d}X_2(s)=(u_1(s)+u_2(s))\mathrm{d}s+\sigma\mathrm{d}C_s, \quad s\in[t_1,1] \end{cases} \\ \quad \boldsymbol{X}(0)=(X_1(0),X_2(0))^\tau=(0,0)^\tau \\ \quad |u_i(s)|\le 1,\ 0\le s\le 1,\ i=1,2. \end{cases} \tag{6.23}$$

We have

$$A_1(s)=\begin{pmatrix}0&0\\1&0\end{pmatrix},\quad B_1(s)=\begin{pmatrix}1&0\\0&1\end{pmatrix},$$

$$A_2(s)=\begin{pmatrix}0&2\\0&0\end{pmatrix},\quad B_2(s)=\begin{pmatrix}-1&0\\1&1\end{pmatrix},\quad S_1=\begin{pmatrix}\frac{1}{2}\\-\frac{2}{3}\end{pmatrix}.$$

It follows from (6.7) that

$$\frac{\mathrm{d}p_2(t)}{\mathrm{d}t}=-\begin{pmatrix}0&0\\2&0\end{pmatrix}p_2(t),\quad p_2(1)=\begin{pmatrix}\frac{1}{2}\\-\frac{2}{3}\end{pmatrix},$$

which has the solution

$$p_2(t)=\begin{pmatrix}\frac{1}{2}\\\frac{1}{3}-t\end{pmatrix}.$$

Hence,

$$\boldsymbol{u}_t^{(2)}=\begin{pmatrix}u_1^{(2)}(t)\\u_2^{(2)}(t)\end{pmatrix}=\begin{pmatrix}sgn(\frac{1}{6}+t)\\sgn(t-\frac{1}{3})\end{pmatrix}.$$

It also follows from (6.7) that

$$\frac{\mathrm{d}p_1(t)}{\mathrm{d}t}=-\begin{pmatrix}0&1\\0&0\end{pmatrix}p_1(t),\quad p_1(t_1)=p_2(t_1)=\begin{pmatrix}\frac{1}{2}\\\frac{1}{3}-t_1\end{pmatrix},$$

and the solution is

$$p_1(t) = \begin{pmatrix} (t_1 - \frac{1}{3})t + (\frac{1}{2} + \frac{1}{3}t_1 - t_1^2) \\ \frac{1}{3} - t_1 \end{pmatrix}.$$

Hence,

$$\boldsymbol{u}_t^{(1)} = \begin{pmatrix} u_1^{(1)}(t) \\ u_2^{(1)}(t) \end{pmatrix} = \begin{pmatrix} sgn[(\frac{1}{3} - t_1)t - (\frac{1}{2} + \frac{1}{3}t_1 - t_1^2)] \\ sgn(t_1 - \frac{1}{3}) \end{pmatrix}.$$

Choose $\varepsilon = 0.01$. By applying Algorithm 6.1, after 10 iterations, we find the optimal switching instant $t_1^* \in [0.592, 0.600]$, and $t_1^* = 0.596$. The corresponding optimal cost is -0.985. The optimal control law and $J(t, \boldsymbol{x})$ are

$$\boldsymbol{u}_t^{(1)*} = \begin{pmatrix} u_1^{(1)*}(t) \\ u_2^{(1)*}(t) \end{pmatrix} = \begin{pmatrix} -1 \\ 1 \end{pmatrix} \quad t \in [0, 0.596),$$

$$\boldsymbol{u}_t^{(2)*} = \begin{pmatrix} u_1^{(2)*}(t) \\ u_2^{(2)*}(t) \end{pmatrix} = \begin{pmatrix} 1 \\ 1 \end{pmatrix} \quad t \in [0.596, 1],$$

$$J(t, \boldsymbol{x}) = \begin{cases} (0.263t + 0.343)x_1 - 0.263x_2 + 0.131t^2 & \\ \quad +0.606t - 0.985, & t \in [0, 0.596) \\ \frac{1}{2}x_1 + (\frac{1}{3} - t)x_2 + t^2 - \frac{1}{6}t - \frac{5}{6}, & t \in [0.596, 1]. \end{cases}$$

6.3 LQ Switched Optimal Control Problem

We consider a kind of special model of switched uncertain systems with a quadratic objective function subject to some linear uncertain differential equations. Then the following uncertain expected value LQ model of switched uncertain systems is considered.

$$\begin{cases} J(0, \boldsymbol{x}_0) = \min\limits_{t_i} \min\limits_{u_s} E\Big[\int_0^T (\frac{1}{2}X_s^\tau Q(t)X_s + X_s^\tau V(t)u_s + \frac{1}{2}u_s^\tau R(t)u_s \\ \qquad +M(t)X_s + N(t)u_s + W(t))\mathrm{d}s + \frac{1}{2}X_T^\tau Q_T X_T + M_T X_T + L_T\Big] \\ \text{subject to} \\ \quad \mathrm{d}X_s = (A_i(s)X_s + B_i(s)u_s)\mathrm{d}s + \sigma(s, u_s, X_s)\mathrm{d}C_s, \\ \qquad s \in [t_{i-1}, t_i),\ i = 1, 2, \ldots, K+1 \\ \quad X_0 = \boldsymbol{x}_0. \end{cases} \tag{6.24}$$

where $T, \boldsymbol{x}_0$ are given, $Q(t) \in R^{n\times n}$, $V(t) \in R^{n\times r}$, $R(t) \in R^{r\times r}$, $M(t) \in R^n$, $N(t) \in R^r$, $W(t) \in R$ are functions of time t and $Q_T \geq 0$, $Q(t) \geq 0$, $R(t) > 0$.

The aim to discuss this model is to find not only an optimal control u_t^* but also an optimal switching law. To begin with we consider the following problem.

$$\begin{cases} J(t,\boldsymbol{x}) = \min\limits_{u_t} E\left[\int_t^T (\frac{1}{2}X_s^\tau Q(t)X_s + X_s^\tau V(t)u_s + \frac{1}{2}u_s^\tau R(t)u_s + M(t)X_s \right. \\ \qquad \left. +N(t)u_s + W(t))\mathrm{d}s + \frac{1}{2}X_T^\tau Q_T X_T + M_T X_T + L_T\right] \\ \text{subject to} \\ \quad \mathrm{d}X_s = (A_i(s)X_s + B_i(s)u_s)\mathrm{d}s + \sigma(s, u_s, X_s)\mathrm{d}C_s, \\ \qquad s \in [t_{i-1}, t_i),\ i = 1, 2, \ldots, K+1 \\ \quad X_t = \boldsymbol{x}. \end{cases} \tag{6.25}$$

By the equation of optimality (2.15) to deal with the model (6.25), the following conclusion can be obtained.

Theorem 6.3 *Assume that $J(t,\boldsymbol{x})$ be twice differentiable on $[t_{i-1}, t_i) \times R^n$. Then we have*

$$-J_t(t,\boldsymbol{x}) = \min_{u_t}\left[\frac{1}{2}\boldsymbol{x}^\tau Q(t)\boldsymbol{x} + \boldsymbol{x}^\tau V(t)u_t + \frac{1}{2}u_t^\tau R(t)u_t + M(t)\boldsymbol{x} + N(t)u_t + W(t)\right.$$
$$\left. +(A_i(t)\boldsymbol{x} + B_i(t)u_t)^\tau \nabla_{\boldsymbol{x}} J(t,\boldsymbol{x})\right] \tag{6.26}$$

where $J_t(t,\boldsymbol{x})$ is the partial derivatives of the function $J(t,\boldsymbol{x})$ in t, and $\nabla_{\boldsymbol{x}} J(t,\boldsymbol{x})$ is the gradient of $J(t,\boldsymbol{x})$ in $\boldsymbol{x}$.

Theorem 6.4 ([8]) *Assume that $J(t,\boldsymbol{x})$ be twice differentiable on $[t_{i-1}, t_i) \times R^n$ $(i = 1, 2, \ldots, K+1)$. Let $Q(t)$, $V(t)$, $R(t)$, $M(t)$, $N(t)$, $W(t)$, $A_i(t)$, $B_i(t)$, $R(t)^{-1}$ be continuous bounded functions of t, and $Q(t) \geq 0$, $Q_T \geq 0$, $R(t) > 0$. The optimal control of model (6.25) when $t \in [t_{i-1}, t_i)$ is that*

$$u_t^{(i)*} = -R(t)^{-1}(B_i(t)^\tau P_i(t) + V(t)^\tau)\boldsymbol{x} - R(t)^{-1}(B_i(t)^\tau S_i^\tau(t) + N(t)^\tau) \tag{6.27}$$

for $i = 1, 2, \ldots, K+1$, where $P_i(t) = P_i^\tau(t)$ and $S_i(t)$ satisfy, respectively,

$$\begin{cases} -\dot{P}_i(t) = Q(t) + P_i(t)A_i(t) + A_i(t)^\tau P_i(t) \\ \qquad -(P_i(t)B_i(t) + V(t))R(t)^{-1}(B_i(t)^\tau P_i(t) + V(t)^\tau) \\ P_{K+1}(T) = Q_T \text{ and } P_i(t_i) = P_{i+1}(t_i) \text{ for } i \leq K, \end{cases} \tag{6.28}$$

and

$$\begin{cases} -\dot{S}_i(t) = M(t) + S_i(t)A_i(t) \\ \qquad -(N(t) + S_i(t)B_i(t))R(t)^{-1}(B_i(t)^\tau P_i(t) + V(t)^\tau) \\ S_{K+1}(T) = M_T \text{ and } S_i(t_i) = S_{i+1}(t_i) \text{ for } i \leq K. \end{cases} \tag{6.29}$$

The optimal value of model (6.25) is

$$J(0,\boldsymbol{x}_0) = \frac{1}{2}\boldsymbol{x}_0^\tau P_1(0)\boldsymbol{x}_0 + S_1(0)\boldsymbol{x}_0 + L_1(0). \tag{6.30}$$

where $L_i(t), t \in [t_{i-1}, t_i)$ satisfies

$$\begin{cases} -\dot{L}_i(t) = W(t) - \frac{1}{2}(S_i(t)B_i(t) + N(t))R(t)^{-1}(B_i(t)^\tau S_i^\tau(t) + N(t)^\tau) \\ L_{K+1}(T) = L_T \text{ and } L_i(t_i) = L_{i+1}(t_i) \text{ for } i \leq K. \end{cases} \tag{6.31}$$

Proof It follows from the equation of optimality (6.26) that

$$\begin{aligned} -J_t(t, \boldsymbol{x}) = \min_{u_t}[&\frac{1}{2}\boldsymbol{x}^\tau Q(t)\boldsymbol{x} + \boldsymbol{x}^\tau V(t)u_t + \frac{1}{2}u_t^\tau R(t)u_t + M(t)\boldsymbol{x} + N(t)u_t + W(t) \\ &+(A_i(t)\boldsymbol{x} + B_i(t)u_t)^\tau \nabla_{\boldsymbol{x}} J(t, \boldsymbol{x})]. \end{aligned} \tag{6.32}$$

Let

$$\begin{aligned} L(u_t^{(i)}) = &\frac{1}{2}\boldsymbol{x}^\tau Q(t)\boldsymbol{x} + \boldsymbol{x}^\tau V(t)u_t^{(i)} + \frac{1}{2}u_t^{(i)\tau} R(t)u_t^{(i)} + M(t)\boldsymbol{x} + N(t)u_t^{(i)} + W(t) \\ &+(A_i(t)\boldsymbol{x} + B_i(t)u_t^{(i)})^\tau \nabla_{\boldsymbol{x}} J(t, \boldsymbol{x}). \end{aligned} \tag{6.33}$$

The optimal control $u_t^{(i)*}$ satisfies

$$\frac{\partial L(u_t^{(i)})}{\partial u_t^{(i)}} = V(t)^\tau \boldsymbol{x} + R(t)u_t^{(i)} + N(t)^\tau + B_i(t)^\tau \nabla_{\boldsymbol{x}} J(t, \boldsymbol{x}) = 0. \tag{6.34}$$

Since

$$\frac{\partial^2 L(u_t^{(i)})}{\partial^2 u_t^{(i)}} = R(t) > 0, \tag{6.35}$$

we have

$$u_t^{(i)*} = -R(t)^{-1}(V(t)^\tau \boldsymbol{x} + N(t)^\tau + B_i(t)^\tau \nabla_{\boldsymbol{x}} J(t, \boldsymbol{x})), \quad t \in [t_{i-1}, t_i). \tag{6.36}$$

Since $J(T, \boldsymbol{x}_T) = \frac{1}{2}X_T^\tau Q_T X_T + M_T X_T + L_T$, we guess

$$J(t, \boldsymbol{x}) = \frac{1}{2}\boldsymbol{x}^\tau P_{K+1}(t)\boldsymbol{x} + S_{K+1}(t)\boldsymbol{x} + L_{K+1}(t), \quad t \in [t_K, T], \tag{6.37}$$

and

$$P_{K+1}(t) = P_{K+1}(t)^\tau, \quad P_{K+1}(T) = Q_T, \quad S_{K+1}(T) = M_T, \quad L_{K+1}(T) = L_T.$$

So

$$J_t(t, \boldsymbol{x}) = \frac{1}{2}\boldsymbol{x}^\tau \dot{P}_{K+1}(t)\boldsymbol{x} + \dot{S}_{K+1}(t)\boldsymbol{x} + \dot{L}_{K+1}(t) \tag{6.38}$$

and

$$\nabla_{\boldsymbol{x}} J(t, \boldsymbol{x}) = P_{K+1}(t)\boldsymbol{x} + S_{K+1}^\tau(t). \tag{6.39}$$

Thus, it follows from (6.36) that

$$u_t^{(K+1)*} = -R(t)^{-1}(B_{K+1}(t)^\tau P_{K+1}(t) + V(t)^\tau)\boldsymbol{x} \\ -R(t)^{-1}(B_{K+1}^\tau S_{K+1}(t)^\tau + N(t)^\tau). \tag{6.40}$$

Substituting (6.38), (6.39), and (6.40) into (6.32) that

$$\begin{aligned} &-\frac{1}{2}\boldsymbol{x}^\tau \dot{P}_{K+1}(t)\boldsymbol{x} - \dot{S}_{K+1}(t)\boldsymbol{x} - \dot{L}_{K+1}(t) \\ &= \frac{1}{2}\boldsymbol{x}^\tau \big[Q(t) + P_{K+1}(t)A_{K+1}(t) + A_{K+1}(t)^\tau P_{K+1}(t) \\ &\quad -(P_{K+1}(t)B_{K+1}(t) + V(t))R(t)^{-1}(B_{K+1}(t)^\tau P_{K+1}(t) + V(t)^\tau)\big]\boldsymbol{x} \\ &\quad + \big[S_{K+1}(t)A_{K+1}(t) - (N(t) + S_{K+1}(t)B_{K+1}(t))\, R(t)^{-1}(B_{K+1}(t)^\tau P_{K+1}(t) \\ &\quad + V(t)^\tau) + M(t)]\boldsymbol{x} \\ &\quad + \left[W(t) - \frac{1}{2}(S_{K+1}(t)B_{K+1}(t) + N(t))R(t)^{-1}(B_{K+1}(t)^\tau S_{K+1}^\tau(t) + N(t)^\tau)\right]. \end{aligned}$$

Therefore, we have

$$\begin{cases} -\dot{P}_{K+1}(t) = Q(t) + P_{K+1}(t)A_{K+1}(t) + A_{K+1}(t)^\tau P_{K+1}(t) \\ \qquad -(P_{K+1}(t)B_{K+1}(t) + V(t))R(t)^{-1} \\ \qquad +(B_{K+1}(t)^\tau P_{K+1}(t) + V(t)^\tau), \\ P_{K+1}(T) = Q_T, \end{cases} \tag{6.41}$$

$$\begin{cases} -\dot{S}_{K+1}(t) = M(t) + S_{K+1}(t)A_{K+1}(t) - (N(t) \\ \qquad +S_{K+1}(t)B_{K+1}(t))R(t)^{-1}(B_{K+1}(t)^\tau P_{K+1}(t) + V(t)^\tau) \\ S_{K+1}(T) = M_T, \end{cases} \tag{6.42}$$

and

$$\begin{cases} -\dot{L}_{K+1}(t) = W(t) - \frac{1}{2}(S_{K+1}(t)B_{K+1}(t) \\ \qquad +N(t))R(t)^{-1}(B_{K+1}(t)^\tau S_{K+1}^\tau(t) + N(t)^\tau) \\ L_{K+1}(T) = L_T \end{cases} \tag{6.43}$$

Hence, $P_{K+1}(t)$, $S_{K+1}(t)$ and $L_{K+1}(t)$ satisfy the Riccati differential equation and boundary condition (6.41), (6.42), and (6.43), respectively.

When $t \in [t_{i-1}, t_i)$ for $i \le K$, assume

$$J(t, \boldsymbol{x}) = \frac{1}{2}\boldsymbol{x}^\tau P_i(t)\boldsymbol{x} + S_i(t)\boldsymbol{x} + L_i(t), \tag{6.44}$$

By the same method as above procedure, we can get

$$u_t^{(i)*} = -R(t)^{-1}(B_i(t)^\tau P_i(t) + V(t)^\tau)\boldsymbol{x} - R(t)^{-1}(B_i(t)^\tau S_i^\tau(t) + N(t)^\tau) \tag{6.45}$$

and

$$J(0, \boldsymbol{x}_0) = \frac{1}{2}\boldsymbol{x}_0^{\tau} P_1(0)\boldsymbol{x}_0 + S_1(0)\boldsymbol{x}_0 + L_1(0). \tag{6.46}$$

where $P_i(t) = P_i^{\tau}(t), S_i(t), L_i(t)$ satisfy, respectively,

$$\begin{cases} -\dot{P}_i(t) = Q(t) + P_i(t)A_i(t) + A_i(t)^{\tau} P_i(t) \\ \qquad\qquad -(P_i(t)B_i(t) + V(t))R(t)^{-1}(B_i(t)^{\tau} P_i(t) + V(t)^{\tau}) \\ P_i(t_i) = P_{i+1}(t_i), \end{cases}$$

$$\begin{cases} -\dot{S}_i(t) = M(t) + S_i(t)A_i(t) - (N(t) + S_i(t)B_i(t))R(t)^{-1}(B_i(t)^{\tau} P_i(t) + V(t)^{\tau}) \\ S_i(t_i) = S_{i+1}(t_i), \end{cases}$$

and

$$\begin{cases} -\dot{L}_i(t) = W(t) - \frac{1}{2}(S_i(t)B_i(t) + N(t))R(t)^{-1}(B_i(t)^{\tau} S_i^{\tau}(t) + N(t)^{\tau}) \\ L_i(t_i) = L_{i+1}(t_i) \end{cases}$$

The theorem is proved.

According to Theorem 6.2, there are $2(K+1)$ matrix Riccati differential equations to be solved in order to solve the model (6.5). Then the optimal cost $J(0, \boldsymbol{x}_0, t_1, \cdots t_K)$ can be obtained by (6.46). Denote $\tilde{J}(t_1, \cdots t_K) = J(0, \boldsymbol{x}_0)$. The next stage is to solve an optimization problem

$$\min_{0 \le t_1 \le t_2 \cdots \le t_K \le T} \tilde{J}(t_1, \cdots, t_K). \tag{6.47}$$

6.4 MACO Algorithm for Optimal Switching Instants

For the model (6.5), we may not obtain the analytical expressions of solutions according to Theorem 6.2. But most optimization algorithms need explicit forms of the first-order derivative of the objective functions. Being presented with such difficulties, evolutionary metaheuristic algorithms may be a good choices to solve Stage (b). An intelligent algorithm combining a mutation ant colony optimization algorithm and a simulated annealing method (MACO) was designed by Zhu [9] to solve continuous optimization models. We will use this algorithm to solve the following optimization problem

$$\begin{cases} \min \tilde{J}(t_1, \cdots, t_K) \\ \text{subject to} \\ 0 \le t_1 \le t_2 \cdots \le t_K \le T \\ t_i \in R(t), i = 1, 2, \ldots, K. \end{cases} \tag{6.48}$$

The vector $t = (t_1, \cdots t_K)$ is a decision vector which is in the feasible set of constrains

$$\Omega = \{t = (t_1, \cdots, t_K) | 0 \le t_1 \le t_2 \cdots \le t_K \le T\}.$$

Assume that $t_i = a_1 a_2 \cdots a_l . a_{l+1} a_{l+2} \cdots a_m$ for $i = 1, 2, \ldots, K$, where l and m ($m \ge l$) are some positive integers and a_k is a natural number which is no less than zero and no more than nine for $k = 1, 2, \ldots, m$. That is

$$t_i = \sum_{k=1}^{m} a_k \times 10^{l-k}, \quad i = 1, 2, \ldots, K. \tag{6.49}$$

where $a_k \in \{0, 1, 2, \ldots, 9\}$ for $k = 1, 2, \ldots, m$. The parameters l and m are selected according to required precision of solutions of problem (6.48).

Let artificial ants walk step by step. Call the numbers $k = 0, 1, \ldots, 9$ to be nodes of each step. For every t_i, each artificial ant is first put on 0 and moves to a node of the 1st step, and then to a node of the 2nd step, until to a node of the mth step. In this movement, an artificial ant walks from a node to the next node according to the strength of the pheromone trails on the latter nodes. If the node of the kth step that an artificial ant selects is j, then equip a_k by j. Once all artificial ants have completed their walk, pheromone trails are updated. Denote the pheromone trail by $\tau_{i;k,j}(s)$ associated to node j of the kth step for the variable t_i at iteration s. The procedures are described as follows.

(1) *Initialization Process*: Randomly generate a feasible solution t as the best solution $\hat{t}$. Set $\tau_{i;k,j}(0) = \tau_0, i = 1, 2, \ldots, K, k = 1, 2, \ldots, m, j = 0, 1, \ldots, 9$, where τ_0 is a parameter.

(2) *Ant Movement*: At step k after building the sequence $\langle a_1, a_2, \cdots, a_k \rangle$, select the next node j of the $(k+1)$th step in the following probability

$$p_{k,k+1} = \frac{\tau_{i;k+1,j}(s)}{\sum_{q=0}^{9} \tau_{i;k+1,q}(s)}. \tag{6.50}$$

After obtaining the sequence $\langle a_1, a_2, \cdots, a_m \rangle$, and form t_i according to Eq. (6.49). The feasible set Ω may be used to check the feasibility of the vector $t = (t_1, \cdots t_K)$.

In order to avoid the premature of the best solution $\hat{t}$ so far, we modify it based on the idea of mutation and Metropolis' acceptance law. Construct a feasible vector t' in the neighbor of $\hat{t}$ as follows: randomly selecting $h_i \in (-1, 1)$, and $l_i \in [0, L)$ for some positive number L, let

$$t' = \hat{t} + (l_1 h_1, l_2 h_2, \cdots l_K h_K)$$

The feasibility of t' may be guaranteed by choosing l_i small enough or $l_i = 0$. If $\Delta f = f(t') - f(\hat{t}) \le 0$, then $\hat{t} \leftarrow t'$. Otherwise, if Metropolis' acceptance law holds, that is, $exp(-\Delta f / T_s) > random(0, 1)$ where $T_s \to 0$ as iteration $s \to \infty$, then denote $\tilde{t} \leftarrow t'$.

(3) *Pheromone Update*: At each moment s, let $\hat{t}$ be the best solution found so far, and t_s be the best solution in the current algorithm iteration s. If $\tilde{J}(t_s) < \tilde{J}(\hat{t})$, then $\hat{t} \leftarrow t_s$. Reinforce the pheromone trails on arcs of $\hat{t}$ and $\tilde{t}$ (if any) and evaporate the pheromone trails on arcs of others:

$$\tau_{i;k,j}(s) = \begin{cases} (1-\rho)\tau_{i;k,j}(s-1) + \rho g(\hat{t}), & \text{if } (k,j) \in \hat{t} \\ (1-\rho)\tau_{i;k,j}(s-1) + \frac{\rho}{2} g(\hat{t}), & \text{if } (k,j) \in \tilde{t} \\ (1-\rho)\tau_{i;k,j}(s-1), & \text{otherwise} \end{cases} \tag{6.51}$$

where ρ, $0 < \rho < 1$, is the evaporation rate, and $g(\boldsymbol{x})$ is a function with that $g(\boldsymbol{x}) \geq g(y)$ if $\tilde{J}(\boldsymbol{x}) < \tilde{J}(y)$.

The algorithm can be summarized as follows.

Algorithm 6.2 (MACO algorithm for solving (6.48))

Step 1. Initialize all pheromone trails with the same amount of pheromone and randomly generate a feasible solution.
Step 2. Ant movement according to the pheromone trails to produce the value of a decision variable.
Step 3. Repeat step 2 to produce $t_1, t_2, \cdots, t_K$ and check them with the feasible set Ω.
Step 4. Repeat step 2 to step 3 for a given number of artificial ants.
Step 5. Update pheromone according to the best feasible solution found so far.
Step 6. Repeat step 2 to step 5 for a given number of cycles.
Step 7. Report the best solution as the optimal solution.

6.4.1 Example

Consider the following example of LQ models for switched uncertain systems

$$\begin{cases} J(0, \boldsymbol{x}_0) = \min\limits_{t_1,t_2} \min\limits_{u(s)} E\left[\int_0^1 (-X(s) - u(s) + \frac{1}{2}u^2(s) + 1)\mathrm{d}s - X^2(1)\right] \\ \text{subject to} \\ \textit{subsystem } 1: \mathrm{d}X(s) = [u(s) - \alpha_1 X(s)]\mathrm{d}s + \sigma X(s)\mathrm{d}C_s, \quad s \in [0, t_1) \\ \textit{subsystem } 2: \mathrm{d}X(s) = [u(s) - \alpha_2 X(s)]\mathrm{d}s + \sigma X(s)\mathrm{d}C_s, \quad s \in [t_1, t_2) \\ \textit{subsystem } 3: \mathrm{d}X(s) = [u(s) - \alpha_3 X(s)]\mathrm{d}s + \sigma X(s)\mathrm{d}C_s, \quad s \in [t_2, 1] \\ \quad X(0) = 1. \end{cases} \tag{6.52}$$

Comparing the example with model (6.24), we have: $Q(t) = 0$, $R(t) = 1$, $V(t) = 0$, $M(t) = -1$, $N(t) = -1$, $W(t) = 1$, $T = 1$, $Q_T = -2$, $M(t)_T = 0$, $L_T = 0$, $A_i(t) = -\alpha_i$, $B_i(t) = 1 (i = 1, 2, 3)$.

Stage (a): Fix t_1, t_2 and formulate $\tilde{J}(t_1, t_2)$ according to Theorem 6.4. It follows from (6.28) and (6.29) that

$$\begin{cases} -\dot{P}_i(t) = -P_i^2(t) - 2\alpha_i P_i(t) \\ P_3(1) = -2, P_3(t_2) = P_2(t_2), P_2(t_1) = P_1(t_1), \end{cases} \tag{6.53}$$

and

$$\begin{cases} -\dot{S}_i(t) = -(P_i(t) + \alpha_i)S_i(t) + P_i(t) - 1 \\ S_3(1) = 0, S_3(t_2) = S_2(t_2), S_2(t_1) = S_1(t_1) \end{cases} \tag{6.54}$$

Then the solutions of Eqs. (6.53) and (6.54) are

$$P_3(t) = \frac{m_3 e^{m_3 t}}{-e^{m_3 t} + n_3}, \quad S_3(t) = \frac{-2(m_3+1)e^{m_3 t} - 2n_3 + c_3 m_3 e^{\frac{1}{2}m_3 t}}{m_3(n_3 - e^{m_3 t})}$$

for $i = 3$, where $m_3 = 2\alpha_3$, $S_{t_3} = -1$, $n_3 = (-S_{t_3} - \alpha_3)e^{m_3}$, $c_3 = (\frac{4}{m_3} + 1)e^{\frac{1}{2}m_3}$.

In addition, we have

$$P_2(t) = \frac{-m_2 S_{t_2} e^{m_2 t}}{S_{t_2} e^{m_2 t} + n_2}, \quad S_2(t) = \frac{2S_{t_2}(m_2+1)e^{m_2 t} - 2n_2 + c_2 m_2 e^{\frac{1}{2}m_2 t}}{m_2(n_2 + S_{t_2} e^{m_2 t})}$$

for $i = 2$, where

$$m_2 = 2\alpha_2, \ S_{t_2} = \frac{1}{2}P_3(t_2), \ n_2 = (-S_{t_2} - \alpha_2)e^{m_2 t_2},$$

$$S'_{t_2} = S_3(t_2), \ c_2 = \left(-\frac{m_2}{2}S'_{t_2} - 2S_{t_2} - 1 - \frac{4S_{t_2}}{m_2}\right)e^{\frac{1}{2}m_2 t_2},$$

and

$$P_1(t) = \frac{-m_1 S_{t_1} e^{m_1 t}}{S_{t_1} e^{m_1 t} + n_1}, \quad S_1(t) = \frac{2S_{t_1}(m_1+1)e^{m_1 t} - 2n_1 + c_1 m_1 e^{\frac{1}{2}m_1 t}}{m_1(n_1 + S_{t_1} e^{m_1 t})}$$

for $i = 1$, where

$$m_1 = 2\alpha_1, \ S_{t_1} = \frac{1}{2}P_2(t_1), \ n_1 = (-S_{t_1} - \alpha_1)e^{m_1 t_1},$$

$$S'_{t_1} = S_2(t_1), \ c_1 = \left(-\frac{m_1}{2}S'_{t_1} - 2S_{t_1} - 1 - \frac{4S_{t_1}}{m_1}\right)e^{\frac{1}{2}m_1 t_1}.$$

According to Theorem 6.4, the optimal value is

$$\tilde{J}(t_1, t_2) = \frac{1}{2}P_1(0) + S_1(0) + L_1(0).$$

where

$$L_1(0) = \int_0^{t_1} \left[-\frac{1}{2} S_1^2(t) + S_1(t) + \frac{1}{2} \right] \mathrm{d}t + \int_{t_1}^{t_2} \left[-\frac{1}{2} S_2^2(t) + S_2(t) + \frac{1}{2} \right] \mathrm{d}t + \int_{t_2}^{1} \left[-\frac{1}{2} S_3^2(t) + S_3(t) + \frac{1}{2} \right] \mathrm{d}t.$$

Stage (b): Find the optimal switching instant t_1^*, t_2^* according to Algorithm 6.2. Choose $\alpha_1 = \frac{1}{3}$, $\alpha_2 = \frac{1}{4}$, $\alpha_3 = \frac{1}{2}$. By applying Algorithm 6.2, we find the optimal switching instant $t_1^* = 0.303$, $t_2^* = 0.462$. The optimal control is

$$u_t^* = \begin{cases} 1 - \dfrac{675.14e^{0.667t} - 1197.62e^{0.333t} - 502.78}{167.68 + 135.07e^{0.667t}} + \dfrac{135.17e^{0.667t}\boldsymbol{x}(t)}{251.39 + 202.5e^{0.667t}}, & t \in [0, 0.303) \\ 1 + \dfrac{10.44e^{0.5t} - 19.79e^{0.25t} + 8.14}{2.04 - 1.74e^{0.5t}} - \dfrac{1.74e^{0.5t}\boldsymbol{x}(t)}{4.07 - 3.48e^{0.5t}}, & t \in [0.303, 0.462) \\ 1 + \dfrac{4e^{t} - 8.24e^{0.5t} + 2.718}{1.359 - e^{t}} - \dfrac{e^{t}\boldsymbol{x}(t)}{1.359 - e^{t}}, & t \in [0.462, 1]. \end{cases}$$

6.5 Optimistic Value Model

Consider an optimistic value model of switched uncertain systems for multidimensional case as follows.

$$\begin{cases} J(0, \boldsymbol{x}_0) = \min\limits_{t_i} \max\limits_{\boldsymbol{u}_s \in [-1,1]^r} F_{\sup}(\alpha) \\ \text{subject to} \\ \quad \mathrm{d}\boldsymbol{X}_s = (\boldsymbol{A}_i(s)\boldsymbol{X}_s + \boldsymbol{B}_i(s)\boldsymbol{u}_s)\mathrm{d}s + \boldsymbol{Q}_i \mathrm{d}\boldsymbol{C}_s, \\ \qquad s \in [t_{i-1}, t_i),\ i = 1, 2, \cdots, K+1 \\ \quad \boldsymbol{X}_t = \boldsymbol{x}. \end{cases} \tag{6.55}$$

where $t_{K+1} = T$, $F = \int_t^T \boldsymbol{f}(s)^\tau \boldsymbol{X}_s \mathrm{d}s + \boldsymbol{S}_T^\tau \boldsymbol{X}_T$, and $F_{\sup}(\alpha) = \sup\{\bar{F} \mid \mathcal{M}\{F \ge \bar{F}\} \ge \alpha\}$ which denotes the α-optimistic value to F. The function $\boldsymbol{f} : [0, T] \to R^n$ is the objective function of dimension n, $\boldsymbol{S}_T \in R^n$. We will use $J(t, \boldsymbol{x})$ to denote the optimal value $\max_{\boldsymbol{u}_s} F_{\sup}(\alpha)$ obtained in $[t, T]$ with the condition that at time t we are in state $\boldsymbol{X}_t = \boldsymbol{x}$.

Applying the equation of optimality (3.4) to deal with model (6.55), the following conclusion can be obtained.

Theorem 6.5 *Let $J(t, \boldsymbol{x})$ be twice differentiable on $[t_{i-1}, t_i) \times R^n$ for $i = 1, 2, \ldots, K+1$. Then we have*

$$-J_t(t, \boldsymbol{x}) = \max_{\boldsymbol{u}_t \in [-1,1]^r} \Big\{\boldsymbol{f}(t)^\tau \boldsymbol{x} + (\boldsymbol{A}_i(t)\boldsymbol{x} + \boldsymbol{B}_i(t)\boldsymbol{u}_t)^\tau \nabla_x J(t, \boldsymbol{x}) + \frac{\sqrt{3}}{\pi} \ln \frac{1-\alpha}{\alpha} \|\nabla_x J(t, \boldsymbol{x})^\tau \boldsymbol{Q}_i\|_1\Big\}, \tag{6.56}$$

where $J_t(t, \boldsymbol{x})$ is the partial derivatives of the function $J(t, \boldsymbol{x})$ in t, $\nabla_x J(t, \boldsymbol{x})$ is the gradient of $J(t, \boldsymbol{x})$ in $\boldsymbol{x}$, and $\|\cdot\|_1$ is the 1-norm for vectors, that is, $\|\boldsymbol{v}\|_1 = \sum_{i=1}^m |v_i|$ for $\boldsymbol{v} = (v_1, v_2 \cdots v_m)$.

6.5.1 Two-Stage Approach

In order to solve problem (6.55), we decompose it into two stages. Stage (a) deals with a conventional uncertain optimal control problem which seeks the optimal value of J with respect to the switching instants. Stage (b) solves an optimization problem in the switching instants.

6.5.2 Stage (a)

Now we fix the switching instants $t_1, t_2, \cdots, t_K$ and handle the following model to find the optimal value:

$$\begin{cases} J(0, \boldsymbol{x}_0, t_1, \cdots, t_K) = \max\limits_{\boldsymbol{u}_s \in [-1,1]^r} \left[\displaystyle\int_0^T \boldsymbol{f}(s)^\tau \boldsymbol{X}_s \mathrm{d}s + \boldsymbol{S}_T^\tau \boldsymbol{X}_T\right]_{\sup} \quad (\alpha) \\ \text{subject to} \\ \quad \mathrm{d}\boldsymbol{X}_s = (\boldsymbol{A}_i(s)\boldsymbol{X}_s + \boldsymbol{B}_i(s)\boldsymbol{u}_s)\mathrm{d}s + \boldsymbol{Q}_i \mathrm{d}\boldsymbol{C}_s \\ \qquad\quad s \in [t_{i-1}, t_i),\ i = 1, 2, \ldots, K+1 \\ \quad \boldsymbol{X}_0 = \boldsymbol{x}_0. \end{cases} \tag{6.57}$$

Applying Eq. (6.56) to model (6.57), we have the following conclusion.

Theorem 6.6 ([10]) *Let $J(t, \boldsymbol{x})$ be twice differentiable on $[t_{i-1}, t_i) \times R^n$ $(i = 1, 2, \ldots, K+1)$. The optimal control $\boldsymbol{u}_t^{(i)*} = (u_1^{(i)*}(t), u_2^{(i)*}(t), \cdots, u_r^{(i)*}(t))^\tau$ of (6.57) is a bang–bang control*

$$u_j^{(i)*}(t) = sgn\{(b_{1j}^{(i)}(t), b_{2j}^{(i)}(t), \cdots, b_{nj}^{(i)}(t))\boldsymbol{p}_i(t)\} \tag{6.58}$$

for $i = 1, 2, \ldots, K+1; j = 1, 2, \ldots, r$, *where* $\boldsymbol{B}_i(t) = (b_{lj}^{(i)}(t))_{n\times r}$ *and* $\boldsymbol{p}_i(t) \in R^n$, $t \in [t_{i-1}, t_i)$, *satisfies*

$$\begin{cases} \dfrac{\mathrm{d}\boldsymbol{p}_i(t)}{\mathrm{d}t} = -\boldsymbol{f}(t) - \boldsymbol{A}_i(t)^\tau \boldsymbol{p}_i(t) \\ \boldsymbol{p}_{K+1}(T) = \boldsymbol{S}_T \text{ and } \boldsymbol{p}_i(t_i) = \boldsymbol{p}_{i+1}(t_i) \text{ for } i \le K. \end{cases} \tag{6.59}$$

The optimal value of model (6.57) *is*

$$\begin{aligned} J(0, \boldsymbol{x}_0, t_1, \cdots, t_K) = \boldsymbol{p}_1(0)^\tau \boldsymbol{x}_0 + \sum_{i=1}^{K+1} \int_{t_{i-1}}^{t_i} \big[\|\boldsymbol{p}_i(t)^\tau \boldsymbol{B}_i(t)\|_1 \\ + \frac{\sqrt{3}}{\pi} \ln \frac{1-\alpha}{\alpha} \|\boldsymbol{p}_i(t)^\tau \boldsymbol{Q}_i\|_1 \big] \mathrm{d}t. \end{aligned} \tag{6.60}$$

Proof First we prove the optimal control of model (6.57) is a bang–bang control. It follows from the equation of optimality (6.56) that

$$\begin{aligned} -J_t(t, \boldsymbol{x}) = \max_{\boldsymbol{u}_t \in [-1,1]^r} \big\{ \boldsymbol{f}(t)^\tau \boldsymbol{x} + (\boldsymbol{A}_i(t)\boldsymbol{x} + \boldsymbol{B}_i(t)\boldsymbol{u}_t)^\tau \nabla_{\boldsymbol{x}} J(t, \boldsymbol{x}) \\ + \frac{\sqrt{3}}{\pi} \ln \frac{1-\alpha}{\alpha} \|\nabla_{\boldsymbol{x}} J(t, \boldsymbol{x})^\tau \boldsymbol{Q}_i\|_1 \big\}. \end{aligned} \tag{6.61}$$

On the right-hand side of (6.61), let $\boldsymbol{u}_t^{(i)*}$ make it the maximum. We have

$$\begin{aligned} &\max_{\boldsymbol{u}_t \in [-1,1]^r} \big\{ \boldsymbol{f}(t)^\tau \boldsymbol{x} + (\boldsymbol{A}_i(t)\boldsymbol{x} + \boldsymbol{B}_i(t)\boldsymbol{u}_t)^\tau \nabla_{\boldsymbol{x}} J(t, \boldsymbol{x}) \\ &\quad + \frac{\sqrt{3}}{\pi} \ln \frac{1-\alpha}{\alpha} \|\nabla_{\boldsymbol{x}} J(t, \boldsymbol{x})^\tau \boldsymbol{Q}_i\|_1 \big\} \\ &= \boldsymbol{f}(t)^\tau \boldsymbol{x} + (\boldsymbol{A}_i(t)\boldsymbol{x} + \boldsymbol{B}_i(t)\boldsymbol{u}_t^{(i)*})^\tau \nabla_{\boldsymbol{x}} J(t, \boldsymbol{x}) + \frac{\sqrt{3}}{\pi} \ln \frac{1-\alpha}{\alpha} \|\nabla_{\boldsymbol{x}} J(t, \boldsymbol{x})^\tau \boldsymbol{Q}_i\|_1. \end{aligned}$$

That is,

$$\max_{\boldsymbol{u}_t \in [-1,1]^r} \{\nabla_{\boldsymbol{x}} J(t, \boldsymbol{x})^\tau \boldsymbol{B}_i(t)\boldsymbol{u}_t\} = \nabla_{\boldsymbol{x}} J(t, \boldsymbol{x})^\tau \boldsymbol{B}_i(t)\boldsymbol{u}_t^{(i)*}. \tag{6.62}$$

Denote

$$\boldsymbol{u}_t^{(i)*} = (u_1^{(i)*}(t), u_2^{(i)*}(t), \cdots, u_r^{(i)*}(t))^\tau$$

and

$$\nabla_{\boldsymbol{x}} J(t, \boldsymbol{x})^\tau \boldsymbol{B}_i(t) = (g_1^{(i)}(t, \boldsymbol{x}), g_2^{(i)}(t, \boldsymbol{x}), \cdots, g_r^{(i)}(t, \boldsymbol{x})).$$

Then,

$$u_j^{(i)*}(t) = \begin{cases} 1, & \text{if } g_j^{(i)}(t, \boldsymbol{x}) > 0 \\ -1, & \text{if } g_j^{(i)}(t, \boldsymbol{x}) < 0 \\ \text{undetermined}, & \text{if } g_j^{(i)}(t, \boldsymbol{x}) = 0 \end{cases} \tag{6.63}$$

for $i = 1, 2, \ldots, K+1; j = 1, 2, \ldots, r$, which is a bang–bang control. The functions $g_j^{(i)}(t, \boldsymbol{x})$ are called switching functions. If at least one switching function equals to zero in some interval, we call it a singular control. But here we only consider switching functions equal to zero at most in some discrete points.

According to (6.61), when $t \in [t_K, T]$, we have

$$-J_t(t, \boldsymbol{x}) = \max_{\boldsymbol{u}_t \in [-1,1]^r} \Big\{ \boldsymbol{f}(t)^\tau \boldsymbol{x} + (\boldsymbol{A}_{K+1}(t)\boldsymbol{x} + \boldsymbol{B}_{K+1}(t)\boldsymbol{u}_t)^\tau \nabla_{\boldsymbol{x}} J(t, \boldsymbol{x}) + \frac{\sqrt{3}}{\pi} \ln \frac{1-\alpha}{\alpha} \|\nabla_{\boldsymbol{x}} J(t, \boldsymbol{x})^\tau \boldsymbol{Q}_{K+1}\|_1 \Big\}. \tag{6.64}$$

Since $J(T, \boldsymbol{x}_T) = \boldsymbol{S}_T^\tau \boldsymbol{x}_T$, we assume

$$J(t, \boldsymbol{x}) = \boldsymbol{p}_{K+1}(t)^\tau \boldsymbol{x} + q_{K+1}(t) \quad \text{and} \quad \boldsymbol{p}_{K+1}(T) = \boldsymbol{S}_T, \ q_{K+1}(T) = 0.$$

So

$$\nabla_{\boldsymbol{x}} J(t, \boldsymbol{x}) = \boldsymbol{p}_{K+1}(t), \ J_t(t, \boldsymbol{x}) = \frac{\mathrm{d}\boldsymbol{p}_{K+1}(t)^\tau}{\mathrm{d}t} \boldsymbol{x} + \frac{\mathrm{d}q_{K+1}(t)}{\mathrm{d}t}. \tag{6.65}$$

Thus, it follows from (6.63) that

$$u_j^{(K+1)*}(t) = \operatorname{sgn}\{(b_{1j}^{(K+1)}(t), b_{2j}^{(K+1)}(t), \cdots, b_{nj}^{(K+1)}(t))\boldsymbol{p}_{K+1}(t)\} \tag{6.66}$$

for $j = 1, 2, \ldots, r$. Substituting (6.65) into (6.64) yields

$$-\frac{\mathrm{d}\boldsymbol{p}_{K+1}(t)^\tau}{\mathrm{d}t}\boldsymbol{x} - \frac{\mathrm{d}q_{K+1}(t)}{\mathrm{d}t} = \boldsymbol{f}(t)^\tau \boldsymbol{x} + (\boldsymbol{A}_{K+1}(t)\boldsymbol{x} + \boldsymbol{B}_{K+1}(t)\boldsymbol{u}_t^{(K+1)*})^\tau \boldsymbol{p}_{K+1}(t) + \frac{\sqrt{3}}{\pi} \ln \frac{1-\alpha}{\alpha} \|\boldsymbol{p}_{K+1}(t)^\tau \boldsymbol{Q}_{K+1}\|_1.$$

Therefore, we have

$$-\frac{\mathrm{d}\boldsymbol{p}_{K+1}(t)^\tau}{\mathrm{d}t} = \boldsymbol{f}(t)^\tau + \boldsymbol{p}_{K+1}(t)^\tau \boldsymbol{A}_{K+1}(t), \quad \boldsymbol{p}_{K+1}(T) = \boldsymbol{S}_T, \tag{6.67}$$

and

$$\begin{cases} -\dfrac{\mathrm{d}q_{K+1}(t)}{\mathrm{d}t} = \boldsymbol{p}_{K+1}(t)^\tau \boldsymbol{B}_{K+1}(t)\boldsymbol{u}_t^{(K+1)*} + \dfrac{\sqrt{3}}{\pi} \ln \dfrac{1-\alpha}{\alpha} \|\boldsymbol{p}_{K+1}(t)^\tau \boldsymbol{Q}_{K+1}\|_1, \\ q_{K+1}(T) = 0. \end{cases} \tag{6.68}$$

Substituting (6.66) into (6.68), we can get

$$q_{K+1}(t) = \int_t^T \left[\|\boldsymbol{p}_{K+1}(s)^\tau \boldsymbol{B}_{K+1}(s)\|_1 + \frac{\sqrt{3}}{\pi} \ln \frac{1-\alpha}{\alpha} \|\boldsymbol{p}_{K+1}(s)^\tau \boldsymbol{Q}_{K+1}\|_1 \right] \mathrm{d}s.$$

So when $t \in [t_K, T]$, we have

$$
\begin{aligned}
J(t, \boldsymbol{x}) &= \boldsymbol{p}_{K+1}(t)^\tau \boldsymbol{x} + q_{K+1}(t) \\
&= \boldsymbol{p}_{K+1}(t)^\tau \boldsymbol{x} + \int_t^T \big[\|\boldsymbol{p}_{K+1}(s)^\tau \boldsymbol{B}_{K+1}(s)\|_1 \\
&\quad + \frac{\sqrt{3}}{\pi} \ln \frac{1-\alpha}{\alpha} \|\boldsymbol{p}_{K+1}(s)^\tau \boldsymbol{Q}_{K+1}\|_1\big] \mathrm{d}s,
\end{aligned}
$$

where $\boldsymbol{p}_{K+1}(t)$ satisfies the Riccati differential equation and boundary condition (6.67).

When $t \in [t_{i-1}, t_i)$ for $i \le K$, assume

$$
J(t, \boldsymbol{x}) = \boldsymbol{p}_i(t)^\tau \boldsymbol{x} + q_i(t), \quad \text{and} \quad \boldsymbol{p}_i(t_i) = \boldsymbol{p}_{i+1}(t_i), \quad q_i(t_i) = q_{i+1}(t_i).
$$

By the same method as the above procedure, we can get

$$
u_j^{(i)*}(t) = \operatorname{sgn}\{(b_{1j}^{(i)}(t), b_{2j}^{(i)}(t), \cdots, b_{nj}^{(i)}(t))\boldsymbol{p}_i(t)\}
$$

for $j = 1, 2, \ldots, r$, where

$$
\begin{cases}
-\dfrac{\mathrm{d}\boldsymbol{p}_i(t)^\tau}{\mathrm{d}t} = \boldsymbol{f}(t)^\tau + \boldsymbol{p}_i(t)^\tau \boldsymbol{A}_i(t), \quad \boldsymbol{p}_i(t_i) = \boldsymbol{p}_{i+1}(t_i) \\
-\dfrac{\mathrm{d}q_i(t)}{\mathrm{d}t} = \|\boldsymbol{p}_i(t)^\tau \boldsymbol{B}_i(t)\|_1 + \dfrac{\sqrt{3}}{\pi} \ln \dfrac{1-\alpha}{\alpha} \|\boldsymbol{p}_i(t)^\tau \boldsymbol{Q}_i\|_1, \quad q_i(t_i) = q_{i+1}(t_i),
\end{cases}
$$

and

$$
\begin{aligned}
J(t, \boldsymbol{x}) &= \boldsymbol{p}_i(t)^\tau \boldsymbol{x} + q_i(t) \\
&= \boldsymbol{p}_i(t)^\tau \boldsymbol{x} + \int_t^{t_i} \left[\|\boldsymbol{p}_i(s)^\tau \boldsymbol{B}_i(s)\|_1 + \frac{\sqrt{3}}{\pi} \ln \frac{1-\alpha}{\alpha} \|\boldsymbol{p}_i(s)^\tau \boldsymbol{Q}_i\|_1 \right] \mathrm{d}s \\
&\quad + q_{i+1}(t_i).
\end{aligned}
$$

Summarily, the optimal value of model (6.57) is

$$
\begin{aligned}
J(0, \boldsymbol{x}_0, t_1, \cdots, t_K) &= \boldsymbol{p}_1(0)^\tau \boldsymbol{x}_0 + \sum_{i=1}^{K+1} \int_{t_{i-1}}^{t_i} \big[\|\boldsymbol{p}_i(t)^\tau \boldsymbol{B}_i(t)\|_1 \\
&\quad + \frac{\sqrt{3}}{\pi} \ln \frac{1-\alpha}{\alpha} \|\boldsymbol{p}_i(t)^\tau \boldsymbol{Q}_i\|_1\big] \mathrm{d}t.
\end{aligned}
$$

The theorem is proved.

6.5.3 Stage (b)

According to Theorem 6.6, there are $(K+1)$ matrix Riccati differential equations to be solved in order to solve the model (6.57). Then the optimal cost $J(0, \boldsymbol{x}_0, t_1, \cdots, t_K)$ can be obtained by (6.60). Denote $\tilde{J}(t_1, \cdots, t_K) = J(0, \boldsymbol{x}_0, t_1, \cdots, t_K)$. The next stage is to solve an optimization problem:

$$\begin{cases} \max \tilde{J}(t_1, \cdots, t_K) \\ \text{subject to} \\ \quad 0 \le t_1 \le t_2 \cdots \le t_K \le T \\ \quad t_i \in R, i = 1, 2, \ldots, K. \end{cases} \tag{6.69}$$

For model (6.57), we may not obtain the analytical expressions and derivative of the optimal reward according to Theorem 6.6. But gradient algorithms need explicit forms of the first-order derivative of the optimal reward. Being presented with such difficulties, evolutionary metaheuristic algorithms such as GA and PSO algorithm are good choices to solve Stage (b) which offer a high degree of flexibility and robustness in dynamic environments.

6.5.4 Example

Consider the following optimal control problem with two uncertain subsystems:

$$\begin{cases} J(0, \boldsymbol{x}_0) = \min\limits_{t_1} \max\limits_{\boldsymbol{u}_s} \left[\frac{1}{2} X_1(1) - \frac{2}{3} X_2(1) \right]_{\sup} \quad (\alpha) \\ \text{subject to} \\ \textit{subsystem } 1: \begin{cases} dX_1(s) = u_1(s) ds + \sigma dC_{s1} \\ dX_2(s) = (X_1(s) + u_2(s)) ds + \sigma dC_{s2}, \quad s \in [0, t_1) \end{cases} \\ \textit{subsystem } 2: \begin{cases} dX_1(s) = (2X_2(s) - u_1(s)) ds + \sigma dC_{s1} \\ dX_2(s) = (u_1(s) + u_2(s)) ds + \sigma dC_{s2}, \quad s \in [t_1, 1] \end{cases} \\ \quad X_1(0) = X_2(0) = 0 \\ \quad |u_i(s)| \le 1, \ 0 \le s \le 1, i = 1, 2. \end{cases}$$

Comparing the example with the model (6.55), we have

$$\boldsymbol{A}_1(s) = \begin{pmatrix} 0 & 0 \\ 1 & 0 \end{pmatrix}, \ \boldsymbol{A}_2(s) = \begin{pmatrix} 0 & 2 \\ 0 & 0 \end{pmatrix}, \ \boldsymbol{B}_1(s) = \begin{pmatrix} 1 & 0 \\ 0 & 1 \end{pmatrix}, \ \boldsymbol{B}_2(s) = \begin{pmatrix} -1 & 0 \\ 1 & 1 \end{pmatrix},$$

$$\boldsymbol{f}(s) = \boldsymbol{0}, \ \boldsymbol{Q}_1 = \boldsymbol{Q}_2 = \begin{pmatrix} \sigma & 0 \\ 0 & \sigma \end{pmatrix}, \ \boldsymbol{S}_1 = \begin{pmatrix} \frac{1}{2} \\ -\frac{2}{3} \end{pmatrix}.$$

Stage (a): Fix t_1 and formulate $\tilde{J}(t_1)$ according to Theorem 6.6. It follows from (6.59) that

$$\frac{\mathrm{d}\boldsymbol{p}_2(t)}{\mathrm{d}t} = -\begin{pmatrix} 0 & 0 \\ 2 & 0 \end{pmatrix}\boldsymbol{p}_2(t), \quad \boldsymbol{p}_2(1) = \begin{pmatrix} \frac{1}{2} \\ -\frac{2}{3} \end{pmatrix}$$

which has the solution

$$\boldsymbol{p}_2(t) = \begin{pmatrix} \frac{1}{2} \\ \frac{1}{3} - t \end{pmatrix}.$$

Hence,

$$\boldsymbol{u}_t^{(2)*} = \begin{pmatrix} u_1^{(2)*}(t) \\ u_2^{(2)*}(t) \end{pmatrix} = \begin{pmatrix} \operatorname{sgn}(-\frac{1}{6} - t) \\ \operatorname{sgn}(\frac{1}{3} - t) \end{pmatrix}.$$

It also follows from (6.59) that

$$\frac{\mathrm{d}\boldsymbol{p}_1(t)}{\mathrm{d}t} = -\begin{pmatrix} 0 & 1 \\ 0 & 0 \end{pmatrix}\boldsymbol{p}_1(t), \quad \boldsymbol{p}_1(t_1) = \boldsymbol{p}_2(t_1) = \begin{pmatrix} \frac{1}{2} \\ \frac{1}{3} - t_1 \end{pmatrix}$$

which has the solution

$$\boldsymbol{p}_1(t) = \begin{pmatrix} (t_1 - \frac{1}{3})t + (\frac{1}{2} + \frac{1}{3}t_1 - t_1^2) \\ \frac{1}{3} - t_1 \end{pmatrix}.$$

Hence

$$\boldsymbol{u}_t^{(1)*} = \begin{pmatrix} u_1^{(1)*}(t) \\ u_2^{(1)*}(t) \end{pmatrix} = \begin{pmatrix} \operatorname{sgn}[(t_1 - \frac{1}{3})t + (\frac{1}{2} + \frac{1}{3}t_1 - t_1^2)] \\ \operatorname{sgn}(\frac{1}{3} - t_1) \end{pmatrix},$$

and

$$\begin{aligned}\tilde{J}(t_1) = \left(\frac{\sqrt{3}}{\pi}\ln\frac{1-\alpha}{\alpha}|\sigma| + 1\right)\int_0^{t_1}\Bigg[\left|\left(t_1 - \frac{1}{3}\right)t + \left(\frac{1}{2} + \frac{1}{3}t_1 - t_1^2\right)\right| \\ + \left|t_1 - \frac{1}{3}\right|\Bigg]\mathrm{d}t + \int_{t_1}^{1}\Bigg[\left|\frac{1}{6} + t\right| + \left|\frac{1}{3} - t\right| \\ + \frac{\sqrt{3}}{\pi}\ln\frac{1-\alpha}{\alpha}|\sigma|\left(\left|\frac{1}{3} - t\right| + \frac{1}{2}\right)\Bigg]\mathrm{d}t\end{aligned}$$

by (6.60).

Stage (b): Find the optimal switching instant t_1^*.

For GA, we keep the parameters as following: population size 40, maximal number of generations 200, crossover probability 0.9, and mutation probability 0.1. For PSO algorithm, the parameters are taken as swarm size 20, maximal number of iterations 300, the first strength of attraction constant 1.49, and the second strength of attraction constant 1.49.

Table 6.1 Results of optimization

Approaches	t_1^*	$\tilde{J}(t_1^*)$
GA-based approach	0.563	0.857
PSO-based approach	0.576	0.856

Let $\sigma = 0.1$, and choose $\alpha = 0.95$. Table 6.1 presents the results by the two approaches.

From this table, we can see that nearly the same results are obtained by GA and PSO approaches. However, compared with GA, PSO algorithm is easier to implement because it has no evolution operators such as crossover and mutation. Therefore, under the condition of about the same result, we are more inclined to use the PSO algorithm for solving the problem.

The optimal control law by PSO is

$$\boldsymbol{u}_t^{(1)*} = \begin{pmatrix} u_1^{(1)*}(t) \\ u_2^{(1)*}(t) \end{pmatrix} = \begin{pmatrix} 1 \\ -1 \end{pmatrix}, \quad t \in [0, 0.576),$$

$$\boldsymbol{u}_t^{(2)*} = \begin{pmatrix} u_1^{(2)*}(t) \\ u_2^{(2)*}(t) \end{pmatrix} = \begin{pmatrix} -1 \\ -1 \end{pmatrix}, \quad t \in [0.576, 1].$$

6.6 Discrete-Time Switched Linear Uncertain System

Considering the following class of discrete-time switched linear uncertain systems consisting of m subsystems.

$$\boldsymbol{x}(k+1) = A_{y(k)}\boldsymbol{x}(k) + B_{y(k)}u(k) + \sigma_{k+1}\xi_{k+1}, \ k = 0, 1, \ldots, N-1 \tag{6.70}$$

where (i) for each $k \in K \triangleq \{0, 1, \cdots, N-1\}$, $\boldsymbol{x}(k) \in R^n$ is the state vector with $\boldsymbol{x}(0)$ given and $u(k) \in R^r$ is the control vector, $y(k) \in M \triangleq \{1, \cdots, m\}$ is the switching control that indicates the active subsystem at stage k; (ii) for each $i \in M$, A_i ,B_i are constant matrices of appropriate dimension; (iii) for each $k \in K$, $\sigma_{k+1} \in R^n$ and $\sigma_{k+1} \neq 0$, ξ_k is the disturbance and $\xi_1, \xi_2, \cdots, \xi_N$ are independent ordinary linear uncertain variables denoted by $\mathcal{L}(-1, 1)$.

The performance of the sequence $u(k)|_{k=0}^{N-1}$ and $y(k)|_{k=0}^{N-1}$ can be measured by the following expected value:

$$E\left[\|\boldsymbol{x}(N)\|_{Q_f}^2 + \sum_{k=0}^{N-1}(\|\boldsymbol{x}(k)\|_{Q_{y(k)}}^2 + \|u(k)\|_{R_{y(k)}}^2)\right] \tag{6.71}$$

where, for any $i \in M$, $Q_i \geq 0, R_i > 0$ and (Q_i, R_i) constitutes the cost-matrix pair of the ith subsystem and $Q_f > 0$ is the terminal penalty matrix. The goal is to solve the following problem.

Problem 6.1 Find $u^*(k)|_{k=0}^{N-1}$ and $y^*(k)|_{k=0}^{N-1}$ to minimize (6.71) subject to the dynamical system (6.70) with initial state $\boldsymbol{x}(0) = \boldsymbol{x}_0$.

By using the dynamic programming approach, we will derive the analytical solution of Problem 6.1. However, we should introduce the recurrence formula first. For any $0 < k < N-1$, let $J(k, \boldsymbol{x}_k)$ be the optimal reward obtainable in $[k, N]$ with the condition that at stage k we are in state $\boldsymbol{x}(k) = \boldsymbol{x}_k$. Then we have

$$\begin{cases} J(k, \boldsymbol{x}_k) = \min\limits_{u(i), y(i), k \leq i \leq N} E\left[\|\boldsymbol{x}_N\|^2_{Q_f} + \sum\limits_{j=k}^{N-1}(\|\boldsymbol{x}(j)\|^2_{Q_{y(j)}} + \|u(j)\|^2_{R_{y(j)}})\right] \\ \text{subject to} \\ \quad \boldsymbol{x}(j+1) = A_{y(j)}\boldsymbol{x}(j) + B_{y(j)}u(j) + \sigma_{j+1}\xi_{j+1}, \\ \quad j = k, \ldots, N-1, \\ \quad \boldsymbol{x}(k) = \boldsymbol{x}_k. \end{cases} \tag{6.72}$$

Theorem 6.7 *For model* (6.72), *we have the following recurrence equation:*

$$J(N, \boldsymbol{x}_N) = \min_{u(N), y(N)} \left[\|\boldsymbol{x}_N\|^2_{Q_f}\right]$$

$$J(k, \boldsymbol{x}_k) = \min_{u(k), y(k)} E\left[\|\boldsymbol{x}_k\|^2_{Q_{y(k)}} + \|u(k)\|^2_{R_{y(k)}} + J(k+1, \boldsymbol{x}_{k+1})\right]$$

Proof The proof is similar to that of Theorem 4.1.

6.6.1 *Analytical Solution*

By using the recurrence equation, the analytical solution of Problem 6.1 can be derived. As in [11], define the following Riccati operator $\rho_i(P) : S_n^+ \to S_n^+$ for given $i \in M$ and $P \in S_n^+$,

$$\rho_i(P) \triangleq Q_i + A_i^\tau P A_i - A_i^\tau P B_i (B_i^\tau P B_i + R_i)^{-1} B_i^\tau P A_i$$

Let $\{H_i\}_{i=0}^N$ denote the set of ordered pairs of matrices defined recursively:

$$H_0 = \{(Q_f, 0)\}, H_{k+1} = \bigcup_{(P,r) \in H_k} \Gamma_k(P, r),$$

with

$$\Gamma_k(P,r) = \bigcup_{i\in M}\{(\rho_i(P), r + \frac{1}{3}\|\sigma_{N-k}\|_P^2)\},\ (P,r)\in H_k$$

for $k = 0, 1, \ldots, N-1$. Suppose that for each $i \in M$, $k = 0, 1, \ldots, N-1$ and $P \geq 0$, the following condition holds

$$|(A_i(k)\boldsymbol{x}(k) + B_i(k)u(k))^\tau P\sigma_{k+1}| \geq \|\sigma_{k+1}\|_P^2 \tag{6.73}$$

which means that at each stage k, the disturbance upon each subsystem is comparatively small.

Next, we will derive the analytical solution of Problem 6.1. First, we have

$$J(N, \boldsymbol{x}_N) = \|\boldsymbol{x}_N\|_{Q_f}^2 = \min_{(P,r)\in H_0}(\|\boldsymbol{x}_N\|_P^2 + r).$$

For $k = N-1$, the following equation holds by Theorem 6.7:

$$\begin{aligned}
&J(N-1, \boldsymbol{x}_{N-1})\\
=&\min_{u(N-1),y(N-1)} E\left[\|\boldsymbol{x}_{N-1}\|_{Q_{y(N-1)}}^2 + \|u(N-1)\|_{R_{y(N-1)}}^2 + J(N, \boldsymbol{x}_N)\right]\\
=&\min_{u(N-1),y(N-1)}\Big\{\|\boldsymbol{x}_{N-1}\|_{Q_{y(N-1)}}^2 + \|u(N-1)\|_{R_{y(N-1)}}^2 + E\big[\big(A_{y(N-1)}\boldsymbol{x}_{N-1}\\
&+B_{y(N-1)}u(N-1) + \sigma_N\xi_N\big)^\tau Q_f\left(A_{y(N-1)}\boldsymbol{x}_{N-1} + B_{y(N-1)}u(N-1) + \sigma_N\xi_N\right)\big]\Big\}\\
=&\min_{u(N-1),y(N-1)}\Big\{\|\boldsymbol{x}_{N-1}\|_{Q_{y(N-1)}+A_{y(N-1)}^\tau Q_f A_{y(N-1)}}^2 + \|u(N-1)\|_{R_{y(N-1)}+B_{y(N-1)}^\tau Q_f B_{y(N-1)}}^2\\
&+2u^\tau(N-1)B_{y(N-1)}^\tau Q_f A_{y(N-1)}\boldsymbol{x}_{N-1} + E\Big[2(A_{y(N-1)}\boldsymbol{x}_{N-1}\\
&+B_{y(N-1)}u(N-1))^\tau Q_f\sigma_N\xi_N + \|\sigma_N\|_{Q_f}^2\xi_N^2\Big]\Big\}.
\end{aligned} \tag{6.74}$$

Denote $a = 2(A_{y(N-1)}\boldsymbol{x}_{N-1} + B_{y(N-1)}u(N-1))^\tau Q_f\sigma_N$, $b = \|\sigma_N\|_{Q_f}^2$ and $s = a/b$. With condition (6.73), we can derive $|s| \geq 2$. Moreover, ξ_N is an ordinary linear uncertain variable and $\xi_N \sim \mathcal{L}(-1, 1)$. According to Example 1.6, the following equations hold

$$E[a\xi_N + b\xi_N^2] = bE[\xi_N^2 + s\xi_N] = \frac{1}{3}b = \frac{1}{3}\|\sigma_N\|_{Q_f}^2. \tag{6.75}$$

Substituting (6.75) into (6.74) yields

$$\begin{aligned} J(N-1,\boldsymbol{x}_{N-1}) = & \min_{u(N-1),y(N-1)} \Big\{ \|\boldsymbol{x}_{N-1}\|^2_{Q_{y(N-1)}+A^{\tau}_{y(N-1)}Q_f A_{y(N-1)}} \\ & + \|u(N-1)\|^2_{R_{y(N-1)}+B^{\tau}_{y(N-1)}Q_f B_{y(N-1)}} \\ & + 2u^{\tau}(N-1)B^{\tau}_{y(N-1)}Q_f A_{y(N-1)}\boldsymbol{x}_{N-1} + \frac{1}{3}\|\sigma_N\|^2_{Q_f} \Big\} \\ \triangleq & \min_{u(N-1),y(N-1)} f(u(N-1), y(N-1)). \end{aligned} \tag{6.76}$$

The optimal control $u^*(N-1)$ satisfies

$$\begin{aligned} & \frac{\partial f}{\partial u(N-1)} \\ = & \ 2(R_{y^*(N-1)} + B^{\tau}_{y^*(N-1)}Q_f B_{y^*(N-1)})u^*(N-1) + 2B^{\tau}_{y^*(N-1)}Q_f A_{y^*(N-1)}\boldsymbol{x}_{N-1} = 0. \end{aligned}$$

Since

$$\frac{\partial^2 f}{\partial u^2(N-1)} = 2(R_{y^*(N-1)} + B^{\tau}_{y^*(N-1)}Q_f B_{y^*(N-1)}) > 0,$$

we have

$$\begin{aligned} & u^*(N-1) \\ = & -(R_{y^*(N-1)} + B^{\tau}_{y^*(N-1)}Q_f B_{y^*(N-1)})^{-1}B^{\tau}_{y^*(N-1)}Q_f A_{y^*(N-1)}\boldsymbol{x}_{N-1}. \end{aligned} \tag{6.77}$$

Substituting (6.77) into (6.76) yields

$$\begin{aligned} & J(N-1,\boldsymbol{x}_{N-1}) \\ = & \min_{y(N-1)} \Big\{ \boldsymbol{x}^{\tau}_{N-1} \Big[Q_{y(N-1)} + A^{\tau}_{y(N-1)}Q_f A_{y(N-1)} - A^{\tau}_{y(N-1)}Q_f B_{y(N-1)}(R_{y(N-1)} \\ & + B^{\tau}_{y(N-1)}Q_f B_{y(N-1)})^{-1}B^{\tau}_{y(N-1)}Q_f A_{y(N-1)} \Big] \boldsymbol{x}_{N-1} + \frac{1}{3}\|\sigma_N\|^2_{Q_f} \Big\}. \end{aligned} \tag{6.78}$$

According to the definition of $\rho_i(P)$ and H_k, Eq. (6.78) can be written as

$$\begin{aligned} J(N-1,\boldsymbol{x}_{N-1}) & = \min_{y(N-1)\in M} \left(\|\boldsymbol{x}_{N-1}\|^2_{\rho_{y(N-1)}(Q_f)} + \frac{1}{3}\|\sigma_N\|^2_{Q_f} \right) \\ & = \min_{(P,r)\in H_1} \left(\|\boldsymbol{x}_{N-1}\|^2_P + r \right). \end{aligned} \tag{6.79}$$

Moreover, according to Eq. (6.79), we have

$$y^*(N-1) = \arg \min_{(P,r)\in H_0} \left\{ \|\boldsymbol{x}_{N-1}\|^2_{\rho_{y(N-1)}(P)} + \frac{1}{3}\|\sigma_N\|^2_P + r \right\}. \tag{6.80}$$

For $k = N-2$, we have

$$\begin{aligned}
&J(N-2, \boldsymbol{x}_{N-2})\\
&= \min_{u(N-2),y(N-2)} E\left[\|\boldsymbol{x}_{N-2}\|^2_{Q_{y(N-2)}} + \|u(N-2)\|^2_{R_{y(N-2)}} + J(N-1, \boldsymbol{x}_{N-1})\right]\\
&= \min_{u(N-2),y(N-2),(P,r)\in H_1} \Big\{\|\boldsymbol{x}_{N-2}\|^2_{Q_{y(N-2)}} + \|u(N-2)\|^2_{R_{y(N-2)}} + E\big[(A_{y(N-2)}\boldsymbol{x}_{N-2}\\
&\quad + B_{y(N-2)}u(N-2) + \sigma_{N-1}\xi_{N-1})^\tau P(A_{y(N-2)}\boldsymbol{x}_{N-2} + B_{y(N-2)}u(N-2)\\
&\quad + \sigma_{N-1}\xi_{N-1})\big] + r\Big\}\\
&= \min_{u(N-2),y(N-2),(P,r)\in H_1} \Big\{\|\boldsymbol{x}_{N-2}\|^2_{Q_{y(N-2)}+A^\tau_{y(N-2)}PA_{y(N-2)}}\\
&\quad + \|u(N-2)\|^2_{R_{y(N-2)}+B^\tau_{y(N-2)}PB_{y(N-2)}} + 2u(N-2)^\tau B^\tau_{y(N-2)}PA_{y(N-2)}\boldsymbol{x}_{N-2}\\
&\quad + E\big[2(A_{y(N-2)}\boldsymbol{x}_{N-2} + B_{y(N-2)}u(N-2))^\tau P\sigma_{N-1}\xi_{N-1}\\
&\quad + \|\sigma_{N-1}\|^2_P\xi^2_{N-1}\big] + r\Big\}.
\end{aligned} \tag{6.81}$$

It follows from a similar computation to (6.75) that

$$\begin{aligned}
&E\big[2(A_{y(N-2)}\boldsymbol{x}_{N-2} + B_{y(N-2)}u(N-2))^\tau P\sigma_{N-1}\xi_{N-1} + \|\sigma_{N-1}\|^2_P\xi^2_{N-1}\big]\\
&= \frac{1}{3}\|\sigma_{N-1}\|^2_P.
\end{aligned}$$

By the similar method to the above process, we can obtain

$$u^*(N-2) = -(R_{y^*(N-2)} + B^\tau_{y^*(N-2)}PB_{y^*(N-2)})^{-1}B^\tau_{y^*(N-2)}PA_{y^*(N-2)}\boldsymbol{x}(N-2),$$

$$\begin{aligned}
J(N-2, \boldsymbol{x}_{N-2}) &= \min_{y(N-2)\in M,(P,r)\in H_1} \left(\|\boldsymbol{x}_{N-2}\|^2_{\rho_{y(N-2)}(P)} + \frac{1}{3}\|\sigma_{N-1}\|^2_P + r\right)\\
&= \min_{(P,r)\in H_2} \left(\|\boldsymbol{x}_{N-2}\|^2_P + r\right),
\end{aligned}$$

and

$$y^*(N-2) = \arg \min_{y(N-2)\in M,(P,r)\in H_1} \left\{\|\boldsymbol{x}_{N-2}\|^2_{\rho_{y(N-2)}(P)} + \frac{1}{3}\|\sigma_{N-1}\|^2_P + r\right\}.$$

By induction, we can obtain the following theorem.

Theorem 6.8 ([12]) *Under condition* (6.73)*, at stage k, for given $\boldsymbol{x}_k$, the optimal switching control and optimal continuous control are*

$$y^*(k) = \arg \min_{y(k)\in M,(P,r)\in H_{N-k-1}} \left\{\|\boldsymbol{x}_k\|^2_{\rho_{y(k)}(P)} + \frac{1}{3}\|\sigma_{k+1}\|^2_P + r\right\}$$

and

$$u^*(k) = -(R_{y^*(k)} + B^{\tau}_{y^*(k)} P^* B_{y^*(k)})^{-1} B^{\tau}_{y^*(k)} P^* A_{y^*(k)} \boldsymbol{x}(k),$$

respectively, where

$$(y^*(k), P^*, r^*) = \arg \min_{y(k) \in M, (P,r) \in H_{N-k-1}} \left\{ \|\boldsymbol{x}_k\|^2_{\rho_{y(k)}(P)} + \frac{1}{3} \|\sigma_{k+1}\|^2_P + r \right\}.$$

The optimal value of Problem 6.1 is

$$J(0, \boldsymbol{x}_0) = \min_{(P,r) \in H_N} (\|\boldsymbol{x}_0\|^2_P + r). \tag{6.82}$$

Remark 6.1 Theorem 6.8 reveals that at iteration k, the optimal value and the optimal control law at all the future iterations only depend on the current set H_k. The above theorem properly transforms the enumeration over the switching sequences in m^N to the enumeration over the pairs of matrices in H_k. It will be shown in the next section that the expression given by (6.82) is more convenient for the analysis and the efficient computation of Problem 6.1.

6.6.2 Two-Step Pruning Scheme

According to Theorem 6.8, at iteration k, the optimal value and the optimal control law at all the future iterations only depend on the current set H_k. However, as k increases, the size of H_k grows exponentially. It becomes unfeasible to compute H_k when k grows large. A natural way of simplifying the computation is to ignore some redundant pairs in H_k. In order to improve computational efficiency, a two-step pruning scheme aiming at removing some redundant pairs will be presented in this section. The first step is a local pruning and the second step is a global pruning. To formalize the above idea, the following definitions are introduced.

Definition 6.1 A pair of matrices $(\hat{P}, \hat{r})$ is called redundant with respect to H if

$$\min_{(P,r) \in H \backslash \{(\hat{P}, \hat{r})\}} \{\|\boldsymbol{x}\|^2_P + r\} = \min_{(P,r) \in H} \{\|\boldsymbol{x}\|^2_P + r\}, \forall \boldsymbol{x} \in R^n.$$

Definition 6.2 The set $\hat{H}$ is called equivalent to H, denoted by $\hat{H} \sim H$ if

$$\min_{(P,r) \in \hat{H}} \{\|\boldsymbol{x}\|^2_P + r\} = \min_{(P,r) \in H} \{\|\boldsymbol{x}\|^2_P + r\}, \forall \boldsymbol{x} \in R^n.$$

Therefore, any equivalent subsets of H_k define the same $J(k, \boldsymbol{x}_k)$. To ease the computation, we shall prune away as many redundant pairs as possible from H_k and obtain an equivalent subset of H_k whose size is as small as possible. In order to remove as many redundant pairs of matrices from H_k as possible, a two-step pruning scheme is applied here. The first step is a local pruning which prunes away some redundant pairs from $\Gamma_k(P, r)$ for any (P, r), and the second step is a global pruning which removing redundant pairs from H_{k+1} after the first step.

6.6.3 Local Pruning Scheme

The goal of local pruning algorithm is removing as many redundant pairs of matrices as possible from $\Gamma_k(P, r)$. However, testing whether a pair is redundant or not is a challenging problem. A sufficient condition for checking pairs redundant or not is given in the following lemma.

Lemma 6.1 ([12]) *A pair $(\hat{P}, \hat{r})$ is redundant in $\Gamma_k(P, r)$ if there exist nonnegative constants $\alpha_1, \alpha_2, \cdots, \alpha_{s-1}$ such that $\sum_{i=1}^{s-1} \alpha_i = 1$ and*

$$\hat{P} \geq \sum_{i=1}^{s-1} \alpha_i P^{(i)} \tag{6.83}$$

where $s = |\Gamma_k(P, r)|$ and $\{(P^{(i)}, r^{(i)})\}_{i=1}^{s-1}$ is an enumeration of $\Gamma_k(P, r)\backslash\{(\hat{P}, \hat{r})\}$.

Proof First, from the definition of $\Gamma_k(P, r) = \bigcup_{i \in M} \{(\rho_i(P), r + \frac{1}{3}\|\sigma_{N-k}\|_P^2)\}$, for any pair $(P^{(i)}, r^{(i)})$ in $\Gamma_k(P, r)$, the second part $r^{(i)}$ is equal to $r + \frac{1}{3}\|\sigma_{N-k}\|_P^2$. Additionally, we know

$$\alpha_1(\hat{P} - P^{(1)}) + \cdots + \alpha_{s-1}(\hat{P} - P^{(s-1)}) \geq 0$$

by the condition (6.83). For any $\boldsymbol{x} \geq 0$, we have

$$\alpha_1 \|\boldsymbol{x}\|_{\hat{P}-P^{(1)}}^2 + \cdots + \alpha_{s-1}\|\boldsymbol{x}\|_{\hat{P}-P^{(s-1)}}^2 \geq 0.$$

So there exists at least one i such that the following formula holds

$$\|\boldsymbol{x}\|_{\hat{P}-P^{(i)}}^2 \geq 0.$$

According to $r^{(i)} = \hat{r}$, we obtain

$$\|\boldsymbol{x}\|_{\hat{P}}^2 + \hat{r} \geq \|\boldsymbol{x}\|_{P^{(i)}}^2 + r^{(i)},$$

which indicates $(\hat{P}, \hat{r})$ is redundant in $\Gamma_k(P, r)$. The proof is completed.

Checking the condition (6.83) in Lemma 6.1 is a LMI feasibility problem which can be solved with MATLAB toolbox LMI. However, Lemma 6.1 cannot remove all the redundant pairs. If the condition in Lemma 6.1 is met, then the pairs under consideration will be discarded; otherwise, the pairs will be kept and get into H_{k+1}. As we know, the size of H_{k+1} is crucial throughout the computational process. So, after this step, we apply a global pruning to H_{k+1}.

6.6.4 Global Pruning Scheme

A pair in H_k being redundant or not can be checked by the following lemma.

Lemma 6.2 ([12]) *A pair* $(\tilde{P}, \tilde{r})$ *is redundant in* H_k *if there exist nonnegative constants* $\alpha_1, \alpha_2, \ldots, \alpha_{l-1}$ *such that* $\sum_{i=1}^{l-1} \alpha_i = 1$ *and*

$$\begin{pmatrix} \tilde{P} & 0 \\ 0 & \tilde{r} \end{pmatrix} \geq \sum_{i=1}^{l-1} \alpha_i \begin{pmatrix} P^{(i)} & 0 \\ 0 & r^{(i)} \end{pmatrix} \tag{6.84}$$

where $l = |H_k|$ *and* $\{(P^{(i)}, r^{(i)})\}_{i=1}^{l-1}$ *is an enumeration of* $H_k \backslash \{(\tilde{P}, \tilde{r})\}$.

The proof of Lemma 6.2 is similar to Lemma 6.1.

A detailed description of the two-step pruning process is given in Algorithm 6.3.

Remark 6.2 Here, after the local pruning, the set H_k is represented by $\tilde{H}_k$. Then $\tilde{H}_k$ is represented by $\hat{H}_k$ after the global pruning. This two-step pruning scheme is different from the approach proposed in [13] which only prunes redundant pairs in H_{k+1}. Because the size of H_{k+1} is much larger than Γ_k, the computation cost of checking whether a pair in H_{k+1} is redundant or not is more complicated than it is in Γ_k whose size is only m. The two-step pruning scheme thus decreases the computational complexity of each round of checking.

Remark 6.3 In order to make our two-step pruning scheme more clearly, we make a metaphor. Image that we have to select several best basketball players from a university with thousands of students. How should we select efficiently? Obviously, one to one competition or one to several competition for all the students in this university is not an efficient method. The global pruning scheme [13] is just like this. Without one step above, two-step pruning scheme is like that, first, we choose some better players from each college or department of the university which can be viewed as a local pruning. Second, the best players are selected by competitions by these better players, and this step can be viewed as a global pruning. From the two-step pruning scheme, we can select the best basketball players from a university efficiently. The similar pruning scheme has been widely used in some influential sport games, such as the regular season and playoffs of NBA, the group phase and knockout round of the Football World Cup.

Remark 6.4 The discrete-time problem is multistage decision-making course. It has obvious difference with continuous-time case [6] not only in the form of solution but also in the methods of solving.

Algorithm 6.3 :(Two-step pruning scheme)

1: Set $\hat{H}_0 = \{(Q_f, 0)\}$;
2: **for** $k = 0$ to $N - 1$ **do**
3:    **for all** $(P, r) \in \hat{H}_k$ **do**
4:       $\Gamma_k(P, r) = \emptyset$;
5:       **for** $i = 1$ to m **do**
6:          $P^{(i)} = \rho_i(P)$,
7:          $r^{(i)} = r + \frac{1}{3}\|\sigma_{N-k}\|_P^2$,
8:          $\Gamma_k(P, r) = \Gamma_k(P, r) \bigcup \{(P^{(i)}, r^{(i)})\}$;
9:       **end for**
10:      **for** $i = 1$ to m **do**
11:         **if** $(P^{(i)}, r^{(i)})$ satisfies the condition in Lemma 6.1, **then**
12:            $\Gamma_k(P, r) = \Gamma_k(P, r) \backslash \{(P^{(i)}, r^{(i)})\}$;
13:         **end if**
14:       **end for**
15:    **end for**
16:    $\hat{H}_{k+1} = \bigcup\limits_{(P,r)\in\hat{H}_k} \Gamma_k(P, r)$;
17:    $\tilde{H}_{k+1} = \hat{H}_{k+1}$;
18:    **for** $i = 1$ to $|\hat{H}_{k+1}|$ **do**
19:       **if** $(\hat{P}^{(i)}, \hat{r}^{(i)})$ satisfies the condition in Lemma 6.2, **then**
20:          $\hat{H}_{k+1} = \hat{H}_{k+1} \backslash \{(\hat{P}^{(i)}, \hat{r}^{(i)})\}$;
21:       **end if**
22:    **end for**
23: **end for**
24: $J(0, x_0) = \min\limits_{(P,r)\in\hat{H}_N} (\|x_0\|_P^2 + r)$.

The sets $\{\hat{H}_k\}_{k=0}^N$ generated by Algorithm 6.3 typically contain much fewer pairs of matrices than $\{H_k\}_{k=0}^N$ and are thus much easier to deal with.

6.6.5   Examples

Example 6.1 Consider the uncertain discrete-time optimal control Problem 6.1 with $N = 10$, $m = 3$ and

$$A_1 = \begin{pmatrix} 2 & 1 \\ 0 & 1 \end{pmatrix}, \ B_1 = \begin{pmatrix} 1 \\ 1 \end{pmatrix}, \ A_2 = \begin{pmatrix} 1 & 1 \\ 1 & 2 \end{pmatrix}, \ B_2 = \begin{pmatrix} 1 \\ 2 \end{pmatrix}, \ A_3 = \begin{pmatrix} 2 & 1 \\ 1 & 2 \end{pmatrix}, \ B_3 = \begin{pmatrix} 2 \\ 1 \end{pmatrix},$$

$$Q_1 = Q_2 = Q_3 = \begin{pmatrix} 1 & 0 \\ 0 & 1 \end{pmatrix}, \ Q_f = \begin{pmatrix} 4 & 1 \\ 1 & 2 \end{pmatrix}, \ R_1 = R_2 = R_3 = 1, \ \sigma_k = \begin{pmatrix} 0.1 \\ 0.1 \end{pmatrix}$$

for $k = 1, 2, \ldots, N$. Algorithm 6.3 is applied to solve this problem. The numbers of elements in $\tilde{H}_k$ and $\hat{H}_k$ at each step is listed in Table 6.2. It turns out that $|\hat{H}_k|$ is

Table 6.2 Size of $\tilde{H}_k$ and $\hat{H}_k$ for Example 6.1

k	1	2	3	4	5	6	7	8	9	10
$\lvert\tilde{H}_k\rvert$	2	5	4	4	7	7	4	7	7	7
$\lvert\hat{H}_k\rvert$	2	2	2	3	3	2	3	3	3	3

Table 6.3 Optimal results of Example 6.1

k	$y^*(k)$	r_k	$\boldsymbol{x}(k)$	$u^*(k)$	$J(k, \boldsymbol{x}_k)$
0	2	–	$(3, -1)^\tau$	−0.7861	12.9774
1	2	0.6294	$(1.2768, -0.5093)^\tau$	−0.2579	2.5122
2	2	0.8116	$(0.5908, -0.1764)^\tau$	−0.1749	0.6456
3	2	−0.7460	$(0.1649, -0.1864)^\tau$	0.0761	0.2084
4	2	0.8268	$(0.1373, 0.0270)^\tau$	−0.1143	0.1582
5	1	0.2647	$(0.0765, -0.0108)^\tau$	−0.0495	0.1137
6	2	−0.8049	$(0.0122, -0.1408)^\tau$	0.1221	0.1113
7	2	−0.4430	$(-0.0503, -0.0685)^\tau$	0.0918	0.0765
8	2	0.0938	$(-0.0716, 0.0057)^\tau$	0.005	0.0443
9	2	0.9150	$(0.0846, 0.0953)^\tau$	−0.1444	0.0678

very small, and the maximum value is 3 as compared to growing exponentially as k increases.

Choose $\boldsymbol{x}_0 = (3, -1)^\tau$, the optimal controls and the optimal values are obtained by Theorem 6.8 and listed in Table 6.3. The data in the fourth column of Table 6.3 are the corresponding states which are derived from $\boldsymbol{x}(k+1) = A_{y^*(k)}\boldsymbol{x}(k) + B_{y^*(k)}u^*(k) + \sigma_{k+1}r_{k+1}$, where r_{k+1} is the realization of uncertain variable $\xi_{k+1} \sim \mathcal{L}(-1, 1)$ and may be generated by $r_{k+1} = \Phi^{-1}_{\xi_{k+1}}(random(0, 1))$ $(k = 0, 1, 2, \ldots, 9)$.

The number of $\tilde{H}_k$ indicates the effect of the local pruning. In order to test the effect of the local pruning further, we increase the number of subsystems and consider the following problem.

Example 6.2 Consider a more complex example with 6 subsystems ($m = 6$). The first three subsystems are the same as Example 6.1 and the other three are chosen as

$$A_4 = \begin{pmatrix} 1 & 2 \\ 0 & 1 \end{pmatrix},\ B_4 = \begin{pmatrix} 0 \\ 1 \end{pmatrix},\ A_5 = \begin{pmatrix} 1 & 2 \\ 1 & 1 \end{pmatrix},\ B_5 = \begin{pmatrix} 1 \\ 3 \end{pmatrix},\ A_6 = \begin{pmatrix} 5 & 1 \\ 1 & 5 \end{pmatrix},\ B_6 = \begin{pmatrix} 2 \\ 1 \end{pmatrix},$$

$$Q_4 = Q_5 = Q_6 = \begin{pmatrix} 1 & 0 \\ 0 & 1 \end{pmatrix},\ R_4 = R_5 = R_6 = 1.$$

The numbers of elements in $\tilde{H}_k$ and $\hat{H}_k$ at each step are listed in Table 6.4. It can be seen that the numbers of $\tilde{H}_k$ and $\hat{H}_k$ does not necessarily increase with the number of

Table 6.4 Size of $\tilde{H}_k$ and $\hat{H}_k$ for Example 6.2

k	1	2	3	4	5	6	7	8	9	10
$\lvert\tilde{H}_k\rvert$	2	5	12	9	9	9	9	9	9	9
$\lvert\hat{H}_k\rvert$	2	4	3	3	3	3	3	3	3	3

Table 6.5 Optimal results of Example 6.2

k	$y^*(k)$	r_k	$\boldsymbol{x}(k)$	$u^*(k)$	$J(k, \boldsymbol{x}_k)$
0	5	–	$(3, -1)^\tau$	−0.7273	11.0263
1	1	0.6294	$(0.3356, -0.1190)^\tau$	−0.1808	0.6251
2	2	0.8116	$(0.4526, -0.2186)^\tau$	−0.0892	0.5116
3	1	−0.7460	$(0.0702, -0.2376)^\tau$	0.1421	0.2063
4	2	0.8268	$(0.1276, -0.0128)^\tau$	−0.0608	0.1428
5	2	0.2647	$(0.0805, 0.0069)^\tau$	−0.0523	0.1121
6	2	−0.8049	$(-0.0454, -0.0908)^\tau$	0.1105	0.1113
7	2	−0.4430	$(-0.0700, -0.0503)^\tau$	0.0931	0.0690
8	1	0.0938	$(-0.0178, 0.0250)^\tau$	−0.0062	0.0426
9	2	0.9150	$(0.0747, 0.1103)^\tau$	−0.1522	0.0615

subsystems. Additionally, with more subsystems, the effectiveness of local pruning becomes more apparent.

Choose $\boldsymbol{x}_0 = (3, -1)^\tau$, the optimal controls and the optimal values are listed in Table 6.5.

References

1. Wang L, Beydoun A, Sun J, Kolmanasovsky I (1997) Optimal hybrid control with application to automotive powertrain systems. Lecture Notes in Control and Information Science 222:190–200
2. Xu X, Antsaklis P (2004) Optimal control of switched systems based on parameterization of the switching instants. IEEE Trans Autom Control 49(1):2–16
3. Benga S, Decarlo R (2005) Optimal control of switching systems. Automatica 41(1):11–27
4. Teo KL, Goh C, Wong K (1991) A unified computational approach to optimal control problems. Longman Scientific and Technical, New York
5. Lee H, Teo K, Rehbock V, Jennings L (1999) Control parametrization enhancing technique for optimal discrete-valued control problems. Automatica 35(8):1401–1407
6. Yan H, Zhu Y (2015) Bang-bang control model for uncertain switched systems. Appl Math Modell 39(10–11):2994–3002
7. Hopfinger E, Luenberger D (1976) On the solution of the unidimensional local minimization problem. J Optim Theory Appl 18(3):425–428
8. Yan H, Sheng L, Zhu Y (2016) Linear quadratic optimization models of uncertain switched systems. ICIC Exp Lett 10(10):2349–2355

9. Zhu Y (2013) An intelligent algorithm: MACO for continuous optimization models. J Int Fuzzy Syst 24(1):31–36
10. Yan H, Zhu Y (2017) Bang-bang control model with optimistic value criterion for uncertain switched systems. J Intell Manuf 28(3):527–534
11. Zhang W, Hu J, Abate A (2009) On the value function of the discrete-time switched LQR problem. IEEE Trans Autom Control 54(11):2669–2674
12. Yan H, Sun Y, Zhu Y (2017) A linear-quadratic control problem of uncertain discrete-time switched systems. J Ind Manag Optim 13(1):267–282
13. Zhang W, Hu J, Lian J (2010) Quadratic optimal control of switched linear stochastic systems. Syst Control Lett 59(11):736–744

Chapter 7
Optimal Control for Time-Delay Uncertain Systems

Assume that an uncertain process X_t $(t \geq -d)$ takes values in a closed set $\mathbf{A} \subset R^n$, which describes the state of a system at time t that started at time $-d < 0$. Here, d describes a constant delay inherent to the system. Let $C_{\mathbf{A}}[-d, 0]$ denote the space of all continuous functions on $[-d, 0]$ taking values in $\mathbf{A}$. For $t \in [-d, 0]$, the process X_t is consistent with a function $\varphi_0 \in C_{\mathbf{A}}[-d, 0]$. For $t \geq 0$, X_{t+s} $(s \in [-d, 0])$ describes the associated segment process of X_t, denoted by

$$\varphi_t(s) = X_{t+s}, \quad s \in [-d, 0].$$

We consider a system whose dynamics may not only depend on the current state but also depend on the segment process through the processes

$$Y_t = \int_{-d}^{0} e^{\lambda s} f(X_{t+s})\mathrm{d}s, \quad \zeta_t = f(X_{t-d}), \quad t \geq 0$$

where $f : R^n \to R^k$ is a differentiable function and $\lambda \in R$ is a constant. The system can be controlled by $u = \{u_t, t \geq 0\}$ taking values in a closed subset U of R^m.

At every time $t \geq 0$, an immediate reward $F(t, X_t, Y_t, u_t)$ is accrued and the terminal state of the system earns a reward $h(X_T, Y_T)$. Then we are looking for a control process u that maximizes the overall expected reward over the horizon $[0, T]$. That is, we consider the following uncertain optimal control problem with time-delay:

Y. Zhu, *Uncertain Optimal Control*, Springer Uncertainty Research,
https://doi.org/10.1007/978-981-13-2134-4_7

$$\begin{cases} J(0, \varphi_0) = \sup\limits_{u \in U} E\left[\int_0^T F(s, X_s, Y_s, u_s)\mathrm{d}s + h(X_T, Y_T)\right] \\ \text{subject to} \\ \quad \mathrm{d}X_s = \mu_1(s, X_s, Y_s, u_s)\mathrm{d}s + \mu_2(X_s, Y_s)\zeta_s \mathrm{d}s \\ \qquad\quad + \sigma(s, X_s, Y_s, u_s)\mathrm{d}C_s, \quad s \in [0, T] \\ \quad X_s = \varphi_0(s), \quad -d \le s \le 0. \end{cases} \tag{7.1}$$

In the above model, X_s is the state vector of n dimension, u_s takes values in a closed subset U of R^m, $F : [0, +\infty) \times R^n \times R^k \times U \to R$ the objective function, and $h : R^n \times R^k \to R$ the function of terminal reward. In addition, $\mu_1 : [0, +\infty) \times R^n \times R^k \times U \to R^n$ is a column-vector function, $\mu_2 : R^n \times R^k \to R^{n\times k}$ a matrix function, $\sigma : [0, +\infty) \times R^n \times R^k \times U \to R^{n\times l}$ a matrix function, and $C_s = (C_{s_1}, C_{s_2}, \ldots C_{s_l})^\tau$, where $C_{s_1}, C_{s_2}, \ldots C_{s_l}$ are independent Liu canonical process. The function $J(0, \varphi_0)$ is the expected optimal reward obtainable in $[0, T]$ with the initial condition that at time 0 we have the state $\varphi_0(s)$ between $-d$ and 0, where $\varphi_0 \in C_{\mathbf{A}}[-d, 0]$ is a given function. The final time $T > 0$ is fixed or free. A feasible control process means that it takes values in the set U.

7.1 Optimal Control Model with Time-Delay

For any $0 < t < T$, $J(t, \varphi_t)$ is the expected optimal reward obtainable in $[t, T]$ with the condition that we have the state $\varphi_t(s)$ between $t - d$ and t. That is, consider the following problem (P):

$$(P)\begin{cases} J(t, \varphi_t) = \sup\limits_{u \in U} E\left[\int_t^T F(s, X_s, Y_s, u_s)\mathrm{d}s + h(X_T, Y_T)\right] \\ \text{subject to} \\ \quad \mathrm{d}X_s = \mu_1(s, X_s, Y_s, u_s)\mathrm{d}s + \mu_2(X_s, Y_s)\zeta_s \mathrm{d}s \\ \qquad\quad +\sigma(s, X_s, Y_s, u_s)\mathrm{d}C_s, \quad s \in [t, T] \\ \quad X_s = \varphi_t(s), \quad s \in [-d, 0]. \end{cases} \tag{7.2}$$

Note that the value function J is defined on the infinite-dimensional space $[0, T] \times C_{\mathbf{A}}[-d, 0]$ so that the equation of optimality (2.15) is not directly applicable. We will formulate an uncertain control problem $(\overline{P})$ with finite-dimensional state space such that an optimal control process for (P) can be constructed from an optimal solution of the problem $(\overline{P})$. In order to transform the uncertain control problem (P), we introduce the following assumption.

Assumption 7.1 There exists an operator $Z : R^n \times R^k \to R^n$ such that

$$e^{\lambda d} D_x Z(x, y)\mu_2(x, y) - D_y Z(x, y) = 0, \quad \forall (x, y) \in R^n \times R^k, \tag{7.3}$$

where $D_x Z(x, y)$ and $D_y Z(x, y)$ denote the Jacobi matrices of Z in x and in y, respectively.

This transformation yields a new state process $Z_t = Z(X_t, Y_t)$. Let $S = \mathbf{A} \times y(C_{\mathbf{A}}[-d, 0])$. For $\psi \in C_{\mathbf{A}}[-d, 0]$, we denote $x(\psi) = \psi(0)$, $y(\psi) = \int_{-d}^{0} e^{\lambda s} f(\psi(s))\mathrm{d}s$ and $\zeta(\psi) = f(\psi(-d))$. Then Z_t take values in $Z(S)$. In order to derive the dynamics of the transformed process Z, we need the following lemma.

Lemma 7.1 ([1]) *Let $G(t, x, y) : [0, +\infty) \times R^n \times R^k \to R^n$ be continuously differentiable function and consider a feasible control process $u_t \in U$. Then the uncertain process $G(t, X_t, Y_t)$ satisfies*

$$\begin{aligned} \mathrm{d}G(t, X_t, Y_t) = \{&G_t(t, X_t, Y_t) + D_x G(t, X_t, Y_t)(\mu_1(t, X_t, Y_t, u_t) + \mu_2(X_t, Y_t)\zeta_t)\}\mathrm{d}t \\ &+ D_x G(t, X_t, Y_t)\sigma(t, X_t, Y_t, u_t)\mathrm{d}C_t + D_y G(t, X_t, Y_t)(f(X_t) \\ &- e^{-\lambda d}\zeta_t - \lambda Y_t)\mathrm{d}t. \end{aligned} \tag{7.4}$$

Proof For a given feasible control process u_t with state process X_t, define a process $\tilde{F}_t$ by

$$\tilde{F}_t = \int_0^t f(X_s)ds.$$

Then the process Y_t has the representation

$$\begin{aligned} Y_t &= \int_{-d}^{0} e^{\lambda s} f(X_{t+s})\mathrm{d}s = \int_{-d}^{0} e^{\lambda s}\mathrm{d}\tilde{F}_{t+s} = e^{\lambda s}\tilde{F}_{t+s}\mid_{-d}^{0} - \int_{-d}^{0} \tilde{F}_{t+s}\mathrm{d}e^{\lambda s} \\ &= \tilde{F}_t - e^{-\lambda d}\tilde{F}_{t-d} - \int_{-d}^{0} \lambda e^{\lambda s}\tilde{F}_{t+s}\mathrm{d}s \\ &= \int_0^t f(X_s)\mathrm{d}s - e^{-\lambda d}\int_0^{t-d} f(X_s)\mathrm{d}s - \lambda\int_{-d}^{0} e^{\lambda s}\int_0^{t+s} f(X_r)\mathrm{d}r\mathrm{d}s. \end{aligned}$$

Thus,

$$\mathrm{d}Y_t = \big(f(X_t) - e^{-\lambda d} f(X_{t-d}) - \lambda Y_t\big)\mathrm{d}t = \big(f(X_t) - e^{-\lambda d}\zeta_t - \lambda Y_t\big)\mathrm{d}t.$$

Applying Theorem 1.14 to $G(t, X_t, Y_t)$, Eq. (7.4) follows.

Now we are able to present the dynamics for $Z_t = Z(X_t, Y_t)$ by using (7.3) and (7.4) as follows. It can be seen that

$$\begin{aligned} \mathrm{d}Z_t &= \mathrm{d}Z(X_t, Y_t) \\ &= D_x Z(X_t, Y_t)(\mu_1(t, X_t, Y_t, u_t) + \mu_2(X_t, Y_t)\zeta_t)\mathrm{d}t \\ &\quad + D_x Z(X_t, Y_t)\sigma(t, X_t, Y_t, u_t)\mathrm{d}C_t + D_y Z(X_t, Y_t)(f(X_t) - e^{-\lambda d}\zeta_t - \lambda Y_t)\mathrm{d}t \\ &= D_x Z(X_t, Y_t)\mu_1(t, X_t, Y_t, u_t)\mathrm{d}t + D_y Z(X_t, Y_t)(f(X_t) - \lambda Y_t)\mathrm{d}t \\ &\quad + D_x Z(X_t, Y_t)\sigma(t, X_t, Y_t, u_t)\mathrm{d}C_t. \end{aligned}$$

Define $\widetilde{\mu} : [0, +\infty) \times R^n \times R^k \times U \to R^n$ by

$$\widetilde{\mu}(t, x, y, u) = D_x Z(x, y)\mu_1(t, x, y, u) + D_y Z(x, y)(f(x) - \lambda y),$$

and $\widetilde{\sigma} : [0, +\infty) \times R^n \times R^k \times U \to R^{n\times l}$ by

$$\widetilde{\sigma}(t, x, y, u) = D_x Z(x, y)\sigma(t, x, y, u).$$

If the functions $\widetilde{\mu}$ and $\widetilde{\sigma}$ as well as h would depend on (x, y) through $Z(x, y)$ only, then the problem (P) could be reduced to a finite-dimensional problem.

Assumption 7.2 There are functions

$$\begin{aligned}&\overline{\mu} : [0, +\infty) \times R^n \times U \to R^n, \quad \overline{\sigma} : [0, +\infty) \times R^n \times U \to R^{n\times l},\\ &\overline{F} : [0, +\infty) \times R^n \times U \to R, \quad \overline{h} : R^n \to R\end{aligned}$$

such that for all $t \in [0, T]$, $u \in U$, $(x, y) \in R^n \times R^k$, we have

$$\begin{aligned}&\overline{\mu}(t, Z(x, y), u) = \widetilde{\mu}(t, x, y, u), \quad \overline{\sigma}(t, Z(x, y), u) = \widetilde{\sigma}(t, x, y, u),\\ &\overline{F}(t, Z(x, y), u) = F(t, x, y, u), \quad \overline{h}(Z(x, y)) = h(x, y).\end{aligned}$$

We introduce a finite-dimensional control problem $(\overline{P})$ associated to (P) via the transformation. For $\varphi_t \in C_{\mathbf{A}}[-d, 0]$, define $z = Z(x(\varphi_t), y(\varphi_t)) \in Z(S)$. Then for $t \in [0, T]$, the problem (P) can be transformed to the problem $(\overline{P})$

$$(\overline{P}) \begin{cases} \overline{J}(t, z) = \sup\limits_{u_t \in U} E\left[\int_t^T \overline{F}(s, Z_s, u_s)\mathrm{d}s + \overline{h}(Z_T)\right] \\ \text{subject to} \\ \quad \mathrm{d}Z_s = \overline{\mu}(s, Z_s, u_s)\mathrm{d}s + \overline{\sigma}(s, Z_s, u_s)\mathrm{d}C_s, \quad s \in [t, T] \\ \quad Z_t = z, \\ \quad u_s \in U, \quad s \in [t, T]. \end{cases} \tag{7.5}$$

The value function $\overline{J}$ of the uncertain optimal control problem $(\overline{P})$ has a finite-dimensional state space. So we can directly use the equation of optimality (2.15) for $(\overline{P})$ and have the main result of this paper.

Theorem 7.1 ([1]) *Suppose that Assumptions 7.1 and 7.2 hold and $\overline{J}_t(t, z)$ is twice differentiable on* $[0, T] \times R^n$. *Then we have*

$$\begin{cases} -\overline{J}_t(t, z) = \sup\limits_{u_t \in U}\{\overline{F}(t, z, u_t) + \nabla_z \overline{J}(t, z)^\tau \overline{\mu}(t, z, u_t)\} \\ \overline{J}(T, Z_T) = \overline{h}(Z_T), \end{cases} \tag{7.6}$$

and $\overline{J}(t, z) = J(t, \varphi_t)$, where $\overline{J}_t(t, z)$ is the partial derivative of the function $\overline{J}(t, z)$ in t, and $\nabla_z \overline{J}(t, z)$ is the gradient of $\overline{J}(t, z)$ in z.

Proof Eq. (7.6) directly follows from the equation of optimality (2.15). In addition, for any $u_t \in U$, we have

$$\begin{aligned}\overline{J}(t,z) &\geq E\left[\int_t^T \overline{F}(s, Z_s, u_s)\mathrm{d}s + \overline{h}(Z_T)\right]\\ &= E\left[\int_t^T F(s, X_s, Y_s, u_s)\mathrm{d}s + h(X_T, Y_T)\right].\end{aligned}$$

Thus,

$$\overline{J}(t,z) \geq \sup_{u_t\in U} E\left[\int_t^T F(s, X_s, Y_s, u_s)\mathrm{d}s + h(X_T, Y_T)\right] = J(t, \varphi_t).$$

Similarly, we can get $J(t, \varphi_t) \geq \overline{J}(t, z)$. Therefore, the theorem is proved.

Remark 7.1 The optimal decision and optimal expected value of problem (P) are determined if Eq. (7.6) has solutions.

7.2 Uncertain Linear Quadratic Model with Time-Delay

In this section, we apply the result obtained in the previous section to study an uncertain LQ problem with time-delay. Let $A_1(t)$, $A_2(t)$, $A_4(t)$, $A_5(t)$, $A_6(t)$, $A_7(t)$, $B(t)$, $H(t)$, $I(t)$, $L(t)$, $M(t)$, $N(t)$, $R(t)$ be continuously differentiable functions of t. What is more, let $A_3 \neq 0$ and a be constants, and $I(t) \leq 0$, $R(t) < 0$. For $\psi \in C_R[-d, 0]$, denote $x(\psi) = \psi(0)$, $y(\psi) = \int_{-d}^0 e^{\lambda s}\psi(s)ds$, $\zeta(\psi) = \psi(-d)$. Then an uncertain LQ problem with time-delay is stated as

$$(LQ)\left\{\begin{array}{l} J(t,\varphi_t) = \sup\limits_{u\in U} E\left[\displaystyle\int_t^T \{I(s)(e^{-\lambda d}X_s + A_3Y_s)^2 + R(s)u_s^2\right.\\ \qquad +H(s)(e^{-\lambda d}X_s + A_3Y_s)u_s + L(s)(e^{-\lambda d}X_s + A_3Y_s)\\ \qquad \left. +M(s)u_s + N(s)\}\,\mathrm{d}s + a(e^{-\lambda d}X_T + A_3Y_T)^2\right]\\ \text{subject to}\\ \quad \mathrm{d}X_s = \{A_1(s)X_s + A_2(s)Y_s + A_3\zeta_s + B(s)u_s + A_4(s)\}\mathrm{d}s + \{A_5(s)X_s\\ \qquad +A_6(s)Y_s + A_7(s)\}\mathrm{d}C_s, \quad s \in [t,T]\\ \quad Y_s = \displaystyle\int_{-d}^0 e^{\lambda r}X_{s+r}\mathrm{d}r, \quad \zeta_s = X_{s-d}, \quad s\in[t,T]\\ \quad X_s = \varphi_t(s), \quad -d \leq s \leq 0\\ \quad u_s \in U, \quad s \in [t,T]. \end{array}\right.$$

where $\varphi_0 \in C_R[-d, 0]$ is a given initial function and $\varphi_t \in C_R[-d, 0]$ is the segment of X_t for $t > 0$, and U is the set of feasible controls. In addition, we are in state $X_t = \boldsymbol{x}$ at time t.

Theorem 7.2 ([1]) *If $A_2(t) = e^{\lambda d} A_3(A_1(t) + e^{\lambda d} A_3 + \lambda)$ and $A_6(t) = e^{\lambda d} A_3 A_5(t)$ hold in the (LQ) model, then the optimal control u_t^* of (LQ) is*

$$u_t^* = -\frac{(H(t) + e^{-\lambda d} B(t)P(t))z + e^{-\lambda d} B(t)Q(t) + M(t)}{2R(t)}, \tag{7.7}$$

where $P(t)$ satisfies

$$\begin{cases} \frac{\mathrm{d}P(t)}{\mathrm{d}t} = \frac{e^{-2\lambda d} B(t)^2}{2R(t)} P(t)^2 + \left(\frac{e^{-\lambda d} H(t)B(t)}{R(t)} - 2A_1(t) - 2A_3 e^{\lambda d} \right) P(t) \\ \quad + \frac{H(t)^2}{2R(t)} - 2I(t) \\ P(T) = 2a, \end{cases} \tag{7.8}$$

and $Q(t)$ is a solution of the following differential equation

$$\begin{cases} \frac{\mathrm{d}Q(t)}{\mathrm{d}t} = \left(\frac{e^{-\lambda d} H(t)B(t) + e^{-2\lambda d} B(t)^2 P(t)}{2R(t)} - A_1(t) - A_3 e^{\lambda d} \right) Q(t) \\ \quad -e^{-\lambda d} P(t)A_4(t) - L(t) + \frac{e^{-\lambda d} M(t)B(t)P(t) + H(t)M(t)}{2R(t)} \\ Q(T) = 0. \end{cases} \tag{7.9}$$

The optimal value of (LQ) is

$$J(t, \varphi_t) = \frac{1}{2} P(t)z^2 + Q(t)z + K(t), \tag{7.10}$$

where $z = e^{-\lambda d}\mathbf{x} + A_3 \int_{-d}^{0} e^{\lambda s} X_{t+s}\mathrm{d}s$, and

$$K(t) = \int_t^T \left\{ \frac{M(s)^2}{4R(s)} + \frac{e^{-2\lambda d} B(t)^2 Q(t)^2}{4R(s)} + \frac{e^{-\lambda d} B(t)M(s)Q(t)}{2R(s)} - N(s) \right. \\ \left. -e^{-\lambda d} Q(s)A_4(s) \right\} \mathrm{d}s. \tag{7.11}$$

Proof The problem (LQ) is a special case of (P). In order to solve (LQ) by employing Theorem 7.1, we need to check Assumptions 7.1 and 7.2 for the (LQ) model. Note that

$$\begin{aligned} \mu_1(t, x, y, u) &= A_1(t)x + A_2(t)y + B(t)u + A_4(t), \quad \mu_2(x, y) = A_3, \\ F(t, x, y, u) &= I(t)(e^{-\lambda d}x + A_3 y)^2 + R(t)u^2 + H(t)(e^{-\lambda d}x + A_3 y)u \\ &\quad + L(t)(e^{-\lambda d}x + A_3 y) + M(t)u + N(t), \\ h(x, y) &= a(e^{-\lambda d}x + A_3 y)^2, \quad \sigma(t, x, y, u) = A_5(t)x + A_6(t)y + A_7(t). \end{aligned}$$

We set $Z(x, y) = e^{-\lambda d}x + A_3 y$ so that Assumption 7.1 is supported in this (LQ) problem. Furthermore, we have

$$\begin{aligned}
\widetilde{\mu}(t, x, y, u) &= Z_x(x, y)\mu_1(t, x, y, u) + Z_y(x, y)(f(x) - \lambda y) \\
&= e^{-\lambda d}(A_1(t)x + A_2(t)y + B(t)u + A_4(t)) + A_3(x - \lambda y) \\
&= (A_1(t) + e^{\lambda d}A_3)Z(x, y) + (e^{-\lambda d}A_2(t) - A_3A_1(t) - e^{\lambda d}A_3^2 - \lambda A_3)y \\
&\quad + e^{-\lambda d}(B(t)u + A_4(t)), \\
\overline{F}(t, x, y, u) &= I(t)Z(x, y)^2 + R(t)u^2 + H(t)Z(x, y)u + L(t)Z(x, y) + M(t)u + N(t), \\
\overline{h}(x, y) &= aZ(x, y)^2, \\
\widetilde{\sigma}(t, x, y, u) &= Z_x(x, y)\sigma(t, x, y, u) \\
&= e^{-\lambda d}(A_5(t)x + A_6(t)y + A_7(t)) \\
&= A_5(t)Z(x, y) - (A_3A_5(t) - A_6(t)e^{-\lambda d})y + e^{-\lambda d}A_7(t).
\end{aligned}$$

Therefore, Assumption 7.2 holds if only if

$$A_2(t) = e^{\lambda d}A_3(A_1(t) + e^{\lambda d}A_3 + \lambda), \;\; A_6(t) = e^{\lambda d}A_3A_5(t).$$

The reduced finite-dimensional uncertain control problem becomes

$$(\overline{LQ}) \begin{cases}
\overline{J}(t, z) = \sup\limits_{u \in U} E\left[\int_t^T \{I(s)Z_s^2 + R(s)u_s^2 + H(s)Z_s u_s + L(s)Z_s \right. \\
\qquad\qquad \left. + M(s)u_s + N(s)\}\,\mathrm{d}s + GZ_T^2\right] \\
\text{subject to} \\
\mathrm{d}Z_s = \left\{(A_1(s) + e^{\lambda d}A_3)Z_s + e^{-\lambda d}(B(s)u_s + A_4(s))\right\}\mathrm{d}s \\
\qquad + \{A_5(s)Z_s + e^{-\lambda d}A_7(s)\}\mathrm{d}C_s, \quad s \in [t, T] \\
Z_t = z, \\
u_s \in U, \quad s \in [t, T]
\end{cases} \tag{7.12}$$

where $z = Z(x(\varphi_t), y(\varphi_t))$. By using Theorem 7.1, we know that $\overline{J}(t, z)$ satisfies

$$-\overline{J}_t(t, z) = \sup_{u_t \in U}\{\overline{F}(t, z, u_t) + \overline{J}_z(t, z)\overline{\mu}(t, z, u_t)\},$$

that is,

$$\begin{aligned}
-\overline{J}_t(t, z) = \sup_{u \in U}\{&I(t)z^2 + R(t)u_t^2 + H(t)zu_t + L(t)z + M(t)u_t + N(t) \\
&+ [(A_1(t) + e^{\lambda d}A_3)z + e^{-\lambda d}(B(t)u_t + A_4(t))]\overline{J}_z\}.
\end{aligned} \tag{7.13}$$

Let

$$\begin{aligned}
g(u_t) = &I(t)z^2 + R(t)u_t^2 + H(t)zu_t + L(t)z + M(t)u_t + N(t) \\
&+ [(A_1(t) + e^{\lambda d}A_3)z + e^{-\lambda d}(B(t)u_t + A_4(t))]\overline{J}_z.
\end{aligned}$$

Setting $\dfrac{\partial g(u_t)}{\partial u_t} = 0$ yields

$$2R(t)u_t + H(t)z + M(t) + e^{-\lambda d} B(t)\overline{J}_z = 0,$$

Hence,

$$u_t^* = -\frac{H(t)z + M(t) + e^{-\lambda d} B(t)\overline{J}_z}{2R(t)}. \tag{7.14}$$

By Eq. (7.13), we have

$$\begin{aligned} -\overline{J}_t(t,z) &= I(t)z^2 + R(t)u_t^{*2} + H(t)zu_t^* + L(t)z + M(t)u_t^* + N(t) \\ &+[(A_1(t) + e^{\lambda d} A_3)z + e^{-\lambda d}(B(t)u_t^* + A_4(t))]\overline{J}_z. \end{aligned} \tag{7.15}$$

Since $\overline{J}(T, Z_T) = GZ_T^2$, we guess

$$\overline{J}(t,z) = \frac{1}{2}P(t)z^2 + Q(t)z + K(t). \tag{7.16}$$

Thus,

$$\overline{J}_t(t,z) = \frac{1}{2}\frac{dP(t)}{dt}z^2 + \frac{dQ(t)}{dt}z + \frac{dK(t)}{dt} \tag{7.17}$$

and

$$\overline{J}_z(t,z) = P(t)z + Q(t). \tag{7.18}$$

Substituting (7.14) and (7.18) into (7.15) yields

$$\begin{aligned} &\overline{J}_t(t,z) \\ &= \left[\frac{H(t)^2}{4R(t)} + \frac{e^{-\lambda d}H(t)B(t)P(t)}{2R(t)} + \frac{e^{-2\lambda d}B(t)^2P(t)^2}{4R(t)} - P(t)A_1(t)\right. \\ &\left. -P(t)A_3e^{\lambda d} - I(t)\right]z^2 + \left[\frac{e^{-\lambda d}H(t)B(t) + e^{-2\lambda d}B(t)^2P(t)}{2R(t)}Q(t)\right. \\ &-A_3e^{\lambda d}Q(t) - A_1(t)Q(t) - L(t) + \frac{e^{-\lambda d}M(t)B(t)P(t) + H(t)M(t)}{2R(t)} \\ &\left.-e^{-\lambda d}P(t)A_4(t)\right]z + \frac{M(t)^2}{4R(t)} + \frac{e^{-2\lambda d}B(t)^2Q(t)^2}{4R(t)} + \frac{e^{-\lambda d}B(t)M(t)Q(t)}{2R(t)} \\ &-N(t) - e^{-\lambda d}Q(t)A_4(t). \end{aligned} \tag{7.19}$$

By Eqs. (7.17) and (7.19), we get

$$\frac{\mathrm{d}P(t)}{\mathrm{d}t} = -2I(t) + \frac{H(t)^2}{2R(t)} + \frac{e^{-\lambda d}H(t)B(t)P(t)}{R(t)} + \frac{e^{-2\lambda d}B(t)^2P(t)^2}{2R(t)}$$
$$-2P(t)(A_1(t) + A_3e^{\lambda d}), \tag{7.20}$$

$$\frac{\mathrm{d}Q(t)}{\mathrm{d}t} = \left(\frac{e^{-\lambda d}H(t)B(t) + e^{-2\lambda d}B(t)^2P(t)}{2R(t)} - A_1(t) - A_3e^{\lambda d}\right)Q(t)$$
$$-e^{-\lambda d}P(t)A_4(t) + \frac{e^{-\lambda d}M(t)B(t)P(t) + H(t)M(t)}{2R(t)} - L(t), \tag{7.21}$$

and

$$\frac{\mathrm{d}K(t)}{\mathrm{d}t} = \frac{M(t)^2}{4R(t)} + \frac{e^{-2\lambda d}B(t)^2Q(t)^2}{4R(t)} + \frac{e^{-\lambda d}B(t)M(t)Q(t)}{2R(t)} - N(t)$$
$$-e^{-\lambda d}Q(t)A_4(t). \tag{7.22}$$

Since $\overline{J}(T, z) = \frac{1}{2}P(T)z^2 + Q(T)z + K(T) = az^2$, we have $P(T) = 2a$, $Q(T) = 0$, and $K(T) = 0$. By Eqs. (7.20) and (7.21), we obtain (7.8) and (7.9). By Eq. (7.22), Eq. (7.11) holds. Therefore,

$$J(t, \varphi_t) = \overline{J}(t, z) = \frac{1}{2}P(t)z^2 + Q(t)z + K(t)$$

is the optimal value of (LQ), and

$$u_t^* = -\frac{(H(t) + e^{-\lambda d}B(t)P(t))z + e^{-\lambda d}B(t)Q(t) + M(t)}{2R(t)}$$

is the optimal control, where

$$z = e^{-\lambda d}x(\varphi_t) + A_3y(\varphi_t) = e^{-\lambda d}\varphi_t(0) + A_3\int_{-d}^{0} e^{\lambda s}\varphi_t(s)\mathrm{d}s$$
$$= e^{-\lambda d}X_t + A_3\int_{-d}^{0} e^{\lambda s}X_{t+s}\mathrm{d}s = e^{-\lambda d}x + A_3\int_{-d}^{0} e^{\lambda s}X_{t+s}\mathrm{d}s.$$

7.2.1 Example

We consider the following example of uncertain optimal control model with time-delay

$$\begin{cases} J(0,\varphi_0) = \sup\limits_{u\in U} E\left[\int_0^2 \left\{-(e^{-1}X_s + Y_s)^2 - u_s^2\right\} \mathrm{d}s + (e^{-1}X_T + Y_T)^2\right] \\ \text{subject to} \\ \quad \mathrm{d}X_t = \{(-e-5)X_t + X_{t-0.2} + u_t\}\mathrm{d}t + \mathrm{d}C_t, \quad t \in [0,2] \\ \quad X_t = \varphi_0(t) = \cos\pi t, \quad -0.2 \le t \le 0 \\ \quad Y_t = \int_{-0.2}^0 e^{5s} X_{t+s}\mathrm{d}s, \quad t \in [0,2] \\ \quad u_t \in R, \quad t \in [0,2]. \end{cases} \tag{7.23}$$

We have $A_1(s) = -(e+5)$, $A_2(s) = 0$, $A_3 = 1$, $A_4(s) = 0$, $B(s) = 1$, $A_5(s) = A_6(s) = 0$, $A_7(s) = 1$, $I(s) = -1$, $R(s) = -1$, $H(s) = L(s) = M(s) = N(s) = 0$, $a = 1$, $\lambda = 5$, $d = 0.2$. Hence, $A_2(t) = eA_3(A_1(t) + eA_3 + 5)$ and $A_6(t) = eA_3A_5(t)$ hold in this model. By Theorem 7.2, the function $Q(t)$ satisfies

$$\begin{cases} \dfrac{\mathrm{d}Q(t)}{\mathrm{d}t} = \left(-\dfrac{1}{2e^2}P(t) + 5\right)Q(t), \quad t \in [0,2] \\ Q(2) = 0. \end{cases}$$

Thus, $Q(t) = 0$ for $t \in [0,2]$, and then $K(t) = 0$ for $t \in [0,2]$. Therefore, the optimal control u_t^* is $u_t^* = \dfrac{e^{-1}P(t)z_t}{2}$, where $z_t = e^{-1}x_t + y_t$, and the optimal value is $J(0,\varphi_0) = \frac{1}{2}P(0)z_0^2$, where $z_0 = e^{-1}x_0 + y_0$, and $P(t)$ satisfies

$$\begin{cases} \dfrac{\mathrm{d}P(t)}{\mathrm{d}t} = -\dfrac{1}{2e^2}P(t)^2 + 10P(t) + 2 \\ P(2) \quad = 2, \end{cases} \tag{7.24}$$

and

$$x_0 = X_0 = 1, \quad y_t = Y_t = \int_{-0.2}^0 e^{5s} X_{t+s}\mathrm{d}s,$$

$$y_0 = Y_0 = \int_{-0.2}^0 e^{5s} X_s ds = \int_{-0.2}^0 e^{5s} \cos\pi s \,\mathrm{d}s = \frac{\pi\sin(0.2\pi) - 5\cos(0.2\pi) + 5e}{e(\pi^2 + 25)}.$$

Since the value of y_t is derived from the value of X_s between $t - 0.2$ and t, the analytical expression of y_t cannot be obtained and so is that of u_t^*.

Now we consider the numerical solutions of the model. Let $\Pi_1 = s_0, s_1, \ldots s_{20}$ be an average partition of $[-0.2, 0]$ (i.e., $-0.2 = s_0 < s_1 < \cdots < s_{20} = 0$), and $\Delta s = 0.01$. Thus,

$$y_t = Y_t = \sum_{i=0}^{20} e^{5s_i} X_{t+s_i}\Delta s.$$

Let $\Pi_2 = t_0, t_1, \ldots t_{200}$ be an average partition of $[0, 2]$ (i.e., $0 = t_0 < t_1 < \cdots < t_{200} = 2$), and $\Delta t = 0.01$. Thus,

$$\Delta X_t = (-(e+5)X_t + X_{t-0.2} + u_t^*)\Delta t + \Delta C_t.$$

Since ΔC_t is a normal uncertain variable with expected value 0 and variance Δt^2, the distribution function of ΔC_t is $\Phi(x) = \left(1 + \exp\left(-\frac{\pi x}{\sqrt{3}\Delta t}\right)\right)^{-1}, x \in R$. We may get a sample point $\widetilde{c}_t$ of ΔC_t from $\widetilde{c}_t = \Phi^{-1}(rand(0,1))$ that $\widetilde{c}_t = \frac{\sqrt{3}\Delta t}{-\pi} \ln\left(\frac{1}{rand(0,1)} - 1\right)$. Thus, x_t, y_t, and u_t may be given by the following iterative equations

$$\begin{aligned} y_{t_j} &= \sum_{i=0}^{20} e^{5s_i} x_{t_j+s_i} \Delta s, \quad u_{t_j} = \frac{e^{-1}}{2} P(t_j)(e^{-1}x_{t_j} + y_{t_j}), \\ x_{t_{j+1}} &= x_{t_j} + \Delta X_t \\ &= x_{t_j} + (-(e+5)x_{t_j} + x_{t_j-0.2} + u_{t_j})\Delta t + \frac{\sqrt{3}\Delta t}{-\pi} \ln\left(\frac{1}{rand(0,1)} - 1\right) \end{aligned}$$

for $j = 0, 1, 2, \ldots, 200$, and $x_{s_i} = \cos \pi s_i$ for $i = 0, 1, \ldots, 20$, where the numerical solution $P(t_j)$ of (7.24) is provided by

$$P(t_{j-1}) = P(t_j) - \left(-\frac{1}{2e^2} P(t_j)^2 + 10P(t_j) + 2\right) \Delta t$$

for $j = 200, 199, \ldots, 2, 1$ with $P(t_{200}) = 2$.

Therefore, the optimal value of the example is $J(0, \varphi_0) = -0.024429$, and the optimal controls and corresponding states are obtained in Table 7.1 for part data.

Table 7.1 Numerical solutions

t	0	0.1	0.2	0.3	0.4	0.5	0.6
x	1.000000	0.472623	0.268005	0.163757	0.103532	0.042242	0.010214
y	0.126709	0.103694	0.063890	0.035297	0.020538	0.012068	0.005922
u	−0.018170	−0.010197	−0.005969	−0.003510	−0.002154	−0.001014	−0.000356
t	0.7	0.8	0.9	1.0	1.1	1.2	1.3
x	−0.008133	−0.028942	−0.002142	−0.014657	−0.002138	−0.046814	−0.030036
y	0.001701	−0.001587	0.000481	0.000381	−0.000242	−0.001997	−0.004128
u	0.000047	0.000449	0.000011	0.000184	0.000038	0.000704	0.000554
t	1.4	1.5	1.6	1.7	1.8	1.9	2.0
x	−0.023194	−0.022888	−0.050667	−0.013823	−0.003976	−0.012452	−0.040871
y	−0.004421	−0.003591	−0.003553	−0.003156	−0.002194	−0.001305	−0.002409
u	0.000466	0.000416	0.000682	0.000160	−0.000047	−0.000621	−0.006417

7.3 Model with Multiple Time-Delays

Consider an uncertain linear systems with multiple time-delays in control input

$$dX_s = \left(a_0(s) + a_1(s)X_s + \sum_{i=1}^{p} B_i(s)u(s-h_i) \right) ds + b(s)dC_s \tag{7.25}$$

with the initial condition $X(t_0) = X_0$, where t_0 is the initial time. Here X_s is the state vector of n dimension, u_s is the control vector of m dimension, $h_i > 0 (i = 1, \ldots, p)$ are positive time-delays, $h = \max\{h_1, \ldots, h_p\}$ is the maximum delay shift, and $C_s = (C_{s_1}, C_{s_2}, \ldots, C_{s_p})$, where $C_{s_1}, C_{s_2}, \ldots, C_{s_p}$ are independent canonical Liu process. And $a_0(s)$, $a_1(s)$, $b(s)$ and $B_i(s)(i = 1, 2, \ldots, p)$ are piecewise continuous matrix functions of appropriate dimensions.

The quadratic cost function to be maximized is defined as follows

$$J(t, \boldsymbol{x}) = \sup_{u_t \in U} E\left(\frac{1}{2} \int_t^T (u_s^{\tau} R(s) u_s + X_s^{\tau} L(s) X_s) ds + X_T^{\tau} \Psi_T X_T \right), \tag{7.26}$$

where $X_t = \boldsymbol{x}$, $R(s)$ is positive, Ψ_T and $L(s)$ are nonnegative definite symmetric matrices and $T > 0$.

Theorem 7.3 ([2]) *Let μ_{1t} be an $n \times n$ integrable uncertain process, μ_{2t} and v_{2t} be two n-dimensional integrable uncertain processes. Then the n-dimensional linear uncertain differential equation*

$$dX_t = (\mu_{1t} X_t + \mu_{2t})dt + v_{2t} dC_t, \tag{7.27}$$

has a solution

$$X_t = U_t \left(X_0 + \int_0^t U_s^{-1} \mu_{2s} ds + \int_0^t U_s^{-1} v_{2s} dC_s \right), \tag{7.28}$$

where

$$U_t = \exp\left(\int_0^t \mu_{1s} ds \right).$$

Proof At first, we define two uncertain processes U_t and V_t via uncertain differential equations,

$$dU_t = \mu_{1t} U_t dt, \ dV_t = U_t^{-1} \mu_{2t} dt + U_t^{-1} v_{2t} dC_t.$$

It follows from the integration by parts that

$$d(U_t V_t) = U_t dV_t + dU_t \cdot V_t = (\mu_{1t} U_t V_t + \mu_{2t})dt + (v_{2t})dC_t.$$

That is, the uncertain process $X_t = U_t V_t$ is a solution of the uncertain differential equation (7.27). The uncertain process U_t can also be written as

$$U_t = \sum_{n=0}^{\infty} \frac{1}{n!} \left(\int_0^t \mu_{1s} \mathrm{d}s \right)^n \cdot U_0.$$

Taking differentiation operations on both sides, we have

$$\mathrm{d}U_t = \mu_{1t} \sum_{n=1}^{\infty} \frac{1}{(n-1)!} \left(\int_0^t \mu_{1s} \mathrm{d}s \right)^{n-1} \cdot U_0 \mathrm{d}t = \mu_{1t} \cdot \exp\left(\int_0^t \mu_{1s} \mathrm{d}s \right) \cdot U_0 \mathrm{d}t.$$

Thus,

$$U_t = \exp\left(\int_0^t u_{1s} \mathrm{d}s \right) \cdot U_0,$$

$$V_t = V_0 + \int_0^t U_s^{-1} \mu_{2s} \mathrm{d}s + \int_0^t U_s^{-1} v_{2s} \mathrm{d}C_s.$$

Taking $U_0 = I$ and $V_0 = X_0$, we get the solution (7.28). The theorem is proved.

Theorem 7.4 ([2]) *For the uncertain linear system with input delay (7.25) and the quadratic criterion (7.26), the optimal control law for $t \geq t_0$ is given by*

$$u^*(t) = -R^{-1}(t) \sum_{i=1}^{p} B_i(t) M_i(t) (P(t)\boldsymbol{x} + Q(t)),$$

where $P(t)$ satisfies

$$\begin{cases} \dot{P}(t) = -\frac{1}{2} \sum_{i=1}^{p} M_i^{\tau}(t) B_i^{\tau}(t) P^{\tau}(t) R^{-1}(t) P(t) \sum_{i=1}^{p} B_i(t) M_i(t) \\ \qquad + L(t) + a_1(t) P(t) \\ P(T) = 2\Psi_T, \end{cases} \tag{7.29}$$

and $Q(t)$ is a solution of the following differential equation

$$\begin{cases} \dot{Q}(t) = -\sum_{i=1}^{p} M_i^{\tau}(t) B_i^{\tau}(t) P(t) R^{-1}(t) Q(t) \sum_{i=1}^{p} B_i(t) M_i(t) \\ \qquad + a_0(t) P(t) + a_1(t) Q(t) \\ Q(T) = 0, \end{cases} \tag{7.30}$$

where $M_i(t) = \exp(-\int_{t-h_i}^{t} a_1(s) \mathrm{d}s)$. The optimal value for $t \geq t_0$ is given by

$$J(t,\boldsymbol{x})=\frac{1}{2}\boldsymbol{x}^{\tau}P(t)\boldsymbol{x}+Q(t)\boldsymbol{x}+K(t),$$

where

$$K(t)=\int_t^T\left\{-\frac{1}{2}\sum_{i=1}^{p}M_i^{\tau}(s)B_i^{\tau}(s)Q(s)^{\tau}R^{-1}(s)Q(s)\sum_{i=1}^{p}B_i(s)M_i(s)\right.$$
$$\left.+a_0(t)Q(s)\right\}\mathrm{d}s. \tag{7.31}$$

Proof For the optimal control problem (7.25) and (7.26), using the equation of optimality (2.15), we get

$$-J_t(t,\boldsymbol{x})=\sup_{u\in U}\left\{\frac{1}{2}(u_t^{\tau}R(t)u_t+\boldsymbol{x}^{\tau}L(t)\boldsymbol{x})+J_x^{\tau}a_0(t)\right.$$
$$\left.J_x^{\tau}a_1(t)\boldsymbol{x}+J_x^{\tau}\sum_{i=1}^{p}B_i(t)u_{t-h_i}\right\}. \tag{7.32}$$

Let

$$g(u_t)=\frac{1}{2}(u_t^{\tau}R(t)u_t+\boldsymbol{x}^{\tau}L(t)\boldsymbol{x})+J_x^{\tau}a_0(t)+J_x^{\tau}a_1(t)\boldsymbol{x}+J_x^{\tau}\sum_{i=1}^{p}B_i(t)u_{t-h_i}.$$

Setting $\frac{\partial g(u_t)}{\partial u_t}=0$ yields

$$R(t)u_t+\sum_{i=1}^{p}M_i^{\tau}(t)B_i^{\tau}(t)J_x=0,$$

where $M_i(t)=\frac{\partial u_{t-h_i}}{\partial u_t}$. Hence,

$$u_t^*=-R^{-1}(t)\sum_{i=1}^{p}M_i^{\tau}(t)B_i^{\tau}(t)J_x. \tag{7.33}$$

By Eq. (7.32), we have

$$-J_t=\frac{1}{2}\left(u_t^{*\tau}R(t)u_t^*+\boldsymbol{x}^{\tau}L(t)\boldsymbol{x}\right)+J_x^{\tau}\left(a_0(t)+a_1(t)\boldsymbol{x}+\sum_{i=1}^{p}B_i(t)u_{t-h_i}^*\right). \tag{7.34}$$

Since $J(T,X_T)=X_T^{\tau}\Psi X_T$, we guess

$$J(t,\boldsymbol{x})=\frac{1}{2}\boldsymbol{x}^{\tau}P(t)\boldsymbol{x}+Q(t)\boldsymbol{x}+K(t).$$

Then

$$J_t = \frac{1}{2}\boldsymbol{x}^\tau \dot{P}(t)\boldsymbol{x} + \dot{Q}(t)\boldsymbol{x} + \dot{K}(t), \tag{7.35}$$

and

$$J_x = P(t)\boldsymbol{x} + Q(t). \tag{7.36}$$

Substituting Eqs. (7.33), (7.36) into Eq. (7.34) yields

$$\begin{aligned}
&-J_t(t, \boldsymbol{x}) \\
&= x^\tau \left\{ -\frac{1}{2}\sum_{i=1}^{p} M_i^\tau(t)B_i^\tau(t)P^\tau(t)R^{-1}(t)P(t)\sum_{i=1}^{p} B_i(t)M_i(t) + L(t) \right. \\
&\quad + a_1(t)P(t)\}\,\boldsymbol{x} + \left\{ -\sum_{i=1}^{p} M_i^\tau(t)B_i^\tau(t)P(t)R^{-1}(t)Q(t)\sum_{i=1}^{p} B_i(t)M_i(t) \right. \\
&\quad + a_0(t)P(t) + a_1(t)Q(t)\}\,\boldsymbol{x} \\
&\quad -\frac{1}{2}\sum_{i=1}^{p} M_i^\tau(t)B_i^\tau(t)Q(t)^\tau R^{-1}(t)Q(t)\sum_{i=1}^{p} B_i(t)M_i(t) + a_0(t)Q(t).
\end{aligned} \tag{7.37}$$

By Eq. (7.35) and Eq. (7.37), we get

$$\begin{aligned}
\dot{P}(t) = &-\frac{1}{2}\sum_{i=1}^{p} M_i^\tau(t)B_i^\tau(t)P^\tau(t)R^{-1}(t)P(t)\sum_{i=1}^{p} B_i(t)M_i(t) \\
&+L(t) + a_1(t)P(t),
\end{aligned} \tag{7.38}$$

$$\begin{aligned}
\dot{Q}(t) = &-\sum_{i=1}^{p} M_i^\tau(t)B_i^\tau(t)P(t)R^{-1}(t)Q(t)\sum_{i=1}^{p} B_i(t)M_i(t) \\
&+a_0(t)P(t) + a_1(t)Q(t),
\end{aligned} \tag{7.39}$$

and

$$\dot{K}(t) = -\frac{1}{2}\sum_{i=1}^{p} M_i^\tau(t)B_i^\tau(t)Q(t)^\tau R^{-1}(t)Q(t)\sum_{i=1}^{p} B_i(t)M_i(t) + a_0(t)Q(t). \tag{7.40}$$

Since $J(T, x) = \frac{1}{2}x^\tau P(T)x + Q(T)x + K(T) = x^\tau \Psi_T x$, we have $P(T) = 2\Psi_T$, $Q(T) = 0$, and $K(T) = 0$. Eqs. (7.29), (7.30) and (7.31) follow directly from Eqs. (7.38), (7.39), and (7.40), respectively. Therefore,

$$J(t, \boldsymbol{x}) = \frac{1}{2}\boldsymbol{x}^\tau P(t)\boldsymbol{x} + Q(t)\boldsymbol{x} + K(t),$$

is the optimal value of the uncertain linear system with input delay equation (7.25) and the quadratic criterion equation (7.26), and

$$u_t^* = -R^{-1}(t)M_i^\tau(t)\sum_{i=1}^{p} B_i^\tau(t)(P(t)\boldsymbol{x} + Q(t)). \tag{7.41}$$

Let us find the value of matrices $M_i(t)$ for this problem. Substituting the optimal control law equation (7.41) into the Eq. (7.25) gives

$$\begin{aligned} \mathrm{d}X_s = \Bigg\{ & -\sum_{i=1}^{p} B_i(s)R^{-1}(s-h_i)M_i^\tau(s-h_i)\sum_{i=1}^{p} B_i^\tau(s-h_i)\big(P(s-h_i)X_{s-h_i} \\ & +Q(s-h_i)\big) + a_0(s) + a_1(s)X_s\Big\}\,\mathrm{d}s + b(s)\mathrm{d}C_s. \end{aligned} \tag{7.42}$$

The multidimensional uncertain differential equation (7.42) has the solution

$$\begin{aligned} X_t = U(r,t)\Bigg\{ & -\int_r^t U(t,s)^{-1}\Bigg\{\sum_{i=1}^{p} B_i(s)R^{-1}(s-h_i)M_i^\tau(s-h_i)\sum_{i=1}^{p} B_i^\tau(s-h_i) \\ & \cdot\big(P(s-h_i)X_{s-h_i} + Q(s-h_i)\big) + a_0(s)\Big\}\,\mathrm{d}s + \int_r^t U(t,s)^{-1}b(s)\mathrm{d}C_s + X_r\Bigg\} \end{aligned} \tag{7.43}$$

where $t, r \geq t_0$, and

$$U(r,t) = \exp\left(-\int_r^t a_1(s)\mathrm{d}s\right),$$

by Theorem 7.3, and we know

$$U(t-h_i,t) = \exp\left(-\int_{t-h_i}^t a_1(s)\mathrm{d}s\right).$$

Since the integral terms in the right-hand side of Eq. (7.43) do not explicitly depend on u_t, we have

$$\partial X_t/\partial u_t = U(r,t)\partial X_r/\partial u_t.$$

It can be converted to

$$\partial u_t/\partial X_t = (\partial u_t/\partial X_r)U(t,r).$$

Hence, the equality

$$Su_t = K_1U(r,t)K_2X_r$$

holds, where $S \in R^{n\times m}$ and $K_1, K_2 \in R^{n\times n}$ can be selected the same for any $t, r \geq t_0$. Writing the last equality for $t + h_i$, $h_i > 0$, we have

$$Su_{t+h_i} = K_1U(r,t+h_i)K_2X_r.$$

Thus,

$$(\partial(Su_t)/\partial Su_{t+h_i}) = U(r,t)(U(r,t+h_i))^{-1} = U(t+h_i,t),$$

which leads to

$$(\partial(Su_t)/\partial u_{t+h_i}) = U(t+h_i,t)S.$$

For any S, using $t-h_i$ instead of t yields

$$S(\partial u_{t-h_i}/\partial u_t) = SM_i(t) = U(t,t-h_i)S = \exp\left(-\int_{t-h_i}^{t} a_1(s)\mathrm{d}s\right)S,$$

for $t \geq t_0 + h_i$. So

$$M_i(t) = \exp\left(-\int_{t-h_i}^{t} a_1(s)\mathrm{d}s\right).$$

The theorem is proved.

7.3.1 Example

Consider the following example of uncertain linear systems with multiple time-delays in control input

$$\begin{cases} J(0,X_0) = \sup\limits_{u\in U} E\left(\dfrac{1}{2}\displaystyle\int_0^2 (u_s^2 + X_s^2)\mathrm{d}s + X_T^2\right), \\ \text{subject to} \\ \mathrm{d}X_t = (X_t + u_{t-0.1} + u_t)\mathrm{d}t + \mathrm{d}C_t, \quad t\in[0,2] \\ u_t = 0, \quad t\in[-0.1,0] \\ X_0 = 1. \end{cases} \tag{7.44}$$

We have $a_0(t)=0$, $a_1(t)=1$, $B(t)=1$, $b_0(t)=0$, $b_1(t)=1$, $R(t)=1$, $L(t)=1$, $\Psi(T)=1$. So we get $M_1(t)=\exp(-0.1)$, and $M_2(t)=1$. By Theorem 7.4, the function $Q(t)$ satisfies

$$\begin{cases} \dfrac{\mathrm{d}Q(t)}{\mathrm{d}t} = -\left(1+\exp(-0.1)\right)^2 P(t)Q(t) + Q(t) \\ Q(2) = 0. \end{cases} \tag{7.45}$$

Thus, $Q(t)=0$ for $t\in[0,2]$, and then $K(t)=0$ for $t\in[0,2]$. So we get the optimal control u_t^* is

$$u_t^* = -\Big(1+\exp(-0.1)\Big)P(t)x,$$

and the optimal value is $J(0, X_0) = \frac{1}{2}P(0)X_0^2$, and $P(t)$ satisfies

$$\begin{cases} \dfrac{\mathrm{d}P(t)}{\mathrm{d}t} = -(1+\exp(-0.1))^2 P(t)^2 + 2P(t) + 2 \\ P(2) = 2. \end{cases} \tag{7.46}$$

Now we consider the numerical solution of this model. Let $S = t_0, t_1, \ldots t_{200}$ be an average partition of $[0, 2]$ (i.e., $0 = t_0 < t_1 < \cdots < t_{200} = 2$), and $\Delta t = 0.01$. Thus,

$$\Delta X_t = (X_t + u^*_{t-0.1} + u^*_t)\Delta t + \Delta C_t.$$

Since ΔC_t is a normal uncertain variable with expected value 0 and variance Δt^2, the distribution function is $\Phi(x) = \left(1 + \exp\left(-\frac{\pi x}{\sqrt{3}\Delta t}\right)\right)^{-1}$, $x \in R$. So we may get a sample point $\tilde{c_t}$ of ΔC_t from $\tilde{c_t} = \Phi^{-1}(rand(0, 1))$ that $\tilde{c_t} = \frac{\sqrt{3}\Delta t}{-\pi} \ln\left(\frac{1}{rand(0,1)} - 1\right)$. Thus, x_t and u_t may be given by the following iterative equations

$$u_{t_j} = -(1 + \exp(-0.1))P(t_j)x_{t_j},$$

$$x_{t_{j+1}} = x_{t_j} + (x_t + u_{t_j-0.1} + u_{t_j})\Delta t + \frac{\sqrt{3}\Delta t}{-\pi} \ln\left(\frac{1}{rand(0, 1)} - 1\right),$$

for $j = 0, 1, 2, \ldots, 200$, and $u_{t_j-0.1} = 0$, where $t_j \in [0, 0.1]$. The numerical solution $P(t_j)$ of (7.46) is provided by

$$P(t_{j-1}) = P(t_j) - \Big(-(1 + \exp(-0.1))^2 P(t_j)^2 + 2P(t_j) + 2\Big)\Delta t,$$

for $j = 200, 199, \ldots, 2, 1$ with $P(t_{200}) = 2$.

Table 7.2 Numerical solutions

t	0	0.1	0.2	0.3	0.4	0.5	0.6
x_t	1.000000	0.996292	1.008533	0.960251	0.968829	0.967966	0.967966
$P(t)$	1.067544	1.067548	1.067554	1.067565	1.067586	1.067624	1.067692
u_t^*	−2.033493	−1.922702	−2.025984	−2.050895	−1.952756	−1.970269	−1.968630
t	0.7	0.8	0.9	1.0	1.1	1.2	1.3
x_t	0.976390	0.964772	0.977926	0.944681	0.940976	0.944992	0.929304
$P(t)$	1.067815	1.068038	1.068441	1.069075	1.070317	1.072567	10.7665
u_t^*	−1.985991	−1.962773	−1.990284	−1.92376	−1.918448	−1.930682	−1.905861
t	1.4	1.5	1.6	1.7	1.8	1.9	2.0
x_t	0.906163	0.86307	0.873138	0.856087	0.784042	0.744824	0.678863
$P(t)$	1.084087	1.097717	1.11969	1.164538	1.252615	1.439701	2.000000
u_t^*	−1.871245	−1.804668	−1.862251	−1.899024	−1.870754	−2.042682	−2.586254

Therefore, the optimal value of the example is $J(0, X_0) = 1.067544$, and the optimal controls and corresponding states are obtained in Table 7.2.

References

1. Chen R, Zhu Y (2013) An optimal control model for uncertain systems with time-delay. J Oper Res Soc Jpn 54(4):243–256
2. Jiang Y, Yan Y, Zhu Y (2016) Optimal control problem for uncertain linear systems with multiple input delays. J Uncertain Anal Appl 4(5):10 pages

Chapter 8
Parametric Optimal Control for Uncertain Systems

As it is well known, the optimal control of linear quadratic model is given in a feedback form, which is determined by the solution of a Riccati differential equation. However, the corresponding Riccati differential equation cannot be solved analytically in many cases. Even if an analytic solution can be obtained, it might be a complex time-oriented function. Then the optimal control is often difficult to be implemented and costly in industrial production. Hence, a practical control in a simplified form should be chosen for overcoming these issues at the precondition of keeping an admissible accuracy of a controller.

This chapter aims at formulating an approximate model with parameter to simplify the form of optimal control for uncertain linear quadratic model and presenting a parametric optimization approach for solving it.

8.1 Parametric Optimization Based on Expected Value

To begin with we consider the following multidimensional uncertain linear quadratic model without control parameter:

$$\begin{cases} J(0, \boldsymbol{x}_0) = \min\limits_{\boldsymbol{u}_s} E\left[\int_0^T \left(\boldsymbol{X}_s^{\tau} Q(s)\boldsymbol{X}_s + \boldsymbol{u}_s^{\tau} R(s)\boldsymbol{u}_s\right) \mathrm{d}s + \boldsymbol{x}_T^{\tau} S_T \boldsymbol{x}_T\right] \\ \text{subject to} \\ \quad \mathrm{d}\boldsymbol{X}_s = (A(s)\boldsymbol{X}_s + B(s)\boldsymbol{u}_s)\mathrm{d}s + (M(s)\boldsymbol{X}_s + N(s)\boldsymbol{u}_s)\mathrm{d}C_s \\ \quad \boldsymbol{X}_0 = \boldsymbol{x}_0, \end{cases} \tag{8.1}$$

where the state $\boldsymbol{X}_s$ is an uncertain vector process of dimension n. The matrix functions $Q(s)$, $R(s)$, $A(s)$, $B(s)$, $M(s)$, $N(s)$ are appropriate size, where $Q(s)$ is symmetric nonnegative definite, $R(s)$ is symmetric positive definite, and S_T is symmetric. For any $0 < t < T$, we use $\boldsymbol{x}$ to denote the state of $\boldsymbol{X}_s$ at time t and $J(t, \boldsymbol{x})$ to denote

Y. Zhu, *Uncertain Optimal Control*, Springer Uncertainty Research,
https://doi.org/10.1007/978-981-13-2134-4_8

the optimal value obtainable in $[t, T]$. Assume that the following two conditions are satisfied.

Assumption 8.1 The elements of $Q(s)$, $R(s)$, $A(s)$, $B(s)$, $M(s)$, $N(s)$, and $R^{-1}(s)$ are continuous and bounded functions on [0, T].

Assumption 8.2 The optimal value $J(t, \boldsymbol{x})$ is a twice differentiable function on $[0, T] \times [a, b]^n$.

Theorem 8.1 ([1]) *A necessary and sufficient condition that $\boldsymbol{u}_t^*$ be an optimal control for model (8.1) is that*

$$\boldsymbol{u}_t^* = -\frac{1}{2}R^{-1}(t)B^{\tau}(t)P(t)\boldsymbol{x}, \tag{8.2}$$

where the function $P(t)$ satisfies the following Riccati differential equation and boundary condition

$$\begin{cases} \dfrac{\mathrm{d}P(t)}{\mathrm{d}t} = -2Q(t) - A^{\tau}(t)P(t) - P(t)A(t) + \frac{1}{2}P(t)B(t)R^{-1}(t)B^{\tau}(t)P(t) \\ P(T) = 2S_T. \end{cases} \tag{8.3}$$

The optimal value of model (8.1) is

$$J(0, \boldsymbol{x}_0) = \frac{1}{2}\boldsymbol{x}_0^{\tau}P(0)\boldsymbol{x}_0. \tag{8.4}$$

Proof Applying Theorem 2.3, we have

$$\min_{\boldsymbol{u}_t}\left\{\boldsymbol{x}^{\tau}Q(t)\boldsymbol{x} + \boldsymbol{u}_t^{\tau}R(t)\boldsymbol{u}_t + (A(t)\boldsymbol{x} + B(t)\boldsymbol{u}_t)^{\tau}\nabla_{\boldsymbol{x}}J(t, \boldsymbol{x}) + J_t(t, \boldsymbol{x})\right\} = 0.$$

Denote $\psi(\boldsymbol{u}_t) = \boldsymbol{x}^{\tau}Q(t)\boldsymbol{x} + \boldsymbol{u}_t^{\tau}R(t)\boldsymbol{u}_t + (A(t)\boldsymbol{x} + B(t)\boldsymbol{u}_t)^{\tau}\nabla_{\boldsymbol{x}}J(t, \boldsymbol{x}) + J_t(t, \boldsymbol{x})$.

First, we verify the necessity. Since $J(T, \boldsymbol{X}_T) = \boldsymbol{x}_T^{\tau}S_T\boldsymbol{x}_T$, we conjecture that $\nabla_{\boldsymbol{x}}J(t, \boldsymbol{x}) = P(t)\boldsymbol{x}$, with boundary condition $P(T) = 2S_T$. Setting $\dfrac{\partial\psi(\boldsymbol{u}_t)}{\partial\boldsymbol{u}_t} = 0$, we have

$$\boldsymbol{u}_t = -\frac{1}{2}R^{-1}(t)B^{\tau}(t)P(t)\boldsymbol{x}.$$

Because $\dfrac{\partial^2\psi(\boldsymbol{u}_t)}{\partial\boldsymbol{u}_t^2} = 2R(t) > 0$, $\boldsymbol{u}_t$ is the optimal control of model (8.1), i.e.,

$$\boldsymbol{u}_t^* = -\frac{1}{2}R^{-1}(t)B^{\tau}(t)P(t)\boldsymbol{x}.$$

Taking the gradient of $\psi(\boldsymbol{u}_t^*)$ with respect to $\boldsymbol{x}$, we have

$$\left(2Q(t) + A^{\tau}(t)P(t) + P(t)A(t) - \frac{1}{2}P(t)B(t)R^{-1}(t)B^{\tau}(t)P(t) + \frac{\mathrm{d}P(t)}{\mathrm{d}t}\right)\boldsymbol{x} = 0.$$

Thus

$$\frac{\mathrm{d}P(t)}{\mathrm{d}t} = -2Q(t) - A^{\tau}(t)P(t) - P(t)A(t) + \frac{1}{2}P(t)B(t)R^{-1}(t)B^{\tau}(t)P(t) \quad (8.5)$$

with $P(T) = 2S_T$. According to the existence and uniqueness theorem of differential equation and Assumption 8.1, we can infer that the solution $P(t)$ is existent and unique. In addition, we have

$$\begin{cases} \left(\frac{\mathrm{d}P(t)}{\mathrm{d}t}\right)^{\tau} = \left[-2Q(t) - A^{\tau}(t)P(t) - P(t)A(t) + \frac{1}{2}P(t)B(t)R^{-1}(t)B^{\tau}(t)P(t)\right]^{\tau} \\ P^{\tau}(T) = 2S_T^{\tau}, \end{cases}$$

That is,

$$\begin{cases} \frac{\mathrm{d}P^{\tau}(t)}{\mathrm{d}t} = -2Q(t) - P^{\tau}(t)A(t) - A^{\tau}(t)P^{\tau}(t) \\ \qquad + \frac{1}{2}P^{\tau}(t)B(t)R^{-1}(t)B^{\tau}(t)P^{\tau}(t) \\ P^{\tau}(T) = 2S_T. \end{cases} \quad (8.6)$$

It follows from Eqs. 8.5 and 8.6 that $P(t)$ and $P^{\tau}(t)$ are solutions of the same Riccati differential equation with the same boundary condition. So, $P(t)$ is symmetric. Further, we have $J(t, \boldsymbol{x}) = \frac{1}{2}\boldsymbol{x}^{\tau}P(t)\boldsymbol{x}$. Then, the optimal value $J(0, \boldsymbol{x}_0)$ is

$$J(0, \boldsymbol{x}_0) = \frac{1}{2}\boldsymbol{x}_0^{\tau}P(0)\boldsymbol{x}_0. \quad (8.7)$$

Then, we prove the sufficient condition. Assume that $J(t, \boldsymbol{x}) = \frac{1}{2}\boldsymbol{x}^{\tau}P(t)\boldsymbol{x}$. Substituting Eqs. (8.2) and (8.3) into $\psi(\boldsymbol{u}_t)$, we have $\psi(\boldsymbol{u}_t^*) = 0$. Because the objective function of model (8.1) is convex, there must be an optimal control solution. Hence, $\boldsymbol{u}_t^*$ is the optimal control. The optimal value $J(0, \boldsymbol{x}_0)$ is

$$J(0, \boldsymbol{x}_0) = \frac{1}{2}\boldsymbol{x}_0^{\tau}P(0)\boldsymbol{x}_0. \quad (8.8)$$

The theorem is proved.

8.1.1 Parametric Optimal Control Model

The parametric optimal control problem we will address here is of the form:

$$\begin{cases} V(0, \boldsymbol{x}_0) = \min\limits_{\boldsymbol{u}_s \in \boldsymbol{U}} E\left[\int_0^T \left(\boldsymbol{X}_s^\tau Q(s)\boldsymbol{X}_s + \boldsymbol{u}_s^\tau R(s)\boldsymbol{u}_s\right) \mathrm{d}s + \boldsymbol{x}_T^\tau S_T \boldsymbol{x}_T\right] \\ \text{subject to} \\ \quad \mathrm{d}\boldsymbol{X}_s = (A(s)\boldsymbol{X}_s + B(s)\boldsymbol{u}_s)\mathrm{d}s + (M(s)\boldsymbol{X}_s + N(s)\boldsymbol{u}_s)\mathrm{d}C_s \\ \quad \boldsymbol{X}_0 = \boldsymbol{x}_0, \end{cases} \tag{8.9}$$

where $\boldsymbol{X}_s$ is a state vector of dimension n with initial condition $\boldsymbol{X}_0 = \boldsymbol{x}_0$ and $\boldsymbol{u}_s$ is a decision vector of dimension r. $\boldsymbol{U} = \{\boldsymbol{K}\boldsymbol{x}_s | \boldsymbol{K} = (k_{ij})_{r\times n} \in R^{r\times n}\}$, where $\boldsymbol{x}_s$ represents the state of $\boldsymbol{X}_s$ at time s with $\boldsymbol{x}_s \in [a, b]^n$. The matrix functions $Q(s)$, $R(s)$, S_T, $A(s)$, $B(s)$, $M(s)$, and $N(s)$ are defined as in model (8.1) and satisfy the Assumption 8.1. For any $0 < t < T$, we use $\boldsymbol{x}$ to denote the state of $\boldsymbol{X}_s$ at time t and $V(t, \boldsymbol{x})$ to denote the optimal value obtainable in $[t, T]$.

Solving an optimal control vector $\boldsymbol{u}_t^*$ of model (8.9) is essentially equivalent to solving an optimal parameter matrix $\boldsymbol{K}^*$. From now on, we assume that $V(t, \boldsymbol{x})$ is a twice differentiable function on $[0, T] \times [a, b]^n$. Applying Eq. (2.15), we obtain

$$\min_{\boldsymbol{K}} \{\boldsymbol{x}^\tau Q(t)\boldsymbol{x} + (\boldsymbol{K}\boldsymbol{x})^\tau R(t)(\boldsymbol{K}\boldsymbol{x}) + (A(t)\boldsymbol{x} + B(t)\boldsymbol{K}\boldsymbol{x})^\tau \nabla_{\boldsymbol{x}} V(t, \boldsymbol{x}) + V_t(t, \boldsymbol{x})\} \\ = 0. \tag{8.10}$$

Note that the $\boldsymbol{u}_t^*$ in Eq. (8.2) can be used to achieve global minimum for model (8.1), and an optimal control $\boldsymbol{u}_t^*$ of model (8.9) can be seen as a local optimal control solution for model (8.1). Therefore, the optimality of optimal parameter matrix $\boldsymbol{K}^*$ means that $V(0, \boldsymbol{x}_0)$ can be close to $J(0, \boldsymbol{x}_0)$ as much as possible. In order to solve an optimal parameter matrix $\boldsymbol{K}^*$, we use $J(t, \boldsymbol{x})$ as a substitute for $V(t, \boldsymbol{x})$, where $J(t, \boldsymbol{x})$ is defined in model (8.1). Hence,

$$\Upsilon(\boldsymbol{K}) = \boldsymbol{x}^\tau Q(t)\boldsymbol{x} + (\boldsymbol{K}\boldsymbol{x})^\tau R(t)(\boldsymbol{K}\boldsymbol{x}) + (A(t)\boldsymbol{x} + B(t)\boldsymbol{K}\boldsymbol{x})^\tau \nabla_{\boldsymbol{x}} J(t, \boldsymbol{x}) + J_t(t, \boldsymbol{x}).$$

Remark 8.1 Because $\boldsymbol{K} \in R^{r\times n}$, we could not obtain the optimal parameter matrix $\boldsymbol{K}^*$ of model (8.9) by taking gradient of $\Upsilon(\boldsymbol{K})$ with respect to $\boldsymbol{K}$.

8.1.2 Parametric Approximation Method

Note that $L([0, T] \times [a, b]^n)$ represents the space of absolutely integrable functions on domain $[0, T] \times [a, b]^n$, where $T > 0$ and $a, b \in R$. For the sake of discussion, we define a norm as

$$\| f(t, \boldsymbol{x}) \| = \int_a^b \cdots \int_a^b \left(\int_0^T |f(t, \boldsymbol{x})|\mathrm{d}t\right) \mathrm{d}x_1 \cdots \mathrm{d}x_n, \tag{8.11}$$

where $f(t, \boldsymbol{x}) \in L([0, T] \times [a, b]^n)$ and $\boldsymbol{x} = (x_1, x_2, \ldots, x_n)^\tau$. The optimal parameter matrix $\boldsymbol{K}^*$ needs to ensure the difference between $\Upsilon(\boldsymbol{K})$ and 0 achieves minimum in the sense of the norm defined above, i.e.,

$$
\begin{aligned}
\boldsymbol{K}^* = \arg \min_{\boldsymbol{K} \in R^{r \times n}} & \parallel \boldsymbol{x}^\tau Q(t)\boldsymbol{x} + (\boldsymbol{K}\boldsymbol{x})^\tau R(t)(\boldsymbol{K}\boldsymbol{x}) + (A(t)\boldsymbol{x} + B(t)\boldsymbol{K}\boldsymbol{x})^\tau \nabla_x J(t, \boldsymbol{x}) \\
& + J_t(t, \boldsymbol{x}) \parallel . \qquad (8.12)
\end{aligned}
$$

We know that $J(t, \boldsymbol{x}) = \frac{1}{2}\boldsymbol{x}^\tau P(t)\boldsymbol{x}$, where the function $P(t)$ satisfies the following matrix Ricatti differential equation and boundary condition

$$
\begin{cases}
\dfrac{\mathrm{d}P(t)}{\mathrm{d}t} = -2Q(t) - A^\tau(t)P(t) - P(t)A(t) + \frac{1}{2}P(t)B(t)R^{-1}(t)B^\tau(t)P(t) \\
P(T) = 2S_T.
\end{cases}
\qquad (8.13)
$$

Remark 8.2 A variety of numerical algorithms have been developed by many researchers for solving the Riccati equation (see Balasubramaniam et al. [2], Caines and Mayne [3], Khan et al. [4]).

Assume that $P(t) = (p_{ij}(t))_{n \times n}$. In solving the matrix Riccati differential equation (8.13), the following system of nonlinear differential equation has occurred:

$$
\dot{p}_{ij}(t) = f_{ij}\left(t, p_{11}(t), \ldots, p_{1n}(t), p_{21}(t), \ldots, p_{2n}(t), \ldots, p_{n1}(t), \ldots, p_{nn}(t)\right) \qquad (8.14)
$$

for $i, j = 1, 2, \ldots, n$. Apparently, matrix Riccati differential equation (8.13) contains n^2 first-order ordinary differential equations with n^2 variables. The Runge–Kutta method is considered as the best tool for the numerical integration of ordinary differential equations. For convenience, the fourth-order Runge–Kutta method is explained for a system of two first-order ordinary differential equations with two variables:

$$
p_{11}(s+1) = p_{11}(s) + \frac{h}{6}(k_1 + 2k_2 + 2k_3 + k_4),
$$

$$
p_{12}(s+1) = p_{12}(s) + \frac{h}{6}(l_1 + 2l_2 + 2l_3 + l_4),
$$

where

$$
\begin{aligned}
k_1 &= f_{11}(t, k_{11}, k_{12}), & l_1 &= f_{12}(t, k_{11}, k_{12}), \\
k_2 &= f_{11}\left(t + \frac{h}{2}, k_{11} + \frac{hk_1}{2}, k_{12} + \frac{hl_1}{2}\right), & l_2 &= f_{12}\left(t + \frac{h}{2}, k_{11} + \frac{hk_1}{2}, k_{12} + \frac{hl_1}{2}\right), \\
k_3 &= f_{11}\left(t + \frac{h}{2}, k_{11} + \frac{hk_2}{2}, k_{12} + \frac{hl_2}{2}\right), & l_3 &= f_{12}\left(t + \frac{h}{2}, k_{11} + \frac{hk_2}{2}, k_{12} + \frac{hl_2}{2}\right), \\
k_4 &= f_{11}(t + h, k_{11} + hk_3, k_{12} + hl_3), & l_4 &= f_{12}(t + h, k_{11} + hk_3, k_{12} + hl_3).
\end{aligned}
$$

In the similar way, the original system (8.13) can be solved for n^2 first-order ordinary differential equations.

Setting

$$L_1 = \left(l_{ij}^{(1)}\right)_{r\times r} = \int_0^T R(t)\mathrm{d}t, \quad L_2 = \left(l_{ij}^{(2)}\right)_{r\times n} = \int_0^T B^\tau(t)P(t)\mathrm{d}t.$$

Then, we have the following theorem to ensure the solvability of optimal control parameter matrix $\boldsymbol{K}^*$.

Theorem 8.2 ([1]) *Denote* $L(\boldsymbol{K}) = \left(L_{ij}(\boldsymbol{K})\right)_{n\times n} = \boldsymbol{K}^\tau L_1 \boldsymbol{K} + \boldsymbol{K}^\tau L_2$. *Then we have*

$$\begin{aligned}\boldsymbol{K}^* = \arg\min_{\boldsymbol{K}\in R^{r\times n}} \Bigg[&\frac{1}{3}(b^2+ba+a^2)(b-a)^n \sum_{i=1}^n L_{ii}(\boldsymbol{K}) \\ &+\frac{1}{4}(b+a)^2(b-a)^n \sum_{i=1}^n \sum_{j=1,j\neq i}^n L_{ij}(\boldsymbol{K}) \Bigg].\end{aligned} \tag{8.15}$$

Proof Applying Eq. (8.12), we have

$$\begin{aligned}\boldsymbol{K}^* = \arg\min_{\boldsymbol{K}\in R^{r\times n}} \parallel &\boldsymbol{x}^\tau Q(t)\boldsymbol{x} + (\boldsymbol{K}\boldsymbol{x})^\tau R(t)(\boldsymbol{K}\boldsymbol{x}) + (A(t)\boldsymbol{x} + B(t)\boldsymbol{K}\boldsymbol{x})^\tau \nabla_{\boldsymbol{x}} J(t,\boldsymbol{x}) \\ &+J_t(t,\boldsymbol{x}) \parallel .\end{aligned}$$

Because $\Upsilon(\boldsymbol{K}) \geq 0$, we have

$$\begin{aligned}\parallel \Upsilon(\boldsymbol{K}) \parallel = \int_a^b \cdots \int_a^b \Bigg(\int_0^T &\big[\boldsymbol{x}^\tau Q(t)\boldsymbol{x} + (\boldsymbol{K}\boldsymbol{x})^\tau R(t)(\boldsymbol{K}\boldsymbol{x}) + (A(t)\boldsymbol{x} \\ &+B(t)(\boldsymbol{K}\boldsymbol{x}))^\tau \nabla_{\boldsymbol{x}} J(t,\boldsymbol{x}) + J_t(t,\boldsymbol{x})\big]\mathrm{d}t\Bigg)\mathrm{d}x_1\cdots\mathrm{d}x_n.\end{aligned}$$

Denote $L(\boldsymbol{K}) = \left(L_{ij}(\boldsymbol{K})\right)_{n\times n} = \boldsymbol{K}^\tau L_1 \boldsymbol{K} + \boldsymbol{K}^\tau L_2$. It holds that

$$\begin{aligned}\boldsymbol{K}^* &= \arg\min_{\boldsymbol{K}\in R^{r\times n}} \int_a^b \cdots \int_a^b \boldsymbol{x}^\tau L(\boldsymbol{K})\boldsymbol{x}\mathrm{d}x_1\cdots\mathrm{d}x_n \\ &= \arg\min_{\boldsymbol{K}\in R^{r\times n}} \Bigg[\frac{1}{3}(b^2+ba+a^2)(b-a)^n \sum_{i=1}^n L_{ii}(\boldsymbol{K}) \\ &\qquad +\frac{1}{4}(b+a)^2(b-a)^n \sum_{i=1}^n \sum_{j=1,j\neq i}^n L_{ij}(\boldsymbol{K}) \Bigg].\end{aligned}$$

The theorem is proved.

Therefore, the optimal control of the model (8.9) is

$$\boldsymbol{u}_t^* = \boldsymbol{K}^*\boldsymbol{x}. \tag{8.16}$$

Assume that $V(t,\boldsymbol{x}) = \frac{1}{2}\boldsymbol{x}^\tau G(t)\boldsymbol{x}$. From Eq. (8.10), we obtain

$$
\begin{aligned}
&Q(t) + \boldsymbol{K}^{*\tau} R(t) \boldsymbol{K}^* + \frac{1}{2} A^\tau(t) G(t) + \frac{1}{2} G(t) A(t) + \frac{1}{2} G(t) B(t) \boldsymbol{K}^* \\
&+ \frac{1}{2} \boldsymbol{K}^{*\tau} B^\tau(t) G(t) + \frac{1}{2} \frac{\mathrm{d}G(t)}{\mathrm{d}t} = 0.
\end{aligned}
$$

Using the fourth-order Runge–Kutta method described above, we can obtain the solution of $G(t)$, where the function $G(t)$ satisfies the following matrix Riccati differential equation and boundary condition

$$
\begin{cases}
\dfrac{\mathrm{d}G(t)}{\mathrm{d}t} = -2Q(t) - 2\boldsymbol{K}^{*\tau} R(t) \boldsymbol{K}^* - A^\tau(t) G(t) - G(t) A(t) - G(t) B(t) \boldsymbol{K}^* \\
\qquad\qquad - \boldsymbol{K}^{*\tau} B^\tau(t) G(t) \\
G(T) = 2S_T.
\end{cases}
\tag{8.17}
$$

Hence, the optimal value of model (8.9) is

$$
V(0, \boldsymbol{x}_0) = \frac{1}{2} \boldsymbol{x}_0^\tau G(0) \boldsymbol{x}_0. \tag{8.18}
$$

8.2 Parametric Optimization Based on Optimistic Value

We will study the following multidimensional uncertain linear quadratic model under optimistic value criterion with control parameter as an approximation of the model (3.24):

$$
\begin{cases}
V(0, \boldsymbol{x}_0) = \inf\limits_{\boldsymbol{u}_s \in U} \left\{ \displaystyle\int_0^T \left(\boldsymbol{X}_s^\tau Q(s) \boldsymbol{X}_s + \boldsymbol{u}_s^\tau R(s) \boldsymbol{u}_s \right) \mathrm{d}s + \boldsymbol{X}_T^\tau S_T \boldsymbol{X}_T \right\}_{\sup} (\alpha) \\
\text{subject to} \\
\quad \mathrm{d}\boldsymbol{X}_s = (A(s) \boldsymbol{X}_s + B(s) \boldsymbol{u}_s) \mathrm{d}s + M(s) \boldsymbol{X}_s \mathrm{d}C_s \\
\quad \boldsymbol{X}_0 = \boldsymbol{x}_0,
\end{cases}
\tag{8.19}
$$

where $\boldsymbol{u}_s$ is a decision vector of dimension r, $\boldsymbol{U} = \{\boldsymbol{K}\boldsymbol{x}_s | \boldsymbol{K} = \boldsymbol{K}_l = (k_{ij}^{(l)})_{r\times n} \in R^{r\times n}, s \in [t_{l-1}, t_l), l = 1, 2, \ldots, m\}$ with $0 = t_0 < t_1 < \cdots < t_{m-1} < t_m = T$ and $\boldsymbol{K}$ is a control parameter matrix. Here, we stipulate the last subinterval $[t_{m-1}, t_m)$ represents the closed interval $[t_{m-1}, t_m]$. For any $0 < t < T$, we use $\boldsymbol{x}$ to denote the state of $\boldsymbol{X}_s$ at time t and $V(t, \boldsymbol{x})$ to denote the optimal value obtainable in $[t, T]$.

Assume that $V(t, \boldsymbol{x})$ is a twice differentiable function on $[0, T] \times [a, b]^n$. According to Theorem 3.2, we have

$$
\begin{aligned}
-V_t(t, \boldsymbol{x}) = \inf_{\boldsymbol{u}_t \in U} \Big\{ & \boldsymbol{x}^\tau Q(t) \boldsymbol{x} + \boldsymbol{u}_t^\tau R(t) \boldsymbol{u}_t + \nabla_{\boldsymbol{x}} V(t, \boldsymbol{x})^\tau (A(t) \boldsymbol{x} + B(t) \boldsymbol{u}_t) \\
& + \frac{\sqrt{3}}{\pi} \ln \frac{1-\alpha}{\alpha} |\nabla_{\boldsymbol{x}} V(t, \boldsymbol{x})^\tau M(t) \boldsymbol{x}| \Big\}.
\end{aligned}
\tag{8.20}
$$

It is noticeable that the optimal control $\boldsymbol{u}_t^*$ of model (8.19) can be seen as a suboptimal control solution for model (3.24). Hence, the optimality of optimal parameter matrix $\boldsymbol{K}^*$ means that the error between $V(0, \boldsymbol{x}_0)$ and $J(0, \boldsymbol{x}_0)$ should be as small as possible. Therefore, we replace $V(t, \boldsymbol{x})$ with $J(t, \boldsymbol{x})$ in Eq. (8.20). For convenience, we denote

$$\begin{aligned}\Gamma(\boldsymbol{K}) &= \boldsymbol{x}^\tau Q(t)\boldsymbol{x} + (\boldsymbol{K}\boldsymbol{x})^\tau R(t)(\boldsymbol{K}\boldsymbol{x}) + \nabla_{\boldsymbol{x}} J(t, \boldsymbol{x})^\tau (A(t)\boldsymbol{x} + B(t)\boldsymbol{K}\boldsymbol{x}) \\ &\quad + \frac{\sqrt{3}}{\pi}\ln\frac{1-\alpha}{\alpha}|\nabla_{\boldsymbol{x}} J(t, \boldsymbol{x})^\tau M(t)\boldsymbol{x}| + J_t(t, \boldsymbol{x}).\end{aligned}$$

Remark 8.3 The optimal parameter matrix $\boldsymbol{K}^*$ cannot be obtained by taking gradient of $\Gamma(\boldsymbol{K})$ because $\boldsymbol{K} \in R^{r\times n}$ is a numerical matrix.

8.2.1 Piecewise Optimization Method

On each subinterval $[t_{l-1}, t_l)$, $l = 1, 2, \ldots, m$, the optimal control parameter matrix $\boldsymbol{K}_l^*$ needs to ensure the difference between $\Gamma\left(\boldsymbol{K}_l^*\right)$ and 0 achieves minimum in the sense of the norm defined by (8.11), i.e.,

$$\begin{aligned}\boldsymbol{K}_l^* = \arg\min_{\boldsymbol{K}_l \in R^{r\times n}} &\parallel \boldsymbol{x}^\tau Q(t)\boldsymbol{x} + (\boldsymbol{K}_l\boldsymbol{x})^\tau R(t)(\boldsymbol{K}_l\boldsymbol{x}) + \nabla_{\boldsymbol{x}} J(t, \boldsymbol{x})^\tau (A(t)\boldsymbol{x} \\ &+ B(t)\boldsymbol{K}_l\boldsymbol{x}) + \frac{\sqrt{3}}{\pi}\ln\frac{1-\alpha}{\alpha}|\nabla_{\boldsymbol{x}} J(t, \boldsymbol{x})^\tau M(t)\boldsymbol{x}| + J_t(t, \boldsymbol{x}) \parallel \quad (8.21)\end{aligned}$$

Assume that $J(t, \boldsymbol{x}) = \frac{1}{2}\boldsymbol{x}^\tau P(t)\boldsymbol{x}$, where the function $P(t)$ satisfies the Riccati differential equation (3.27) and boundary condition $P(T) = 2S_T$.

Theorem 8.3 ([5]) *Denote* $\boldsymbol{W} = \int_{t_{l-1}}^{t_l} R(t)\mathrm{d}t$, $\boldsymbol{Y} = \int_{t_{l-1}}^{t_l} P(t)B(t)\mathrm{d}t$. *Then we have*

$$\begin{aligned}\boldsymbol{K}_l^* &= \arg\min_{\boldsymbol{K}_l \in R^{r\times n}} \int_{\boldsymbol{x}\in[a,b]^n} \boldsymbol{x}^\tau(\boldsymbol{K}_l^\tau \boldsymbol{W}\boldsymbol{K}_l + \boldsymbol{Y}\boldsymbol{K}_l)\boldsymbol{x}\mathrm{d}\boldsymbol{x} \\ &= \arg\min_{\boldsymbol{K}_l \in R^{r\times n}} \left[\frac{1}{3}(b^2 + ba + a^2)(b-a)^n \sum_{i=1}^n Z_{ii}(\boldsymbol{K}_l)\right. \\ &\qquad \left. + \frac{1}{4}(b+a)^2(b-a)^n \sum_{i=1}^n \sum_{j=1, j\neq i}^n Z_{ij}(\boldsymbol{K}_l)\right], \quad (8.22)\end{aligned}$$

where

$$\boldsymbol{Z}(\boldsymbol{K}_l) = \left(Z_{ij}(\boldsymbol{K}_l)\right)_{n\times n} = \boldsymbol{K}_l^\tau \boldsymbol{W}\boldsymbol{K}_l + \boldsymbol{Y}\boldsymbol{K}_l.$$

Proof It follows from Eq. (8.21) that

$$
\begin{aligned}
\boldsymbol{K}_l^* = & \arg\min_{\boldsymbol{K}_l \in R^{r\times n}} \parallel \boldsymbol{x}^\tau Q(t)\boldsymbol{x} + (\boldsymbol{K}_l\boldsymbol{x})^\tau R(t)(\boldsymbol{K}_l\boldsymbol{x}) + \nabla_{\boldsymbol{x}} J(t,\boldsymbol{x})^\tau (A(t)\boldsymbol{x} \\
& + B(t)\boldsymbol{K}_l\boldsymbol{x}) + \tfrac{\sqrt{3}}{\pi}\ln\tfrac{1-\alpha}{\alpha}|\nabla_{\boldsymbol{x}} J(t,\boldsymbol{x})^\tau M(t)\boldsymbol{x}| + J_t(t,\boldsymbol{x}) \parallel \\
= & \arg\min_{\boldsymbol{K}_l \in R^{r\times n}} \parallel (\boldsymbol{K}_l\boldsymbol{x})^\tau R(t)(\boldsymbol{K}_l\boldsymbol{x}) + \nabla_{\boldsymbol{x}} J(t,\boldsymbol{x})^\tau B(t)\boldsymbol{K}_l\boldsymbol{x} \parallel \\
= & \arg\min_{\boldsymbol{K}_l \in R^{r\times n}} \int_{\boldsymbol{x}\in[a,b]^n} \int_{t_{l-1}}^{t_l} [(\boldsymbol{K}_l\boldsymbol{x})^\tau R(t)(\boldsymbol{K}_l\boldsymbol{x}) + \nabla_{\boldsymbol{x}} J(t,\boldsymbol{x})^\tau B(t)\boldsymbol{K}_l\boldsymbol{x}]\,\mathrm{d}t\mathrm{d}\boldsymbol{x} \\
= & \arg\min_{\boldsymbol{K}_l \in R^{r\times n}} \int_{\boldsymbol{x}\in[a,b]^n} \int_{t_{l-1}}^{t_l} \left[\boldsymbol{x}^\tau \boldsymbol{K}_l^\tau R(t)\boldsymbol{K}_l\boldsymbol{x} + \boldsymbol{x}^\tau P(t)B(t)\boldsymbol{K}_l\boldsymbol{x}\right]\mathrm{d}t\mathrm{d}\boldsymbol{x}.
\end{aligned}
$$

Denote $\boldsymbol{W} = \int_{t_{l-1}}^{t_l} R(t)\mathrm{d}t$, $\boldsymbol{Y} = \int_{t_{l-1}}^{t_l} P(t)B(t)\mathrm{d}t$. Then

$$
\begin{aligned}
\boldsymbol{K}_l^* = & \arg\min_{\boldsymbol{K}_l \in R^{r\times n}} \int_{\boldsymbol{x}\in[a,b]^n} \boldsymbol{x}^\tau(\boldsymbol{K}_l^\tau \boldsymbol{W}\boldsymbol{K}_l + \boldsymbol{Y}\boldsymbol{K}_l)\boldsymbol{x}\mathrm{d}\boldsymbol{x} \\
= & \arg\min_{\boldsymbol{K}_l \in R^{r\times n}} \left[\tfrac{1}{3}(b^2+ba+a^2)(b-a)^n \sum_{i=1}^{n} Z_{ii}(\boldsymbol{K}_l)\right. \\
& \left. + \tfrac{1}{4}(b+a)^2(b-a)^n \sum_{i=1}^{n} \sum_{j=1, j\neq i}^{n} Z_{ij}(\boldsymbol{K}_l)\right].
\end{aligned}
$$

The theorem is proved.

Here, we use the fourth-order Runge–Kutta method to reversely calculate the numerical value of $P(t)$ on each subinterval. In the first step, we calculate $P(t)$ on interval $[t_{m-1}, t_m)$ with the boundary value $P(m) = P(T)$. Then, in the ith ($i = 2, \ldots, m$) step, we calculate $P(t)$ on interval $[t_{m-i}, t_{m-i+1})$, where the boundary value $P(t_{m-i+1})$ is obtained in $(i-1)$th step. At last, we calculate the integral value of $P(t)B(t)$ on each subinterval $[t_{m-1}, t_m)$, $l = 1, 2, \ldots, m$.

It follows from Eq. (8.22) that the optimal parameter matrix $\boldsymbol{K}_l^*$ can be obtained by the method of derivation. Hence, the optimal control of model (8.19) is

$$
\boldsymbol{u}_t^* = \boldsymbol{K}_l^*\boldsymbol{x}, \quad l = 1, 2, \ldots, m, \ t_{l-1} \le t < t_l. \tag{8.23}
$$

Assume that $V(t, \boldsymbol{x}) = \frac{1}{2}\boldsymbol{x}^\tau G(t)\boldsymbol{x}$. Let

$$
\begin{aligned}
\Omega_3 = \{(t,\boldsymbol{x}) |\ & \boldsymbol{x}^\tau G(t)M(t)\boldsymbol{x} \ge 0, (t,\boldsymbol{x}) \in [t_{l-1}, t_l) \times [a,b]^n, l = 1, 2, \ldots, m\}, \\
\Omega_4 = \{(t,\boldsymbol{x}) |\ & \boldsymbol{x}^\tau G(t)M(t)\boldsymbol{x} < 0, (t,\boldsymbol{x}) \in [t_{l-1}, t_l) \times [a,b]^n, l = 1, 2, \ldots, m\}.
\end{aligned}
$$

Substituting the piecewise continuous control $\boldsymbol{u}_t^*$ into Eq. (8.20), we have

$$
\begin{aligned}
& \boldsymbol{x}^\tau \left(Q(t) + \boldsymbol{K}_l^{*\tau} R(t)\boldsymbol{K}_l^* + \frac{1}{2}A^\tau(t)G(t) + \frac{1}{2}G(t)A(t) + \frac{1}{2}G(t)B(t)\boldsymbol{K}_l^* \right. \\
& \left. + \frac{1}{2}\boldsymbol{K}_l^{*\tau} B^\tau(t)G(t) + \frac{1}{2}\frac{\mathrm{d}G(t)}{\mathrm{d}t} \right)\boldsymbol{x} + \frac{\sqrt{3}}{\pi}\ln\frac{1-\alpha}{\alpha}|\boldsymbol{x}^\tau G(t)M(t)\boldsymbol{x}| = 0.
\end{aligned}
$$

Then, the function $G(t)$ satisfies the following matrix Riccati differential equation

$$\frac{dG(t)}{dt} = \begin{cases} -2Q(t) - 2\boldsymbol{K}_l^{*\tau} R(t)\boldsymbol{K}_l^* - A^\tau(t)G(t) - G(t)A(t) - G(t)B(t)\boldsymbol{K}_l^* \\ \quad -\boldsymbol{K}_l^{*\tau} B^\tau(t)G(t) - \frac{\sqrt{3}}{\pi}\ln\frac{1-\alpha}{\alpha}G(t)M(t) \\ \quad -\frac{\sqrt{3}}{\pi}\ln\frac{1-\alpha}{\alpha}M(t)^\tau G(t), \qquad \text{if } (t, \boldsymbol{x}) \in \Omega_3, \\ -2Q(t) - 2\boldsymbol{K}_l^{*\tau} R(t)\boldsymbol{K}_l^* - A^\tau(t)G(t) - G(t)A(t) - G(t)B(t)\boldsymbol{K}_l^* \\ \quad -\boldsymbol{K}_l^{*\tau} B^\tau(t)G(t) + \frac{\sqrt{3}}{\pi}\ln\frac{1-\alpha}{\alpha}G(t)M(t) \\ \quad +\frac{\sqrt{3}}{\pi}\ln\frac{1-\alpha}{\alpha}M(t)^\tau G(t), \qquad \text{if } (t, \boldsymbol{x}) \in \Omega_4 \end{cases} \tag{8.24}$$

and boundary condition $G(T) = 2S_T$.

Similar to the solving procedure of $P(t)$, we can also calculate the numerical value of $G(t)$ at each point t_{l-1}, $l = 1, 2, \ldots, m$, with $G(T) = 2S_T$. Thus, the optimal value of model (8.19) is

$$V(0, \boldsymbol{x}_0) = \frac{1}{2}\boldsymbol{x}_0^\tau G(0)\boldsymbol{x}_0. \tag{8.25}$$

References

1. Li B, Zhu Y (2017) Parametric optimal control for uncertain linear quadratic models. Appl Soft Comput 56:543–550
2. Balasubramaniam P, Samath J, Kumaresan N, Kumar A (2006) Solution of matrix Riccati differential equation for the linear quadratic singular system using neural networks. Appl Math Comput 182(2):1832–1839
3. Caines P, Mayne D (2007) On the discrete time matrix Riccati equation of optimal control. Int J Control 12(5):785–794
4. Khan N, Ara A, Jamil M (2011) An efficient approach for solving the Riccati equation with fractional orders. Comput Math Appl 61(9):2683–2689
5. Li B, Zhu Y, Chen Y (2017) The piecewise optimisation method for approximating uncertain optimal control problems under optimistic value criterion. Int J Syst Sci 48(8):1766–1774

Chapter 9
Applications

9.1 Portfolio Selection Models

9.1.1 Expected Value Model

Portfolio selection problem is a classical problem in financial economics of allocating personal wealth between investment in a risk-free security and investment in a single risk asset. Under the assumption that the risk asset earns a random return, Merton [1] studied a portfolio selection model by stochastic optimal control, and Kao [2] considered a generalized Merton's model. If we assume that the risk asset earns an uncertain return, this generalized Merton's model may be solved by uncertain optimal control.

Let X_t be the wealth of an investor at time t. The investor allocates a fraction w of the wealth in a sure asset and remainder in a risk asset. The sure asset produces a rate of return b. The risk asset is assumed to earn an uncertain return and yields a mean rate of return μ ($\mu > b$) along with a variance of σ^2 per unit time. That is to say, the risk asset earns a return $\mathrm{d}r_t$ in time interval $(t, t + \mathrm{d}t)$, where $\mathrm{d}r_t = \mu \mathrm{d}t + \sigma \mathrm{d}C_t$, and C_t is a canonical Liu process. Thus

$$\begin{aligned} X_{t+\mathrm{d}t} &= X_t + bwX_t\mathrm{d}t + \mathrm{d}r_t(1-w)X_t \\ &= X_t + bwX_t\mathrm{d}t + (\mu\mathrm{d}t + \sigma\mathrm{d}C_t)(1-w)X_t \\ &= X_t + [bw + \mu(1-w)]X_t\mathrm{d}t + \sigma(1-w)X_t\mathrm{d}C_t. \end{aligned} \tag{9.1}$$

Assume that an investor is interested in maximizing the expected utility over an infinite time horizon. Then, a portfolio selection model [3] is provided by

$$\begin{cases} J(t,x) \equiv \max\limits_{w} E\left[\displaystyle\int_0^{\infty} e^{-\beta t}\, \frac{(wX_t)^{\lambda}}{\lambda}\, \mathrm{d}t\right] \\ \text{subject to} \\ \qquad \mathrm{d}X_t = [bwX_t + \mu(1-w)X_t]\mathrm{d}t + \sigma(1-w)X_t\mathrm{d}C_t, \end{cases}$$

Y. Zhu, *Uncertain Optimal Control*, Springer Uncertainty Research,
https://doi.org/10.1007/978-981-13-2134-4_9

where $\beta > 0,\ 0 < \lambda < 1$. By the equation of optimality (2.7), we have

$$-J_t = \max_w \left\{ e^{-\beta t} \frac{(wx)^\lambda}{\lambda} + (b - \mu) wxJ_x + \mu xJ_x \right\} = \max_w L(w),$$

where $L(w)$ represents the term in the braces. The optimal w satisfies

$$\frac{\partial L(w)}{\partial w} = e^{-\beta t}(wx)^{\lambda-1}x + (b - \mu)xJ_x = 0,$$

or

$$w = \frac{1}{x}\left[(\mu - b)J_x e^{\beta t}\right]^{\frac{1}{\lambda-1}}.$$

Hence

$$-J_t = \frac{1}{\lambda}e^{-\beta t}\left[(\mu - b)J_x e^{\beta t}\right]^{\frac{\lambda}{\lambda-1}} + (b - \mu)\left[(\mu - b)J_x e^{\beta t}\right]^{\frac{1}{\lambda-1}} J_x + \mu xJ_x,$$

or

$$-J_t e^{\beta t} = \left(\frac{1}{\lambda} - 1\right)\left[(\mu - b)J_x e^{\beta t}\right]^{\frac{\lambda}{\lambda-1}} + \mu xJ_x e^{\beta t}. \tag{9.2}$$

We conjecture that $J(t, x) = kx^\lambda e^{-\beta t}$. Then

$$J_t = -k\beta x^\lambda e^{-\beta t}, \quad J_x = k\lambda x^{\lambda-1}e^{-\beta t}.$$

Substituting them into Eq. (9.2) yields

$$k\beta x^\lambda = \left(\frac{1}{\lambda} - 1\right)(\mu - b)^{\frac{\lambda}{\lambda-1}}(k\lambda)^{\frac{\lambda}{\lambda-1}}x^\lambda + \mu k\lambda x^\lambda,$$

or

$$(k\lambda)^{\frac{1}{\lambda-1}} = \frac{\beta - \mu\lambda}{(1 - \lambda)(\mu - b)^{\frac{\lambda}{\lambda-1}}}.$$

So we get

$$k\lambda = \left(\frac{\beta - \mu\lambda}{1 - \lambda}\right)^{\lambda-1} \frac{1}{(\mu - b)^\lambda}.$$

Therefore, the optimal fraction of investment on sure asset is determined by

$$w = (\mu - b)^{\frac{1}{\lambda-1}}(k\lambda)^{\frac{1}{\lambda-1}} = \frac{\beta - \mu\lambda}{(1 - \lambda)(\mu - b)}.$$

Remark 9.1 Note that the optimal fraction of investment on sure asset or risk asset is independent of total wealth. This conclusion is similar to that in the case of randomness [2].

9.1.2 Optimistic Value Model

Consider the following optimistic value model [4] provided by

$$\begin{cases} J(t,x) \equiv \max\limits_{\omega} \left[\int_0^{\infty} e^{-\beta t} \frac{(\omega X_t)^{\lambda}}{\lambda} \mathrm{d}t \right]_{\sup} (\alpha) \\ \text{subject to} \\ \quad \mathrm{d}X_t = [b\omega + \mu(1-\omega)]X_t \mathrm{d}t + \sigma(1-\omega)X_t \mathrm{d}C_t \end{cases} \tag{9.3}$$

where $\alpha \in (0, 1)$ is a specified confidence level, $\beta > 0$ and $0 < \lambda < 1$. Conjecture that $J_x(t, x) \geq 0$. Then by the equation of optimality (3.12), we have

$$\begin{aligned} -J_t = \max_{\omega} & \left\{ e^{-\beta t} \frac{(\omega x)^{\lambda}}{\lambda} + J_x \omega x \left(b - \mu - \sigma \frac{\sqrt{3}}{\pi} \ln \frac{1-\alpha}{\alpha} \right) \right. \\ & \left. + J_x x \left(\mu + \sigma \frac{\sqrt{3}}{\pi} \ln \frac{1-\alpha}{\alpha} \right) \right\} \triangleq \max_{\omega} L(\omega) \end{aligned}$$

where $L(\omega)$ represents the term enclosed by the braces. The optimal ω satisfies

$$\frac{\partial L(\omega)}{\partial \omega} = e^{-\beta t}(x\omega)^{\lambda-1}x + J_x \left(b - \mu - \sigma \frac{\sqrt{3}}{\pi} \ln \frac{1-\alpha}{\alpha} \right) x = 0,$$

or

$$\omega = \frac{1}{x} \left[\left(\mu + \sigma \frac{\sqrt{3}}{\pi} \ln \frac{1-\alpha}{\alpha} - b \right) J_x e^{\beta t} \right]^{\frac{1}{\lambda-1}}.$$

Substituting the preceding result into $\max\limits_{\omega} L(\omega)$, we obtain

$$\begin{aligned} -J_t = & \frac{1}{\lambda} e^{-\beta t} \left[\left(\mu + \sigma \frac{\sqrt{3}}{\pi} \ln \frac{1-\alpha}{\alpha} - b \right) J_x e^{\beta t} \right]^{\frac{\lambda}{\lambda-1}} + \left(\mu + \sigma \frac{\sqrt{3}}{\pi} \ln \frac{1-\alpha}{\alpha} \right) x J_x \\ & + J_x \left(b - \mu - \sigma \frac{\sqrt{3}}{\pi} \ln \frac{1-\alpha}{\alpha} \right) \left[\left(\mu + \sigma \frac{\sqrt{3}}{\pi} \ln \frac{1-\alpha}{\alpha} - b \right) J_x e^{\beta t} \right]^{\frac{1}{\lambda-1}} \end{aligned}$$

which may be rewritten as

$$-e^{\beta t}J_t = \left(\frac{1}{\lambda}-1\right)\left[\left(\mu+\sigma\frac{\sqrt{3}}{\pi}\ln\frac{1-\alpha}{\alpha}-b\right)J_x e^{\beta t}\right]^{\frac{\lambda}{\lambda-1}} + \left(\mu+\sigma\frac{\sqrt{3}}{\pi}\ln\frac{1-\alpha}{\alpha}\right)xJ_x e^{\beta t}. \tag{9.4}$$

We conjecture that $J(t, x) = kx^\lambda e^{-\beta t}$. Then $J_t = -k\beta x^\lambda e^{-\beta t}$, $J_x = k\lambda x^{\lambda-1}e^{-\beta t}$. Substituting them into (9.4) yields

$$k\beta = \left(\frac{1}{\lambda}-1\right)\left[\left(\mu+\sigma\frac{\sqrt{3}}{\pi}\ln\frac{1-\alpha}{\alpha}-b\right)k\lambda\right]^{\frac{\lambda}{\lambda-1}} + \left(\mu+\sigma\frac{\sqrt{3}}{\pi}\ln\frac{1-\alpha}{\alpha}\right)k\lambda.$$

So we get

$$k\lambda = \left[\frac{\beta-\left(\mu+\sigma\frac{\sqrt{3}}{\pi}\ln\frac{1-\alpha}{\alpha}\right)\lambda}{1-\lambda}\right]^{\lambda-1}\frac{1}{\left(\mu+\sigma\frac{\sqrt{3}}{\pi}\ln\frac{1-\alpha}{\alpha}-b\right)^{\lambda}}.$$

Therefore, the optimal ω is

$$\omega = \frac{\beta-\left(\mu+\sigma\frac{\sqrt{3}}{\pi}\ln\frac{1-\alpha}{\alpha}\right)\lambda}{(1-\lambda)\left(\mu+\sigma\frac{\sqrt{3}}{\pi}\ln\frac{1-\alpha}{\alpha}-b\right)} \tag{9.5}$$

Remark 9.2 The conclusions obtained here are different from that in the case of expected value model of uncertain optimal control studied in Sect. 2.6. Here, the optimal fraction and the optimal reward depend on all the parameters β, λ, b, μ, and σ, while the conclusions in Sect. 2.6 depend on the parameters β, λ, b, and μ. However, there are still some similar conclusions. First, in both cases, the optimal fraction of investment on risk-free asset or risky asset is independent of total wealth. Second, the optimal reward $J(t, x)$ of both two cases can be expressed as the product of a power function with respect to x and an exponential function with respect to t.

9.2 Manufacturing Technology Diffusion Problem

There are three phases in the life cycle of any new technology: research and development, transfer and commercialization, and operation and regeneration [5]. Investigations on the technology diffusion originated in some researches of marketing diffusion, such as Bass [6], Horsky and Simon [7]. Technology diffusion refers to the transition of technology's economic value during the transfer and operation phases of a technology life cycle. Modeling of technology diffusion must address two aspects: regularity due to the mean depletion rate of the technology's economic value and uncertainty owing to the disturbances occurring in technological evolution and innovation. Liu [8] studied a flexible manufacturing technology diffusion problem in a stochastic environment. If we employ uncertain differential equations as a framework to model technology diffusion problems, this flexible manufacturing technology diffusion in [8] may be solved by uncertain optimal control model with the Hurwicz criterion.

Let X_t be the potential market share at time t (state variable) and u be the proportional production level (control variable). An annual production rate can be determined as uX_t. The selling price has been fairly stable at p per unit time. The unit production cost has been a function of the annual production rate and can be calculated as cuX_t, where c is a cost conversion coefficient. With constant β as a fixed learning percentage, the learning effect can be expressed as βX_t. Thus, the typical drift is

$$b(t, X_t, u) = -\frac{uX_t}{1+\beta X_t}.$$

The diffusion is $\sigma(t, X_t, u) = \sqrt{aX_t}$, where $a > 0$ is scaling factor. Since the unit profit is $(p - cuX_t)$, and the production rate is $\frac{uX_t}{1+\beta X_t}$, the unit profit function f is expressed as

$$f(t, u, X_t) = (p - cuX_t)\frac{uX_t}{1+\beta X_t}.$$

Let $k > 0$ be the discount rate and $e^{-kT}h_0(k - \mu^{X_T})$ be the salvage value at the end time, with $\mu > 1,\ k \geq 1$. Then, a manufacturing technology diffusion problem can be defined as to choose an appropriate control $\hat{u}$, so that the Hurwicz weighted average total profit is maximized. The model [9] is provided by

$$\begin{cases} J(0, x_0) \equiv \max\limits_u H_\alpha^\rho \left\{ \displaystyle\int_0^T e^{-kt}\left[(p - cuX_t)\frac{uX_t}{1+\beta X_t}\right] \mathrm{d}t + e^{-kT}h_0(k - \mu^{X_T}) \right\} \\ \text{subject to} \\ \mathrm{d}X_t = \left[-\dfrac{uX_t}{1+\beta X_t}\right]\mathrm{d}t + \sqrt{aX_t}\mathrm{d}C_t\,. \end{cases}$$

Conjecture that $J_x(t, x) \geq 0$. Then applying the equation of optimality (3.12), we have

$$-J_t = \max_u \left\{ e^{-kt}(p - cux)\frac{ux}{1+\beta x} - J_x\frac{ux}{1+\beta x} \right.$$
$$\left. + J_x\sqrt{ax}(2\rho - 1)\left(\sigma\frac{\sqrt{3}}{\pi}\ln\frac{1-\alpha}{\alpha}\right)\right\} \triangleq \max_u L(u)$$

where $L(u)$ represents the term enclosed by the braces. The optimal u satisfies

$$\frac{\partial L(u)}{\partial u} = -e^{-kt}cx \cdot \frac{ux}{1+\beta x} + e^{-kt}(p - cux) \cdot \frac{x}{1+\beta x} - J_x \frac{x}{1+\beta x} = 0,$$

or

$$u = \frac{1}{2cx}(p - e^{kt}J_x).$$

Substituting the above result into $\max_u L(u)$, we obtain

$$-J_t = \frac{(e^{-kt}p - J_x)(p - e^{kt}J_x)}{4c(1+\beta x)} + J_x\sqrt{ax}(2\rho - 1)\left(\sigma\frac{\sqrt{3}}{\pi}\ln\frac{1-\alpha}{\alpha}\right).$$

Conjecture that $J(t,x) = e^{-kt}y(x)$, and this gives $J_t = -ke^{-kt}y(x)$, $J_x = e^{-kt}$ $y'(x)$. Using the last expression, denoting $(2\rho - 1)\left(\sigma\frac{\sqrt{3}}{\pi}\ln\frac{1-\alpha}{\alpha}\right)$ by parameter q, we find

$$ky(x) = \frac{(p - y'(x))^2}{4c(1+\beta x)} + q\sqrt{ax}y'(x)\,.$$

Letting $\lambda(x) = y'$, then we have

$$y = \frac{\lambda^2 + 2\left[2cq(1+\beta x)\sqrt{ax} - p\right]\lambda + p^2}{4kc(1+\beta x)}\,,\quad y' = \lambda\,.$$

The derivative of the right side of the first expression should be equal to the right side of the second expression. So we get

$$\begin{aligned}&\left[\lambda + 2cq(1+\beta x)\sqrt{ax} - p(1+\beta x)\right]\frac{\mathrm{d}\lambda}{\mathrm{d}x}\\ &= \frac{\beta}{2(1+\beta x)}\lambda^2 + 2kc(1+\beta x)\lambda + \left[\frac{\beta p^2}{2(1+\beta x)} - \frac{cqa(1+\beta x)}{2\sqrt{ax}} - \frac{\beta p}{(1+\beta x)}\right].\end{aligned}$$

This differential equation is a second type of Abelian equation with respect to $\lambda(x)$ with the following form

$$[\lambda + g(x)]\frac{\mathrm{d}\lambda}{\mathrm{d}x} = f_2(x)\lambda^2 + f_1(x)\lambda + f_0\,. \tag{9.6}$$

Then, solving the ordinary differential equation (9.6) with the terminal condition

$$J_{X_T} = \frac{\partial[e^{-kT}h_0(k - \mu^{X_T})]}{\partial X_T} = e^{-kT}h_0(-\ln\mu \cdot \mu^{X_T}) = e^{-kT}y'(X_T),$$

we get

$$
\begin{aligned}
y' &= \lambda(x)\\
&= -g(x) + L^{-1}\left[I + \left(\int (f_1 + g' - 2f_2 g)L dx\right)^2 + 2\int (f_0 - f_1 g + f_2 g^2)L^2 \mathrm{d}x\right]^{\frac{1}{2}}\\
&= p(\beta x + 1) - 2cq\sqrt{ax}(\beta x + 1) + (2\beta x + 2)^{\frac{1}{2}}\Bigg\{ I_0 + \frac{1}{2}(2\beta c x^2\left(acq^2 + \mu p\right)\\
&\quad - \frac{2}{15}cq\sqrt{ax}(8c\mu x(3\beta x + 5) + 20\beta p x + 15) + px(4c\mu + \beta p) - \frac{(p-2)p}{\beta x + 1}\Bigg)\\
&\quad + \frac{c^2(\sqrt{\beta x + 1}(3a\beta q x + 4\mu\sqrt{ax}(\beta x + 1))}{18 a \beta^2 x}\\
&\quad \frac{+3aq\sqrt{\beta x}\,\mathrm{arc\,sinh}^{-1}\left(\sqrt{\beta x}\right))^2}{}\Bigg\}^{\frac{1}{2}}
\end{aligned}
$$

where $L = \exp(-\int f_2\, \mathrm{d}x)$, and I_0 satisfies the equation $\lambda(X_T) = h_0(-\ln\mu\cdot\mu^{X_T})$.

The optimal proportional production level is determined by

$$u = \frac{1}{2cx}(p - \lambda(x)).$$

And the $J_x = e^{-kt}\lambda(x)$ denotes the rate of the current value function.

9.3 Mitigation Policies for Uncertain Carbon Dioxide Emissions

Climate change is accelerating and has become one of the most troublesome pollution issues to the whole society. Over the past 20 years, a lot of effort has been toward evaluating policies to control the accumulative greenhouse gases (GHG) that rise in the earth's atmosphere to lead global warming and ocean acidification. The major subject has been studied is to stabilize greenhouse gases concentration level, chiefly carbon dioxide (CO_2). Besides the emissions from natural systems, more emissions that increase the atmospheric carbon dioxide are generated by human activities, and the determinate mathematical model describing a climate economy dynamic system is presented in DeLara and Doyen [10], Doyen et al. [11], Nordhaus [12]. Inspired by this work, we plug uncertain variables into the dynamic system and deal with the management of the interaction between economic growth and greenhouse gas emissions. In order to formulate mathematical models, we use the following notations:

- $M(t)$: the atmospheric CO_2 concentration level measured in mg/kg at time t (state variable);

- $Q(t)$: the aggregated economic production level such as gross world product (GWP) at time t, measured in trillion US dollar (state variable);
- $u(t)$: the abatement rate reduction of CO_2 emissions, $0 \le u(t) \le 1$ (control variable);
- $M_{-\infty}$: the preindustrial equilibrium atmospheric concentration;
- δ: the parameter that stands for the natural rate of removal of atmospheric CO_2 to unspecified sinks;
- $\mathscr{E}(Q(t))$: the CO_2 emissions released by the economic production $Q(t)$;
- ξ^e_{t+1}: the uncertain rate of growth for the production level (uncertain variable);
- ξ^p_{t+1}: the conversion factor from emissions to concentration, it sums up highly complex physical mechanisms which are denoted by an uncertain variable. And then $\xi^p_{t+1} \cdot E(Q(t))$ stands for the CO_2 retention in the atmosphere (uncertain variable);
- T: a positive integer denotes the number of managing time stages.

We present the dynamics of the carbon cycle and global economic production described by uncertain differential equations

$$M(t+1) = M(t) - \delta(M(t) - M_{-\infty}) + \xi^p_{t+1} \cdot \mathscr{E}(Q(t)) \cdot (1 - u(t)), \quad (9.7)$$
$$Q(t+1) = (1 + \xi^e_{t+1}) \cdot Q(t), \quad (9.8)$$

where time t vary in $\{0, 1, \ldots, T-1\}$. M_0 and Q_0 denote the initial CO_2 concentration level and initial production level, respectively. This carbon cycle dynamics (9.7) can be rewrote as $(M(t+1) - M_{-\infty}) = (1-\delta)(M(t) - M_{-\infty}) + \xi^p_{t+1} \cdot \mathscr{E}(Q(t)) \cdot (1 - u(t))$, which represents the anthropogenic perturbation of a natural system from a preindustrial equilibrium atmospheric concentration. While dynamics (9.8) indicates that abatement policies or costs do not directly influence the economy, assuredly it is a restrictive assumption but this is normally used in modeling for GHG reduction policies. In addition, we suppose that the uncertain variables are stage-by-stage independent.

Consider a physical or environmental requirement as a constraint through the limitation of CO_2 concentrations below a tolerable threshold at the specific final horizon T. This concentration target is pursued to avoid danger

$$M(T) \le M_{lim}. \quad (9.9)$$

And now, we add $C(Q(t), u(t))$ to specify the abatement costs function, and the parameter $\rho \in (0, 1)$ denotes the discount factor. If the total cost is to be minimized, the controller has to balance his desire to minimize the cost due to the current decision against his desire to avoid future situations where high cost is inevitable.

We study the following pessimistic value model of uncertain optimal control problem:

$$\begin{cases} J(M_0, Q_0, 0) = \min\limits_{u(0),\cdots,u(T-1)} \left[\sum\limits_{t=0}^{T-1} \rho^t C(Q(t), u(t)) \right]_{\inf} (\alpha) \\ \text{subject to} \\ \quad (9.7), (9.8)\text{and}(9.9) \end{cases} \tag{9.10}$$

where the parameter $\alpha \in (0, 1]$ denotes the predetermined confidence level. It is similar to the literature [11], assumed the abatement costs function C having the following multiplicative form

$$C(Q(t), u(t)) = \mathscr{E}(Q(t))\cdot \left(\frac{Q(t)}{Q_0}\right)^{-\mu} \cdot L(u(t)),$$

and in this work, we set $C(Q(t), u(t))$ is linear or quadratic with respect to abatement rate $u(t)$ via, respectively, designing $L(u(t)) = \eta u(t)$ or $L(u(t)) = \eta u^2(t)/2$, where coefficient μ interrelates with the technical progress rate and η relies on the price of the backstop technology.

The problem is solved at 1-year intervals, and $T = 40$. Initial CO_2 concentration level and initial production level are set according to the data from Web site http://co2now.org/ and http://data.worldbank.org.cn/ in 2013, respectively. So we have $M_0 = 396.48$ ppm and $Q_0 = 75.62$ trillion US\$. The concentration target is fixed to $M_{lim} = 450$ ppm, while preindustrial level $M_{-\infty} = 274$ ppm.

We give the confidence level $\alpha = 0.90$, natural removal rate $\delta = 0.017$, parameters of the abatement cost functions $\eta = 100$, $\mu = 1.03$, and the discount factor $\rho = 1/1.08$.

Indeterminate factors $\xi_1^p, \ldots, \xi_T^p$ are specified as independent normal uncertain variables whose uncertainty distribution is

$$\Phi^p(x) = \left(1 + \exp\left(\frac{\pi(e-x)}{\sigma\sqrt{3}}\right)\right)^{-1},$$

with $e = 0.64$ and $\sigma = 0.02$. Additionally, $\xi_1^e, \ldots, \xi_T^e$ are independent uncertain variables following a linear distribution which is denoted by $\mathscr{L}(a, b)$. That is

$$\Phi^e(x) = \begin{cases} 0, & \text{if } x \le a \\ (x-a)/(b-a), & \text{if } a \le x \le b \\ 1, & \text{if } x \ge b \end{cases}$$

where $a = 0.00$, $b = 0.06$.

The feasible solutions can be illustrated by Fig. 9.1 drawn support from uncertain simulation algorithm in Sect. 1.4. It shows several CO_2 concentrations trajectories, and the concentrations $M(t)$ sometimes are larger than terminal target M_{lim}. This indicates that even though the final state is restricted to a target sets, it allows for exceeding the boundary during the time. And it should point out that an uncertain

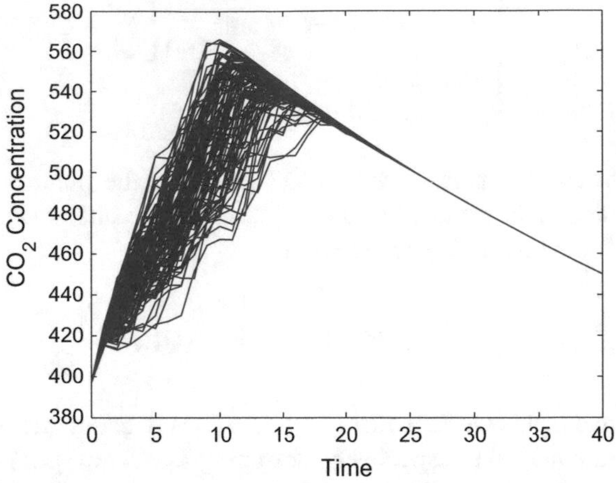

Fig. 9.1 Feasible CO_2 concentrations trajectories

variable ξ_{t+1} denoted by c_{t+1} could be realized by $\Phi(c_{t+1}) = r_{t+1}$, where r_{t+1} is a arbitrarily generated number from the interval [0.000012, 0.999988].

Using the recursion equation [13] for pessimistic value model, we give the numerical results and simulations. Referring to Tables 9.1 and 9.2, two different cost functions are used to obtain the minimal pessimistic discounted intertemporal costs $J(M_0, Q_0, 0)$. Since the CO_2 concentrations $M(t)$ and the productions $Q(t)$ are both uncertain processes, that is for each fixed time t, $M(t)$ and $Q(t)$ are uncertain variables. We can only realize a typical sample path; the state path and its interrelated optimal control sequence are shown in two tables, respectively. In both cases, the optimal abatement rates $u^*(t)$ always increase along the time. Additionally, an occurrence of a jump in $u^*(t)$ appears for the linear cost case while it will vanish for the quadratic case and replaced by a gentle change of slope.

Obviously, the minimal total abatement cost is an uncertain variable that is why we measure it with the pessimistic criterion. As displayed in Fig. 9.2, we compare these realization points of uncertain cost with its 0.9-pessimistic value and its expected value. It can be observed that the minimal total abatement cost is larger in pessimistic criterion than in an expected criterion; although the minimal expected cost is optimal, the realizations can be far from it while the pessimistic one may be not hard to reach. Minimizing the pessimistic cost is cautious to some extent, and it actually provides the least bad cost with belief degree 0.90. However in this problem, the target is finding mitigation policies to stabilize CO_2 concentration. From this perspective, using the pessimistic criterion probably does not strongly support the costs of mitigation but prevents the damages to come.

From the numerical results presented in Table 9.1, we minimized the pessimistic cost subject to the 0.90-level belief degree and 450 ppm concentration limit. And we obtained an optimal objective value of about \$28433.85 per ton of carbon. Now, we make α vary from 0.50-level to 0.95-level and make M_{lim} vary from 425 to 475 ppm and then plot the minimal total abatement cost J as a function of α and M_{lim}.

Table 9.1 Numerical results of a linear cost function

Stage t	$M(t)$	$Q(t)$	$u^*(t)$	$J(M_t, Q_t, t)$
0	396.480000	75.620000	0.000000	28433.846959
1	418.029777	79.296437	0.000000	28366.159557
2	441.607240	79.745692	0.000000	28335.740263
3	466.099984	84.209178	0.000000	28289.196715
4	490.725457	87.184703	0.000000	28220.606863
5	517.275456	90.652666	0.000000	28137.879774
6	544.278491	93.030048	0.000000	27917.934252
7	569.971202	96.442804	0.653296	27462.278899
8	575.226580	97.529059	0.915863	25798.165196
9	572.776084	101.886388	0.983667	23810.618218
10	568.256459	106.739505	0.997243	21862.785776
11	563.351325	112.726579	0.999502	20035.616317
12	558.451074	117.806237	0.999912	18349.632558
13	553.618711	120.989779	0.999981	16798.841164
14	548.865964	126.507147	0.999997	15356.200292
15	544.193363	127.831488	0.999999	14036.783455
16	539.600099	135.096188	1.000000	12797.900309
17	535.084901	141.857216	1.000000	11655.264841
18	530.646458	143.318092	1.000000	10611.685584
19	526.283469	143.483696	1.000000	9648.214678
20	521.994650	147.381889	1.000000	8749.371183
21	517.778741	149.816954	1.000000	7920.416150
22	513.634502	150.239983	1.000000	7156.410683
23	509.560715	150.675485	1.000000	6448.720878
24	505.556183	152.547221	1.000000	5792.073685
25	501.619728	159.547804	1.000000	5179.228183
26	497.750193	161.015280	1.000000	4617.739125
27	493.946439	169.021565	1.000000	4093.112362
28	490.207350	171.964579	1.000000	3611.794660
29	486.531825	180.365734	1.000000	3163.452958
30	482.918784	182.501925	1.000000	2752.106697
31	479.367165	189.489651	1.000000	2369.648177
32	475.875923	194.352181	1.000000	2016.915617
33	472.444032	205.914720	1.000000	1688.948335
34	469.070484	213.858083	1.000000	1386.980634
35	465.754285	224.283190	1.000000	1107.577394
36	462.494463	225.884472	1.000000	850.554526
37	459.290057	226.983886	1.000000	612.667432
38	456.140126	229.043773	1.000000	392.406883
39	453.043744	238.916966	1.000000	188.415427
40	450.000000			

Table 9.2 Numerical results of a quadratic cost function

Stage t	$M(t)$	$Q(t)$	$u^*(t)$	$J(M_t, Q_t, t)$
0	396.480000	75.620000	0.230734	13954.616663
1	414.102111	76.232243	0.251515	13175.250168
2	430.182082	78.385833	0.273516	12816.208493
3	446.207244	79.471808	0.305302	12414.119242
4	461.849058	82.594846	0.336776	12111.566258
5	477.722490	84.641280	0.370666	11777.661671
6	491.924979	89.373509	0.420572	11489.982335
7	505.997931	94.079504	0.480506	11177.812447
8	518.423756	99.213143	0.546959	10818.000990
9	529.421357	100.355853	0.602629	10377.812551
10	537.188036	103.637119	0.678483	9898.460554
11	543.111163	107.635592	0.767154	9388.686952
12	546.305648	110.711027	0.786680	8827.377163
13	549.227256	113.848791	0.907502	8289.543869
14	548.030187	116.489896	0.982796	7673.014121
15	544.030791	119.356515	0.996734	7025.972836
16	539.569191	119.783798	0.999381	6419.302248
17	535.080001	122.759330	0.999904	5850.597982
18	530.645342	124.811180	0.999979	5325.662378
19	526.283249	124.954256	0.999996	4842.848051
20	521.994611	126.668240	0.999999	4393.131094
21	517.778733	128.235256	1.000000	3978.088757
22	513.634500	132.522000	1.000000	3590.909522
23	509.560715	134.056360	1.000000	3234.957235
24	505.556183	140.074593	1.000000	2903.122840
25	501.619728	140.544752	1.000000	2599.272444
26	497.750193	148.707979	1.000000	2314.235888
27	493.946439	153.958208	1.000000	2052.181650
28	490.207350	154.300744	1.000000	1811.510753
29	486.531825	157.089153	1.000000	1588.268494
30	482.918784	163.553166	1.000000	1380.572322
31	479.367165	173.089621	1.000000	1188.047631
32	475.875923	180.872230	1.000000	1010.608430
33	472.444032	187.917450	1.000000	846.783978
34	469.070484	193.086141	1.000000	695.600510
35	465.754285	199.105181	1.000000	555.752822
36	462.494463	204.859040	1.000000	426.525938
37	459.290057	211.492653	1.000000	306.991229
38	456.140126	222.616870	1.000000	196.383043
39	453.043744	225.880005	1.000000	94.373121
40	450.000000			

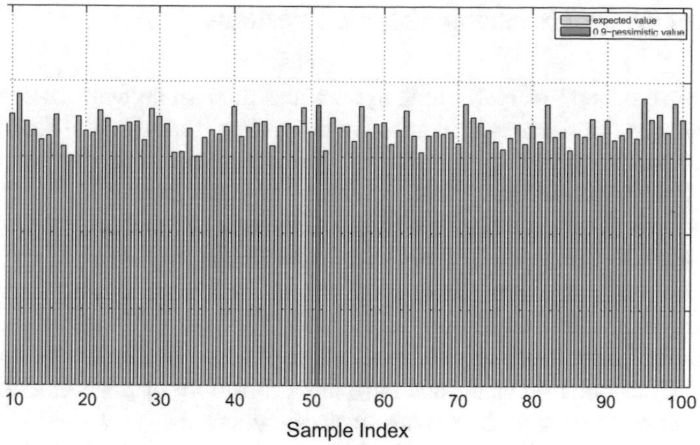

Fig. 9.2 Intertemporal discounted costs realizations

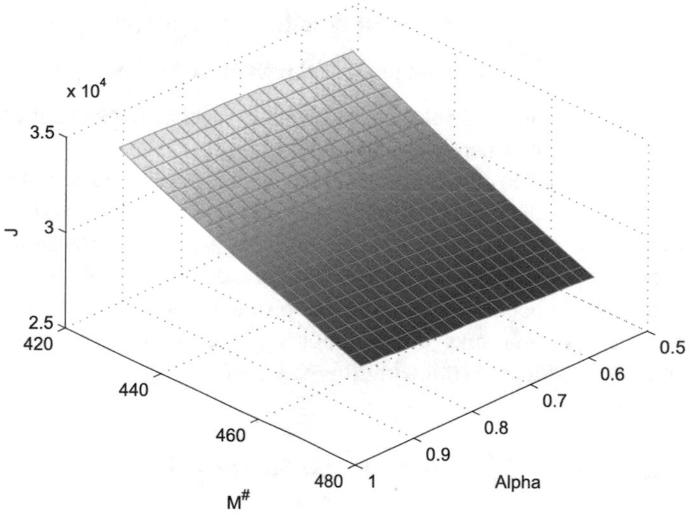

Fig. 9.3 Optimal objective value respected to confidence level and concentrations tolerable threshold

As shown in Fig. 9.3, the color is deeper, and the value J is larger. It turns out that if concentration limit is fixed, the minimal total abatement cost increases with respect to belief degree, this can be interpreted as if the higher belief degree is set which means the lower risk we can bear, and greater cost may be needed to satisfy the given target constraint. Simultaneously, the minimal total abatement cost decreases with respect to concentration limit; that is, if we relaxed the target constraint, the corresponding abatement cost can be cut under the same belief degree. It well displays the trade-offs between sustainability thresholds and risk.

9.4 Four-Wheel Steering Vehicle Problem

With the development of road traffic system and modern automotive engineering, the vehicle safety problem is increasingly outstanding. Based on the control theory, the vehicle performance has been studied by many researchers. For instance, Dirk and John [14] addressed the lateral control of vehicles during high-g emergency maneuvers. Marzbanrad et al. [15] studied a stochastic optimal control problem of a vehicle suspension on a random road. March and Shim [16] developed an integrated control system of active front steering and normal force control using fuzzy reasoning to enhance the vehicle-handling performance. Chen et al. [17] proposed a search scheme for finding robust and reliable solutions that are less sensitive to stochastic behaviors of customer demands and have low probability of route failures, respectively, in vehicle routing problem with stochastic demands.

We will consider an optimal control problem of four-wheel steering vehicle. When the vehicle involved in lane changing, overtaking, or turning behavior in high speed, the sideslip angle is expected to be close to zero and the yaw angular velocity's steady-state gain remains stable such as to ensure the vehicle security and operational stability. Let $\boldsymbol{X}_s = \begin{pmatrix} X_1(s) \\ X_2(s) \end{pmatrix}$ be the state variable, where $X_1(s)$ and $X_2(s)$ represent the sideslip angle and yaw angular velocity, respectively, and $\boldsymbol{u}_s$ be the control variable, which represent the front wheel steering angle.

Because of the influence of some uncertain factors (such as road waterlogging and drivers mind), it is unreasonable to describe the evolution of state variable in simple deterministic or stochastic analysis. Here, we employ uncertain differential equation as a framework for modeling the event and then the optimal control problem of four-wheel steering vehicle becomes an uncertain optimal control problem. The purpose of decision maker is to minimize the vehicle energy consumption at confidence level α. Therefore, we consider the following four-wheel steering vehicle optimal control model:

$$\begin{cases} J(0, \boldsymbol{x}_0) = \inf\limits_{\boldsymbol{u}_s} \left\{ \int_0^3 \left(\boldsymbol{X}_s^\tau Q(s) \boldsymbol{X}_s + \boldsymbol{u}_s^\tau R(s) \boldsymbol{u}_s \right) \mathrm{d}s + \boldsymbol{X}_3^\tau S_3 \boldsymbol{X}_3 \right\}_{\sup} (\alpha) \\ \text{subject to} \\ \mathrm{d}\boldsymbol{X}_s = (A(s)\boldsymbol{X}_s + B(s)\boldsymbol{u}_s)\mathrm{d}s + M(s)\boldsymbol{X}_s \mathrm{d}C_s \\ \boldsymbol{X}_0 = \boldsymbol{x}_0, \end{cases} \tag{9.11}$$

where

$$Q(s) = \begin{pmatrix} \frac{s}{2} & \frac{s}{4} \\ \frac{s}{4} & \frac{s}{2} \end{pmatrix}, \quad R(s) = \frac{s+2}{4}, \quad S_3 = \begin{pmatrix} \frac{1}{2} & 1 \\ 1 & \frac{1}{2} \end{pmatrix}, \quad A(s) = \begin{pmatrix} -\frac{s}{4} & 0 \\ 0 & -\frac{s}{4} \end{pmatrix},$$

$$B(s) = \begin{pmatrix} \frac{1}{10} \\ \frac{1}{5} \end{pmatrix}, \quad M(s) = \begin{pmatrix} \frac{s}{20} & \frac{s}{30} \\ \frac{s}{20} & \frac{s}{10} \end{pmatrix}, \quad \boldsymbol{x}_0 = \left(\frac{\pi}{6} \; \frac{1}{2} \right)^\tau, \quad \alpha = 0.9,$$

and $\boldsymbol{x}_s \in [0, 2]$. Here, we note that the matrix functions $Q(s)$ and $R(s)$ are weighting matrices, matrix function S_3 is terminal deviation, and matrix function $M(s)$ is perturbation coefficient. The value $\frac{s}{4}$ in $A(s)$ represents the reciprocal of first-order inertial element. The elements $\frac{1}{10}$ and $\frac{1}{5}$ in $B(s)$ are ratios of the steady-state gain of sideslip angle and yaw angular velocity to first-order inertial element, respectively.

According to the analyses in Sect. 9.3, the optimal control of model (9.11) will be a complex time-oriented function such that it is impractical or undesirable to implement. Then, we introduce a control parameter vector $\boldsymbol{K}$ into model (9.11) for simplifying the expression of optimal control. The corresponding model with control parameter can be formulated as follows:

$$\begin{cases} V(0, \boldsymbol{x}_0) = \inf\limits_{\boldsymbol{u}_s \in U} \left\{ \int_0^3 \left(\boldsymbol{X}_s^\tau Q(s)\boldsymbol{X}_s + \boldsymbol{u}_s^\tau R(s)\boldsymbol{u}_s\right) \mathrm{d}s + \boldsymbol{X}_3^\tau S_3 \boldsymbol{X}_3 \right\}_{\sup} (\alpha) \\ \text{subject to} \\ \mathrm{d}\boldsymbol{X}_s = (A(s)\boldsymbol{X}_s + B(s)\boldsymbol{u}_s)\mathrm{d}s + M(s)\boldsymbol{X}_s \mathrm{d}C_s \\ \boldsymbol{X}_0 = \boldsymbol{x}_0, \end{cases} \tag{9.12}$$

where $U = \{\boldsymbol{K}\boldsymbol{x}_s | \boldsymbol{K} = (K_1, K_2) = (K_1^{(l)}, K_2^{(l)}) \in R^{1\times 2}, s \in [t_{l-1}, t_l), l = 1, 2, 3\}$ with $t_0 = 0, t_1 = 1, t_2 = 2$ and $t_3 = 3$.

Assume that $J(t, \boldsymbol{x}) = \frac{1}{2}\boldsymbol{x}^\tau P(t)\boldsymbol{x}$, where $P(t) = \left(p_{ij}(t)\right)_{2\times 2}$. Applying Eqs. (3.26) and (3.27), we know the optimal control of model (9.11) is

$$\boldsymbol{u}_t^* = -\frac{1}{2} R^{-1}(t) B^\tau(t) P(t)\boldsymbol{x}, \tag{9.13}$$

where the function $P(t)$ satisfies the following matrix Riccati differential equation and boundary condition

$$\begin{cases} \dfrac{\mathrm{d}P(t)}{\mathrm{d}t} = -2Q(t) - A^\tau(t)P(t) - P(t)A(t) - \frac{\sqrt{3}}{\pi}\ln\frac{1-\alpha}{\alpha}P(t)M(t) - \frac{\sqrt{3}}{\pi}\ln\frac{1-\alpha}{\alpha}M^\tau(t)P(t) \\ \qquad\qquad + \frac{1}{2}P(t)B(t)R^{-1}(t)B^\tau(t)P(t) \\ P(3) = 2S_3. \end{cases} \tag{9.14}$$

Using the fourth-order Runge–Kutta method, the solutions of $p_{11}(t), p_{12}(t), p_{21}(t)$, and $p_{22}(t)$ can be obtained as shown in Figs. 9.4, 9.5, 9.6, and 9.7, respectively, with

$$P(2) = \begin{pmatrix} 1.2837 & 0.7997 \\ 0.7997 & 1.1696 \end{pmatrix}, P(1) = \begin{pmatrix} 1.3727 & 0.6065 \\ 0.6065 & 1.2043 \end{pmatrix}, P(0) = \begin{pmatrix} 1.3595 & 0.5348 \\ 0.5348 & 1.1659 \end{pmatrix}. \tag{9.15}$$

From Eq. (8.22), we have the following results:

$$K_1^* = \begin{cases} -0.1586, & \text{if } t \in [0, 1) \\ -0.1543, & \text{if } t \in [1, 2) \\ -0.2009, & \text{if } t \in [2, 3], \end{cases} \quad K_2^* = \begin{cases} -0.1512, & \text{if } t \in [0, 1) \\ -0.1754, & \text{if } t \in [1, 2) \\ -0.2362, & \text{if } t \in [2, 3]. \end{cases} \tag{9.16}$$

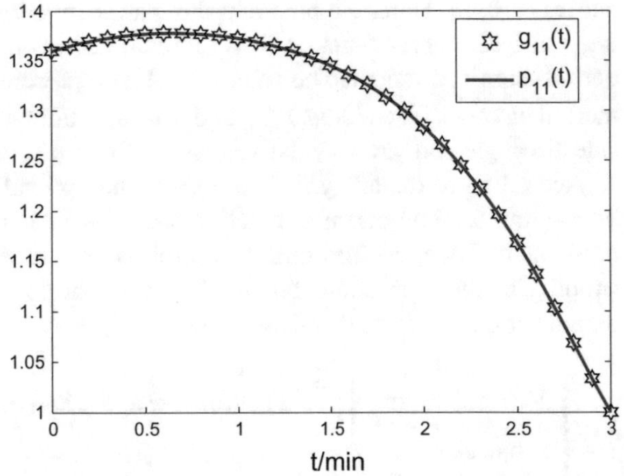

Fig. 9.4 Comparison of $g_{11}(t)$ and $p_{11}(t)$

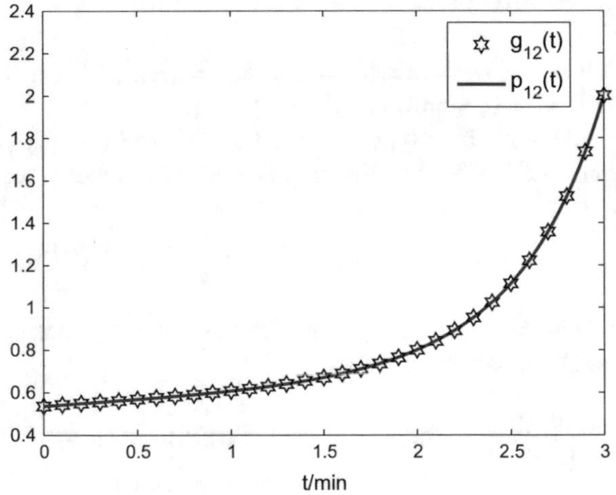

Fig. 9.5 Comparison of $g_{12}(t)$ and $p_{12}(t)$

In view of Eq. (8.17), the solutions of $g_{11}(t)$, $g_{12}(t)$, $g_{21}(t)$, and $g_{22}(t)$ can be obtained as shown in Figs. 9.4, 9.5, 9.6, and 9.7, respectively, with

$$G(2) = \begin{pmatrix} 1.2839 & 0.7996 \\ 0.7996 & 1.1696 \end{pmatrix}, \ G(1) = \begin{pmatrix} 1.3728 & 0.6066 \\ 0.6066 & 1.2045 \end{pmatrix}, \ G(0) = \begin{pmatrix} 1.3599 & 0.5353 \\ 0.5353 & 1.1667 \end{pmatrix}. \tag{9.17}$$

The optimal front wheel steering angle of model (9.12) is

$$\boldsymbol{u}_t^* = K_1^* X_1(t) + K_2^* X_2(t). \tag{9.18}$$

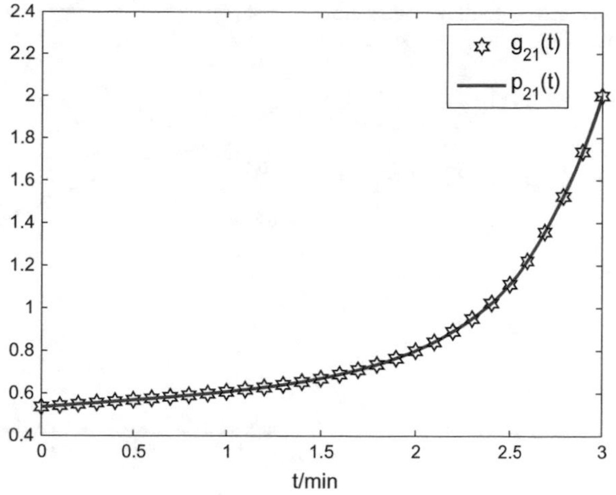

Fig. 9.6 Comparison of $g_{21}(t)$ and $p_{21}(t)$

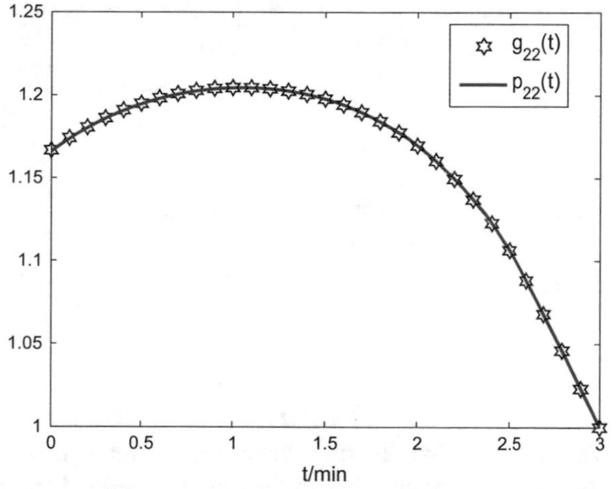

Fig. 9.7 Comparison of $g_{22}(t)$ and $p_{22}(t)$

And, the optimal value is

$$V(0, \boldsymbol{x}_0) = \frac{1}{2}\boldsymbol{x}_0^{\tau} G(0)\boldsymbol{x}_0 = 0.4724.$$

The optimal front wheel steering angle can be translated into an optimal steering wheel angle in a certain steering gear ratio. When controller implements the optimal steering wheel angle on steering wheel, the vehicle energy consumption will be minimized at confidence level 0.9 and the minimum energy consumption is 0.4724. We find that the coefficient of state variable $\boldsymbol{x}$ in Eq. (9.13) is a complex time-varying function such that the steering wheel has to make an uninterrupted and complex change, which will increase the design cost and complexity of controller.

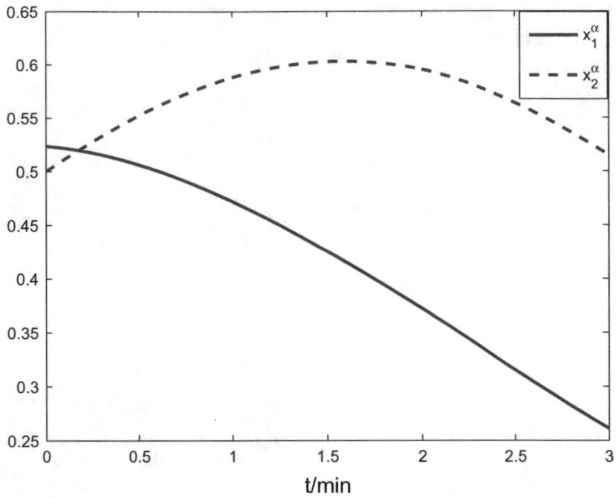

Fig. 9.8 α-paths of two state variables with $\alpha = 0.9$

Table 9.3 Optimal values and errors at different confidence level α

α	$V(0, \boldsymbol{x}_0)$	$J(0, \boldsymbol{x}_0)$	The absolute error	The relative error (%)
0.9000	0.4724	0.4721	$2.8573e - 004$	0.0605
0.8000	0.5343	0.5339	$3.7801e - 004$	0.0708
0.7000	0.5841	0.5837	$4.5539e - 004$	0.0780
0.6500	0.6078	0.6073	$5.0778e - 004$	0.0836
0.6000	0.6314	0.6308	$5.4646e - 004$	0.0866
0.5500	0.6555	0.6549	$6.2506e - 004$	0.0954

But the coefficients K_1^* and K_2^* in Eq. (9.18) are two piecewise constant functions, which shows that steering wheel can be manipulated more smoothly and we can use a switching system to implement the optimal control $\boldsymbol{u}_t^*$ such that the controller can be designed more easily and economically. The α-paths x_1^α and x_2^α of two state variables x_1 and x_2 with $\alpha = 0.9$ are shown in Fig. 9.8. So, the optimal control $\boldsymbol{u}_t^*$ is acceptable and more practical. Moreover, the absolute error of two optimal values is $|V(0, \boldsymbol{x}_0) - J(0, \boldsymbol{x}_0)| = |0.4724 - 0.4721| = 0.0003$ and the relative error is 0.0605%. In addition, the optimal values and errors at different confidence levels α are given in Table 9.3, which means the higher confidence level is, the lower error will become.

References

1. Merton R (1971) Optimal consumption and portfolio rules in a continuous time model. J Econ Theory 3:373–413
2. Kao E (1997) An introduction to stochastic processes. Wadsworth Publishing Company
3. Zhu Y (2010) Uncertain optimal control with application to a portfolio selection model. Cybern Syst 41(7):535–547
4. Sheng L, Zhu Y (2013) Optimistic value model of uncertain optimal control. Int J Uncertain Fuzziness Knowl Based Syst 21:75–83 (Suppl 1)
5. Yong J, Zhou X (1999) Stochastic controls: hamiltonian systems and HJB equations. Springer, New York
6. Bass F (1969) A new product growth for model consumer durables. Manag Sci 15(3):215–227
7. Horsky D, Simon L (1983) Advertising and the diffusion of new products. Mark Sci 2(1):1–17
8. Liu J (2000) On the dynamics of stochastic diffusion of manufacturing technology. Eur J Oper Res 124(5):601–614
9. Sheng L, Zhu Y, Hamalainen T (2013) An uncertain optimal control model with Hurwicz criterion. Appl Math Comput 224:412–421
10. DeLara M, Doyen L (2008) Sustainable management of natural resources: mathematical models and methods. Springer, Berlin
11. Doyen L, Dumas P, Ambrosi P (2008) Optimal timing of CO_2 mitigation policies for a cost-effectiveness model. Math Comput Model 48(5–6):882–897
12. Nordhaus W (1994) Managing the global commons: the economics of climate change. MIT Press, Cambridge
13. Sheng L, Zhu Y, Yan H, Wang K (2017) Uncertain optimal control approach for CO_2 mitigation problem. Asian J Control 19(6):1931–1942
14. Dirk E, John M (1995) Effects of model complexity on the performance of automated vehicle steering controllers: model development, validation and comparison. Veh Syst Dyn 24(2):163–181
15. Marzbanrad J, Ahmadi G, Zohoor H, Hojjat Y (2004) Stochastic optimal preview control of a vehicle suspension. J Sound Vib 275(3):973–990
16. March C, Shim T (2007) Integrated control of suspension and front steering to enhance vehicle handling. Proc Inst Mech Eng Part D J Automob Eng 221(D4):377–391
17. Chen X, Feng L, Soon O (2012) A self-adaptive memeplexes robust search scheme for solving stochastic demands vehicle routing problem. Int J Syst Sci 43(7):1347–1366

Index

A
α-path, 24

B
Bang bang, 99

C
Carbon dioxide emission, 193

E
Empirical uncertainty distribution, 3
Equation of optimality, 29, 31, 41, 49, 53, 61, 111
Event, 1
Expected value, 4

F
Four-wheel steering vehicle, 200

G
Global pruning scheme, 151

H
Hurwicz criterion, 52
Hybrid intelligent algorithm, 75

I
Impulse-free, 39
Indefinite LQ optimal control, 83
Independent, 4
Independent increment, 22

L
Linear quadratic model, 71
Linear uncertainty distribution, 3
Liu integral, 23
Liu process, 23
Local pruning scheme, 150

M
MACO algorithm, 133
Manufacturing technology diffusion, 190
Multistage uncertain system, 69
Multiple time delay, 168

N
Normal uncertainty distribution, 3

O
Optimistic value, 16
Ordinary uncertain variable, 10

P
Parametric approximation method, 180
Parametric optimal control, 177
Pessimistic value, 16
Piecewise optimization method, 184
Portfolio selection, 187
Principle of optimality, 28, 48, 53
Product uncertain measure, 2
Product uncertainty space, 2

Y. Zhu, *Uncertain Optimal Control*, Springer Uncertainty Research,
https://doi.org/10.1007/978-981-13-2134-4

R
Recurrence equation, 70, 145
Regular, 39
Regular uncertainty distribution, 5
Riccati differential equation, 34, 58, 132

S
Saddle point, 111
Singular uncertain system, 39, 60
Stationary independent increment process, 22
Switched uncertain system, 122

T
Time-delay, 157
Two-stage algorithm, 123
Two-stage approach, 138
Two-step pruning scheme, 149

U
Uncertain differential equation, 23
Uncertain expected value optimal control, 27
Uncertain linear quadratic model, 33, 57
Uncertain measure, 1
Uncertain optimistic value optimal control, 47
Uncertain process, 21
Uncertain simulation, 20
Uncertainty distribution, 3
Uncertainty space, 2
Uncertain variable, 3

V
Variance, 5

W
Well-posed, 85

Z
Zigzag uncertainty distribution, 3

Printed in the United States
By Bookmasters